P9-EDU-805

The Newtonian revolution

The Newtonian revolution

With illustrations of the transformation
of scientific ideas

I. BERNARD COHEN
Harvard University

CAMBRIDGE UNIVERSITY PRESS
Cambridge
London New York New Rochelle
Melbourne Sydney

Published by the Press Syndicate of the University of Cambridge
The Pitt Building, Trumpington Street, Cambridge CB2 1RP
32 East 57th Street, New York, NY 10022, USA
296 Beaconsfield Parade, Middle Park, Melbourne 3206, Australia

© Cambridge University Press 1980
First published 1980

Printed in the United States of America
Typeset and printed by Heritage Printers, Inc., Charlotte, North Carolina

Library of Congress Cataloging in Publication Data

Cohen, I Bernard, 1914–

The Newtonian revolution.

1. Physics–History. 2. Science–History.
3. Newton, Sir Isaac, 1642–1727. Principia.
I. Title.
QC7.C66   509   79-18637
ISBN 0 521 22964 2

35,193

CAMROSE LUTHERAN COLLEGE
LIBRARY

To
W. Sidney Allen (Cambridge)
Marshall Clagett (Princeton)
Tullio Gregory (Rome)
and my colleagues
Erwin N. Hiebert
A. I. Sabra

CONTENTS

Contents

Contents

PREFACE

The origins of this book go back to 1966, when I had the honor of giving the Wiles Lectures in the Queen's University of Belfast, sponsored by the foundation established by Mrs. Janet P. Boyd in memory of her father. This foundation is remarkable in its conception. It not only provides a lecturer on an aspect of history, but ensures that each lecture will be discussed by the Belfast historians and research students and an invited group of historians from other universities. The evening discussions, following each afternoon's lecture, were of great value in helping me to make more precise certain basic issues. I am especially grateful for having thus been able to test certain primary viewpoints in an audience of colleagues and of general historians, and to profit by the reactions of Rupert and Marie Boas Hall, John Herivel, Michael Hoskin, George Huxley, D. T. Whiteside, and W. P. D. Wightman. I am indebted to my academic host, Professor J. C. Becket, to Mrs. Janet P. Boyd, and to Vice-Chancellor and Mrs. Michael Grant for much personal kindness.

The completion of a published version of these lectures occurs a decade or so later than had been expected. This delay has been caused, in the first instance, by the consuming labor of completing the *Introduction to Newton's 'Principia'* and of the editing of Newton's *Principia* with variant readings (undertaken in concert with Alexandre Koyré and with the assistance of Anne Whitman). The preparation of this edition became a far more onerous undertaking than had been originally conceived, owing to the untimely death of Professor Koyré, which deprived us of his direct aid, wisdom, and experience during the final stages of that task. It was not until after the publication of that edition (1971, 1972) that I was free to return to the assignment of preparing the Wiles Lectures for publication.

In the meanwhile, however, I had published a redaction of two of the lectures in a privately distributed edition, and versions of each of them were printed in scholarly journals. In particular, a central theme of the lectures and of this book based upon them— the "transformation" of scientific ideas—was further developed in articles and tried out in my university lectures and seminars at Harvard University. This bit of personal history is recounted as part of the supplement at the end of Ch. 5.

This book, like the lectures themselves, centers about the scientific life of Isaac Newton, but does so as a key to the understanding of an aspect of Newtonian science and as a means of understanding scientific change generally. The book thus deals with the Newtonian revolution in science, in the ways in which I believe Newton's contemporaries and immediate predecessors in the exact sciences conceived him to have made a "revolution". In this expression, I am not imposing an anachronistic historical judgment based upon twentieth-century concepts of scientific change, but I am rather going back to the actual expression used by creative scientists and analysts of scientific change in the age of Newton. This book is thus part of a series of general studies I have been making of the history and concept of revolution in the sciences and of some of the main features of Newton's *Principia*.

I have concentrated my attention on Newton's *Principia*, because it was in this work that there was fully developed what I have called the "Newtonian style". The essence of this style was an ability to separate the study of the exact sciences into two parts: the development of the mathematical consequences of imaginative constructs or systems and the subsequent application of the mathematically derived results to the explanation of phenomenological reality. I have called this aspect of the science of the *Principia* the "Newtonian style", fully cognizant that it was not invented by Newton out of whole cloth and that it is very similar to what has been called the Galilean style.

The Newtonian style has three phases. Phase one usually begins with nature simplified and idealized, leading to an imaginative construct in the mathematical domain, a system in geometric space in which mathematical entities move in mathematical time according to certain set conditions which tend to be expressible as mathematical laws or relations. Consequences are deduced by means of mathematical techniques and are then transferred to the observed world of physical nature, where in phase two a comparison and contrast is made with experiential data and the laws or rules derived from such data. This usually results in an alteration of the original mathematical construct or system, or a new phase one, which in turn leads to a new phase two. Thus Newton starts out with a mass point in a central force field and deduces a law of areas. Later he will add conditions of a second body mutually interacting with the original one, then still other such bodies. He will eventually consider bodies of finite sizes and specified shapes and constitutions rather than essentially mass points, and will even pursue the possibilities of various types of resisting mediums through which they may move. In phase three Newton applies the results obtained in phases one and two (roughly corresponding to bks. one and two of the *Principia*) to natural philosophy, so as to elaborate his 'System of the World' (bk. three). For Newton there was a sequel, not part of the published *Principia*, the attempt to find out how such a force as universal gravity can exist and act according to the laws he had discovered. One of the explanations he put forth invoked a model in which an aether with a variation in density related to the distribution of matter could produce effects like those of gravity.

The great power of the Newtonian style was that it made possible the study of forces of different sorts in relation to motions in general, and in relation to those motions observed in the external world, without any inhibiting considerations as to whether such forces can actually (or do actually) exist in nature. The Newtonian style was successful in the *Principia*, even though there were conspicuous failures to achieve complete solutions (as in the case of the motion of the moon). In his studies of optics Newton attempted to follow a similar line of development, but the subject matter did not prove fully amenable to the Newtonian style. Hence to find out how Newton attempted to develop the subject of optics in the Newtonian style, it is necessary to turn to the posthumously published *Optical Lectures* or *Lectiones opticae* and to certain optical MSS (as D. T. Whiteside has done in the *Mathematical Papers*; see, notably, Newton, 1967–, vol. 3, pp. 450–454, vol. 6, pp. 422–434). Only bare traces of the Newtonian style are discernible in Newton's published *Opticks*, which he cast in a different mold, so that it became a book of experiment in a popular style, rather than an illustration of the method of elaboration by mathematical techniques of the properties of imaginative constructs. And, in fact, it is not always clear which experiments were actually made, or were made exactly as reported. Just as the Newtonian style was not really successful in optics in the sense that it had been in dynamics and celestial mechanics, so it was barren in relation to Newton's theory of matter. It is at once evident, I believe, from this analysis that nothing resembling the Newtonian revolution in science could have occurred in the seventeenth and eighteenth centuries in the biological or life sciences.

Although this book is concentrated on the science of the *Principia*, some examples are brought in from other aspects of Newtonian science, from the science of other periods, and from branches of physics other than dynamics and celestial mechanics. For it is my belief that the analysis of scientific change as a sequence of transformations is universally applicable and may help us to understand the detailed and individual steps which in their totality comprise great revolutions in scientific thought.

The second part of the book deals with transformations in the history of scientific thought. This aspect of scientific change is then illustrated by an examination of the treatment of Kepler's laws by Newton. This second part of the book is related to the first in a number of ways. The discussion of revolutions produced in science by successive transformations illuminates the prior discussion of revolutions in science. The elaboration of the role of Kepler's laws in the formation of Newton's celestial dynamics and of his system of the world based on universal gravity completes the earlier presentation of the Newtonian style. I have devoted a wholly separate book to the cognate general theme of *Revolution in Science: History, Analysis, and Significance of a Name and a Concept*.

I have divided each chapter into sections, so that the reader who may not wish to follow every step of the argument may find those parts which may satisfy his or her interests and needs. Although there are a series of cross-references which link the chapters together, I have also tried (even at the expense of occasionally repeating in summary an idea fully developed in another chapter) to have each chapter be an independent statement, so that it may be read without over-dependence upon what has gone before.

I am grateful to many friends, colleagues, and students, with whom I have discussed these ideas. I owe a special debt of gratitude to those who have looked at the typescript and given me the advantage of their helpful suggestions: Lorraine J. Daston, Joel Genuth, Ernan McMullin, Simon Schaffer, Michael Shank, and especially D. T. Whiteside. I am grateful to the National Science Foundation, which has supported the research on the scientific thought of Isaac Newton and his *Principia* on which this book is centrally based. I am equally grateful to the Spencer Foundation (Chicago), which has supported my research on the historical relations between the natural and physical sciences and the social and behavioral sciences; for it is this research which has led me to understand the history and the nature of revolutions in science in general and hence the revolution in science associated with Isaac Newton.

Cambridge, Mass.   I.B.C.
July 1980

PART ONE

The Newtonian revolution and the
Newtonian style

1

The Newtonian revolution in science

1.1 Some basic features of the Scientific Revolution

A study of the Newtonian revolution in science rests on the fundamental assumption that revolutions actually occur in science. A correlative assumption must be that the achievements of Isaac Newton were of such a kind or magnitude as to constitute a revolution that may be set apart from other scientific revolutions of the sixteenth and seventeenth centuries. At once we are apt to be plunged deep into controversy. Although few expressions are more commonly used in writing about science than "scientific revolution", there is a continuing debate as to the propriety of applying the concept and term "revolution" to scientific change.[1] There is, furthermore, a wide difference of opinion as to what may constitute a revolution. And although almost all historians would agree that a genuine alteration of an exceptionally radical nature (*the* Scientific Revolution[2]) occurred in the sciences at some time between the late fifteenth (or early sixteenth) century and the end of the seventeenth century, the question of exactly when this revolution occurred arouses as much scholarly disagreement as the cognate question of precisely what it was. Some scholars would place its origins in 1543, the year of publication of both Vesalius's great work on the fabric of the human body and Copernicus's treatise on the revolutions of the celestial spheres (Copernicus, 1543; Vesalius, 1543). Others would have the revolution be inaugurated by Galileo, possibly in concert with Kepler, while yet others would see Descartes as the true prime revolutionary. Contrariwise, a whole school of historians declare that many of the most significant features of the so-called Galilean revolution had emerged during the late Middle Ages.[3]

A historical analysis of the Newtonian revolution in science does not, however, require participation in the current philosophical and sociological debates on these issues. For the fact of the matter is that the concept of revolution in science—in the sense in which we would understand this term nowadays—arose during Newton's day and was applied (see §2.2) first to a part of mathematics in which he made his greatest contribution, the calculus, and then to his work in celestial mechanics. Accordingly, the historian's task may legitimately be restricted to determining what features of Newton's science seemed so extraordinary in the age of Newton as to earn the designation of revolution. There is no necessity to inquire here into the various meanings of the term "revolution" and to adjudge on the basis of each such meaning the correctness of referring to a Newtonian revolution in the sciences.

The new science that took form during the seventeenth century may be distinguished by both external and internal criteria from the science and the philosophical study or contemplation of nature of the antecedent periods. Such an external criterion is the emergence in the seventeenth century of a scientific "community": individuals linked together by more or less common aims and methods, and dedicated to the finding of new knowledge about the external world of nature and of man that would be consonant with—and, accordingly, testable by—experience in the form of direct experiment and controlled observation. The existence of such a scientific community was characterized by the organization of scientific men into permanent formal societies, chiefly along national lines, with some degree of patronage or support by the state.[4] The primary goal of such societies was the improvement of "natural knowledge".[5] One way by which they sought to gain that end was through communication; thus the seventeenth century witnessed the establishment of scientific and learned journals, often the organs of scientific societies, including the *Philosophical Transactions* of the Royal Society of London, the *Journal des Sçavans*, and the *Acta eruditorum* of Leipzig.[6] Another visible sign of the existence of a "new science" was the founding of research institutions, such as the Royal Greenwich Observatory, which celebrated its three-hundredth birthday in 1975. Newton's scientific career exhibits aspects of these several manifestations of the new science and the scientific community. He depended on the Astronomer Royal, John Flamsteed, for observational evidence that Jupiter might perturb the orbital motion of Saturn near conjunction and later needed lunar positions from Flamsteed at the Greenwich Observatory in order to test and to advance his lunar theory, especially in the 1690s. His first scientific publication was his famous article on light and colors, which appeared in the pages of the *Philosophical Transactions;* his *Principia* was officially published by the Royal Society, of which he became president in 1703 (an office he kept until his death in 1727). While the Royal Society was thus of great importance in Newton's scientific life, it cannot be said that his activities in relation to that organization or its journal were in any way revolutionary.

The signs of the revolution can also be seen in internal aspects of science: aims, methods, results. Bacon and Descartes agreed on one aim of the new science, that the fruits of scientific investigation would be the improvement of man's condition here on earth:[7] agriculture, medicine, navigation and transportation, communication, warfare, manufacturing, mining.[8] Many scientists of the seventeenth century held to an older point of view, that the pursuit of scientific understanding of nature was practical insofar as it might advance man's comprehension of the divine wisdom and power. Science was traditionally practical in serving the cause of religion; but a revolutionary feature of the new science was the additional pragmatic goal of bettering everyday life here and now through applied science. The conviction that had been developing in the sixteenth and seventeenth centuries, that a true goal of the search for scientific truth must be to affect the material conditions of life, was then strong and widely shared, and constituted a novel and even a characteristic feature of the new science.

Newton often declared his conviction as to the older of these practicalities, as when he wrote to Bentley about his satisfaction in having advanced the cause of true religion by his scientific discoveries. Five years after the publication of his *Principia*, he wrote to Bentley that while composing the *Principia* ('my Treatise about our system'), 'I had an eye upon such Principles as might work with considering Men, for the Belief of a Deity' (Newton, 1958, p. 280; 1959–1977, vol. 3, p. 233). About two decades later, in 1713, he declared in the concluding general scholium to the *Principia* that the system of the world 'could not have arisen without the design and dominion of an intelligent and powerful being'. Newton was probably committed to some degree to the new practicality; at least he served as advisor to the official group concerned with the problem of finding methods of determining the longitude at sea. Yet it was not Newton himself, but other scientists such as Halley, who attempted to link the Newtonian lunar theory with the needs of navigators, and the only major practical innovation that he produced was an instrument for science (the reflecting telescope) rather than inventions for man's more mundane needs.[9]

Another feature of the revolution was the attention to method. The attempts to codify method–by such diverse figures as Descartes, Bacon, Huygens, Hooke, Boyle, and Newton–signify that discoveries were to be made by applying a new tool of inquiry (a *novum organum*, as Bacon put it) that would direct the mind unerringly to the uncovering of nature's secrets. The new method was largely experimental, and has been said to have been based on induction; it also was quantitative and not merely observational and so could lead to mathematical laws and principles. I believe that the seventeenth-century evaluation of the importance of method was directly related to the role of experience (experiment and observation) in the new science. For it seems to have been a tacit postulate that any reasonably skilled man or woman should be able to reproduce an experiment or observation, provided that the report of that experiment or observation was given honestly and in sufficient detail. A consequence of this postulate was that anyone who understood the true methods of scientific enquiry and had acquired the necessary skill to make experiments and observations could have made the discovery in the first instance–provided, of course, that he had been gifted with the wit and insight to ask the right questions.[10]

This experimental or experiential feature of the new science shows itself also in the habit that arose of beginning an enquiry by repeating or reproducing an experiment or observation that had come to one's attention through a rumor or an oral or written report. When Galileo heard of a Dutch optical invention that enabled an observer to see distant objects as clearly as if they were close at hand, he at once set himself to reconstructing such an instrument.[11] Newton relates how he had bought a prism 'to try therewith the celebrated *Phaenomena* of *Colours*'.[12] From that day to this, woe betide any investigator whose experiments and observations could not be reproduced, or which were reported falsely; this attitude was based upon a fundamental conviction that nature's occurrences are constant and reproducible, thus subject to universal laws. This twin requirement of performability and reproducibility imposed a code of honesty and integrity upon the scientific community that is itself yet another distinguishing feature of the new science.

The empirical aspect of the new science was just as significant with respect to the results achieved as with respect to the aims and methods. The law of falling bodies, put forth by Galileo, describes how real bodies actually fall on this earth—due consideration being given to the difference between the ideal case of a vacuum and the realities of an air-filled world, with winds, air resistance, and the effects of spin. Some of the laws of uniform and accelerated motion announced by Galileo can be found in the writings of certain late medieval philosopher-scientists, but the latter (with a single known exception of no real importance[13]) never even asked whether these laws might possibly correspond to any real or observable motions in the external world. In the new science, laws which do not apply to the world of observation and experiment could have no real significance, save as mathematical exercises. This point of view is clearly enunciated by Galileo in the introduction of the subject of 'naturally accelerated motion', in his *Two New Sciences* (1638). Galileo states the aim of his research to have been 'to seek out and clarify the definition that best agrees with that [accelerated motion] which nature employs' (Galileo, 1974, p. 153; 1890–1909, vol. 8, p. 197). From his point of view, there is nothing 'wrong with inventing at pleasure some kind of motion and theorizing about its consequent properties, in the way that some men have derived spiral and conchoidal lines from certain motions, though nature makes no use of these [paths]'. But this is different from studying motion in nature, for in exploring phenomena of the real external world, a definition is to be sought that accords with nature as revealed by experience:

> But since nature does employ a certain kind of accelera-
> tion for descending heavy things, we decided to look
> into their properties so that we might be sure that the
> definition of accelerated motion which we are about
> to adduce agrees with the essence of naturally accelerated
> motion. And at length, after continual agitation of mind,
> we are confident that this has been found, chiefly for the
> very powerful reason that the essentials successively
> demonstrated by us correspond to, and are seen to be in

agreement with, that which physical experiments [*naturalia experimenta*] show forth to the senses [ibid.]. Galileo's procedure is likened by him to having 'been led by the hand to the investigation of naturally accelerated motion by consideration of the custom and procedure of nature herself'.

Like Galileo, Newton the physicist saw the primary importance of concepts and rules or laws that relate to (or arise directly from) experience. But Newton the mathematician could not help but be interested in other possibilities. Recognizing that certain relations are of physical significance (as that 'the periodic times are as the 3/2 power of the radii', or Kepler's third law), his mind leaped at once to the more universal condition (as that 'the periodic time is as any power R^n of the radius R').[14] Though Newton was willing to explore the mathematical consequences of attractions of spheres according to any rational function of the distance, he concentrated on the powers of index 1 and −2 since they are the ones that occur in nature: the power of index 1 of the distance from the center applies to a particle within a solid sphere and the power of index −2 to a particle outside either a hollow or solid sphere.[15] It was his aim, in the *Principia*, to show that the abstract or 'mathematical principles' of the first two books could be applied to the phenomenologically revealed world, an assignment which he undertook in the third book. To do so, after Galileo, Kepler, Descartes, and Huygens, was not in itself revolutionary, although the scope of the *Principia* and the degree of confirmed application could well be so designated and thus be integral to the Newtonian revolution in science.

An excessive insistence on an out-and-out empirical foundation of seventeenth-century science has often led scholars to exaggerations.[16] The scientists of that age did not demand that each and every statement be put to the test of experiment or observation, or even have such a capability, a condition that would effectively have blocked the production of scientific knowledge as we know it. But there was an insistence that the goal of science was to understand the real external world, and that this required the possibility of predicting testable results and retrodicting the data of actual experience: the accumulated results of experiment and controlled observation. This continual growth of factual knowledge garnered from the researches and observations made all over the world, paralleled by an equal and continual advance of understanding, was another major aspect of the new science, and has been a distinguishing characteristic of the whole scientific enterprise ever since. Newton certainly made great additions to the stock of knowledge. In the variety and fundamental quality of these contributions we may see the distinguishing mark of his great creative genius, but this is something distinct from having created a revolution.

1.2   *A Newtonian revolution in science: the varieties of Newtonian science*

In the sciences, Newton is known for his contributions to pure and applied mathematics, his work in the general area of optics, his experiments and speculations relating to theory of matter and chemistry (including alchemy), and his systematization of rational mechanics (dynamics) and his celestial dynamics (including the Newtonian "system of the world"). Even a modest portion of these achievements would have sufficed to earn him an unquestioned place among the scientific immortals. In his own day (as we shall see below in Ch. 2), the word "revolution" began to be applied to the sciences in the sense of a radical change; one of the first areas in which such a revolution was seen to have occurred was in the discovery or invention of the calculus: a revolution in mathematics.[1] There is also evidence aplenty that in the age of Newton and afterwards, his *Principia* was conceived to have ushered in a revolution in the physical sciences. And it is precisely this revolution whose characteristic features I aim to elucidate.

Newton's studies of chemistry and theory of matter yielded certain useful results[2] and numerous speculations. The latter were chiefly revealed in the queries at the end of the *Opticks*, especially the later ones,[3] and in such a tract as the *De natura acidorum*.[4] The significance of these writings and their influence have been aggrandized (from Newton's day to ours) by the extraordinary place in science held by their author. At best, they are incomplete and programmatic and—in a sense—they chart out a possible revolution, but a revolution never achieved by Newton nor ever realized along the lines that he set down. Newton's program and suggestions had a notable influence on the science of the eighteenth century, particularly the development of theories of heat and electricity (with their subtle elastic fluids) (cf. Cohen, 1956, Ch. 7, 8). Newton had a number of brilliant insights into the structure of matter and the process of chemical reaction, but the true revolution in chemistry did not come into being until the work of Lavoisier, which was not directly Newtonian (see Guerlac, 1975).

The main thrust of Newton's views on matter was the hope of deriving 'the rest of the phenomena of nature by the same kind of reasoning from mechanical principles' that had served in deducing 'the motions of the planets, the comets, the moon, and the sea'. He was convinced that all such phenomena, as he said in the preface (1686) to the first edition of the *Principia*, 'may depend upon certain forces by which the particles of bodies . . . are either mutually impelled [attracted] toward one another so as to cohere in regular figures' or 'are repelled and recede from one another'.[5] In this way, as he put it on another occasion, the analogy of nature would be complete: 'Whatever reasoning holds for greater motions, should hold for lesser ones as well. The former depend upon the greater attractive forces of larger bodies, and I suspect that the latter depend upon the lesser forces, as yet unobserved, of insensible particles'. In short, Newton would have nature be thus 'exceedingly simple and conformable to herself'.[6] This particular program was a conspicuous failure. Yet it was novel and can even be said to have had revolutionary features, so that it may at best represent a failed (or at least a never-achieved) revolution. But since we are concerned here with a positive Newtonian revolution, Newton's hope to develop a micro-mechanics analogous to his successful macro-mechanics is not our main concern. We cannot wholly neglect this topic, however, since it has been alleged that Newton's mode of attack on the physics of gross bodies and his supreme success in celestial mechanics was the product of his investigations of short-range forces, despite the fact that Newton himself said (and said repeatedly) that it was his success in the area of gravitation that led him to believe that the forces of particles could be developed in the same style. R. S. Westfall (1972, 1975) would not even stop there, but would have the 'forces of attraction between particles of matter', and also 'gravitational attraction which was probably the last one [of such forces] to appear', be 'primarily the offspring of alchemical active principles'. This particular thesis is intriguing in that it would give a unity to Newton's intellectual endeavor; but I do not believe it can be established by direct evidence (see Whiteside, 1977). In any event, Newton's unpublished papers on alchemy and his published (and unpublished) papers on chemistry and theory of matter hardly merit the appellation of "revolution", in the sense of the radical influence on the advance of science that was exerted by the *Principia*.

In optics, the science of light and colors, Newton's contributions were outstanding. But his published work on 'The Reflections, Refractions, Inflexions [i.e., diffraction] & Colours of Light', as the *Opticks* was subtitled, was not revolutionary in the sense that the *Principia* was. Perhaps this was a result of the fact that the papers and book on optics published by Newton in his lifetime do not boldly display the mathematical properties of forces acting (as he thought) in the production of dispersion and other optical phenomena, although a hint of a mathematical model in the Newtonian style is given in passing in the *Opticks* (see §3.11) and a model is developed more fully in sect. 14 of bk. one of the *Principia*. Newton's first published paper was on optics, specifically on his prismatic experiments relating to dispersion and the composition of sunlight and the nature of color. These results were expanded in his *Opticks* (1704; Latin ed. 1706; second English ed. 1717/1718), which also contains his experiments and conclusions on other aspects of optics, including a large variety of what are known today as diffraction and interference phenomena (some of which Newton called the "inflexion" of light). By quantitative experiment and measurement he explored the cause of the rainbow, the formation of "Newton's rings" in sunlight and in monochromatic light, the colors and other phenomena produced by thin and thick "plates", and a host of other optical effects.[7] He explained how bodies exhibit colors in relation to the type of illumination and their selective powers of absorption and transmission or reflection of different colors. The *Opticks*, even apart from the queries, is a brilliant display of the experimenter's art, where (as Andrade, 1947, p. 12, put it so well) we may see Newton's 'pleasure in shaping'. Some of his measurements were so precise that a century later they yielded to Thomas Young the correct values, to within less than 1 percent, of the wavelengths of light of different colors.[8] Often cited as a model of how to perform quantitative experiments and how to analyze a difficult problem by experiment,[9] Newton's studies of light and color and his *Opticks* nevertheless did not create a revolution and were not ever considered as revolutionary in the age of Newton or afterwards. In this sense, the *Opticks* was not epochal.

From the point of view of the Newtonian revolution in science, however, there is one very significant aspect of the *Opticks*: the fact that in it Newton developed the most complete public statement he ever made of his philosophy of science or of his conception of the experimental scientific method. This methodological declaration has, in fact, been a source of some confusion ever since, because it has been read as if it applies to all of Newton's work, including the *Principia*.[10] The final paragraph of qu. 28 of the *Opticks* begins by discussing the rejection of any 'dense Fluid' supposed to fill all space, and then castigates 'Later Philosophers' (i.e., Cartesians and Leibnizians) for 'feigning Hypotheses for explaining all things mechanically, and referring other Causes to Metaphysicks'. Newton asserts, however, that 'the main Business of natural Philosophy is to argue from Phaenomena without feigning Hypotheses, and to deduce Causes from Effects, till we come to the very first Cause, which certainly is not mechanical'.[11] Not only is the main assignment 'to unfold the Mechanism of the World', but it is to 'resolve' such questions as: 'What is there in places almost empty of Matter . . . ?' 'Whence is it that Nature doth nothing in vain; and whence arises all that Order and Beauty which we see in the World?' What 'hinders the fix'd Stars from falling upon one another?' 'Was the Eye contrived without Skill in Opticks, and the Ear without Knowledge of Sounds?' or 'How do the Motions of the Body follow from the Will, and whence is the Instinct in Animals?'

In qu. 31, Newton states his general principles of analysis and synthesis, or resolution and composition, and the method of induction:

> As in Mathematicks, so in Natural Philosophy, the
> Investigation of difficult Things by the Method of Analysis,
> ought ever to precede the Method of Composition. This
> Analysis consists in making Experiments and Observa-
> tions, and in drawing general Conclusions from them
> by Induction, and admitting of no Objections against the
> Conclusions, but such as are taken from Experiments,
> or other certain Truths. For Hypotheses are not to be
> regarded in experimental Philosophy. And although the
> arguing from Experiments and Observations by Induc-
> tion be no Demonstration of general Conclusions; yet it
> is the best way of arguing which the Nature of Things
> admits of, and may be looked upon as so much the
> stronger, by how much the Induction is more general.

Analysis thus enables us to

> proceed from Compounds to Ingredients, and from
> Motions to the Forces producing them; and in general,
> from Effects to their Causes, and from particular Causes to
> more general ones, till the Argument end in the
> most general.

This method of analysis is then compared to synthesis or composition:

> And the Synthesis consists in assuming the Causes
> discover'd, and establish'd as Principles, and by them
> explaining the Phaenomena proceeding from them,
> and proving the Explanations.[12]

The lengthy paragraph embodying the foregoing three extracts is one of the most often quoted statements made by Newton, rivaled only by the concluding General Scholium of the *Principia*, with its noted expression: *Hypotheses non fingo*.

Newton would have us believe that he had himself followed this "scenario":[13] first, to reveal by "analysis" some simple results that were generalized by induction, thus proceeding from effects to causes and from particular causes to general causes; then, on the basis of these causes considered as principles, to explain by "synthesis" the phenomena of observation and experiment that may be derived or deduced from them, 'proving the Explanations'. Of the latter, Newton says that he has given an 'Instance . . . in the End of the first Book' where the 'Discoveries being proved [by experiment] may be assumed in the Method of Composition for explaining the Phaenomena arising from them'. An example, occurring at the end of bk. one, pt. 2, is props. 8–11, with which pt. 2 concludes. Prop. 8 reads: 'By the discovered Properties of Light to explain the Colours made by Prisms'. Props. 9–10 also begin: 'By the discovered Properties of Light to explain . . .', followed (prop. 9) by 'the Rain-bow' and (prop. 10) by 'the permanent Colours of Natural Bodies'. Then, the concluding prop. 11 reads: 'By mixing coloured Lights to compound a beam of Light of the same Colour and Nature with a beam of the Sun's direct Light'.

The formal appearance of the *Opticks* might have suggested that it was a book of synthesis, rather than analysis, since it begins (bk. one, pt. 1) with a set of eight 'definitions' followed by eight 'axioms'. But the elucidation of the propositions that follow does not make explicit reference to these axioms, and many of the individual propositions are established by a method plainly labeled 'The PROOF by Experiments'. Newton himself states clearly at the end of the final qu. 31 that in bks. one and two he has 'proceeded by . . . Analysis' and that in bk. three (apart from the queries) he has 'only begun the Analysis'. The structure of the *Opticks* is superficially similar to that of the *Principia*, for the *Principia* also starts out with a set of 'definitions' (again eight in number), followed by three 'axioms' (three 'axiomata sive leges motus'), upon which the propositions of the first two books are to be constructed (as in the model of Euclid's geometry). But then, in bk. three of the *Principia*, on the system of the world, an ancillary set of so-called 'phenomena' mediate the application of the mathematical results of bks. one and two to the motions and properties of the physical universe.[14] Unlike the *Opticks*, the *Principia* does make use of the axioms and definitions.[15] The confusing aspect of Newton's stated method of analysis and synthesis (or composition) in qu. 31 of the *Opticks* is that it is introduced by the sentence 'As in Mathematicks, so in Natural Philosophy . . .', which was present when this query first appeared (as qu. 23) in the Latin *Optice* in 1706, 'Quemadmodum in Mathematica, ita etiam in Physica . . .'. A careful study, however, shows that Newton's usage in experimental natural philosophy is just the reverse of the way "analysis" and "synthesis" (or "resolution" and "composition") have been traditionally used in relation to mathematics, and hence in the *Principia*–an aspect of Newton's philosophy of science that was fully understood by Dugald Stewart a century and a half ago but that has not been grasped by present-day commentators on Newton's scientific method, who would even see in the *Opticks* the same style that is to be found in the *Principia*[16] (this point is discussed further in §3.11).

Newton's "method", as extracted from his *dicta* rather than his *opera*, has been summarized as follows: 'The main features of Newton's method, it seems, are: The rejection of hypotheses, the stress upon induction, the working sequence (induction precedes deduction), and the inclusion of metaphysical arguments in physics' (Turbayne, 1962, p. 45). Thus Colin Turbayne would have 'the deductive procedure' be a defining feature of Newton's 'mathematical way' and Descartes's '*more geometrico*' respectively: 'Descartes's "long chains of reasoning" were deductively linked. Newton's demonstrations were reduced to "the form of propositions in the mathematical way" '. He would criticize those analysts who have not recognized that the defining property of 'the Cartesian "geometrical method" or the Newtonian "mathematical way" '– 'paradoxical as it may seem–need be neither geometrical nor mathematical. Its defining property is demonstration, not the nature of the terms used in it'.[17] It may be observed that the phrase used here, 'the Newtonian "mathematical way" ', or 'Newton's "mathematical way" ', so often quoted in philosophical or methodological accounts of Newton's science, comes from the English translation of Newton's *System of the World*[18] but is not to be found in any of the manuscript versions of that tract, including the one that is still preserved among Newton's papers (see Dundon, 1969; Cohen, 1969*a*, 1969*c*).

The Newtonian revolution in the sciences, however, did not consist of his use of deductive reasoning, nor of a merely external form of argument that was presented as a series of demonstrations from first principles or axioms. Newton's outstanding achievement was to show how to introduce mathematical analysis into the study of nature in a rather new and particularly fruitful way, so as to disclose *Mathematical Principles of Natural Philosophy*, as the *Principia* was titled in full: *Philosophiae naturalis principia mathematica*. Not only did Newton exhibit a powerful means of applying mathematics to nature, he made use of a new mathematics which he himself had been forging and which may be hidden from a superficial observer by the external mask of what appears to be an example of geometry in the traditional Greek style (see n. 10 to §1.3).

In the *Principia* the science of motion is developed in a way that I have characterized as the Newtonian style. In Ch. 3 it shall be seen that this style consists of an interplay between the simplification and idealization of situations occurring in nature and their analogues in the mathematical domain. In this manner Newton was able to produce a mathematical system and mathematical principles that could then be applied to natural philosophy, that is, to the system of the world and its rules and data as determined by experience. This style permitted Newton to treat problems in the exact sciences as if they were exercises in pure mathematics and to link experiment and observation to mathematics in a notably fruitful manner. The Newtonian style also made it possible to put to one side, and to treat as an independent question, the problem of the cause of universal gravity and the manner of its action and transmission.

The Newtonian revolution in the sciences was wrought by and revealed in the *Principia*. For more than two centuries, this book set the standard against which all other science was measured; it became the goal toward which scientists in such diverse fields as paleontology, statistics, and biochemistry would strive in order to bring their own fields to a desired high estate.[19] Accordingly, I have striven in the following pages to explore and to make precise the qualities of Newton's *Principia* that made it so revolutionary. Chief among them, as I see it, is the Newtonian style, a clearly thought out procedure for combining mathematical methods with the results of experiment and observation in a way that has been more or less followed by exact scientists ever since. This study concentrates mainly on the *Principia*, because of the supreme and unique importance of that treatise in the Scientific Revolution and in the intellectual history of man. In the *Principia* the role of induction is minimal and there is hardly a trace of that analysis which Newton said should always precede synthesis.[20] Nor is there any real evidence whatever that Newton first discovered the major propositions of the *Principia* in any way significantly different from the way in which they are published with their demonstrations.[21] Newton's studies of optical phenomena, chemistry, theory of matter, physiological and sensational psychology, and other areas of experimental philosophy did not successfully exhibit the Newtonian style. Of course, whatever Newton said about method, or induction, or analysis and synthesis, or the proper role of hypotheses, took on an added significance because of the commanding scientific position of the author. This position was attained as a result of the revolution in science that, in the age of Newton (and afterwards), was conceived to be centered in his mathematical principles of natural philosophy and his system of the world (see Ch. 2). The general philosophical issues of induction, and of analysis and synthesis, gained their importance after Newton had displayed the system of the world governed by universal gravity, but they played no significant role in the way the Newtonian style is used in the elaboration of that system or in the disclosure of that universal force.

1.3   *Mathematics in the new science (1): a world of numbers*

After modern science had emerged from the crucible of the Scientific Revolution, a characteristic expression of one aspect of it was given by Stephen Hales, often called the founder of plant physiology.[1] An Anglican clergyman and an ardent Newtonian, Hales wrote (1727) that 'we are assured that the all-wise Creator has observed the most exact proportions, of *number, weight and measure*, in the make of all things'; accordingly, 'the most likely way . . . to get any insight into the nature of those parts of the creation, which come within our observation, must in all reason be to number, weigh and measure' (Hales, 1969, p. xxxi). The two major subjects to which Hales applied this rule were plant and animal physiology: specifically the measurement of root and sap pressures in different plants under a variety of conditions and the measurement of blood pressure in animals. Hales called his method of enquiry 'statical', from the Latin version of the Greek word for weighing—in the sense that appears to have been introduced into the scientific thought of the West by Nicolaus Cusanus in the fifteenth century, in a treatise entitled *De staticis experimentis* (cf. Guerlac, 1972, p. 37; and Viets, 1922).

In the seventeenth century two famous examples of this 'statical' method were Santorio's experiments on the changes in weight that occurred in the daily life cycle of man (Grmek, 1975), and Van Helmont's experiment on the willow tree. The latter consisted of filling an earthen pot with a weighed quantity of soil that had been dried in a furnace, in which Helmont planted a previously weighed 'Trunk or Stem' of a willow tree. He 'covered the lip or mouth of the Vessel with an Iron plate covered with Tin', so that the dust flying about should not be 'co-mingled with the Earth' inside the vessel. He watered the earth regularly with rain water or distilled water for five years, and found that the original tree, weighing 5 pounds, now had grown to a weight of '169 pounds, and about three ounces' (ignoring 'the weight of the leaves that fell off in the four Automnes'). Since the earth in the vessel, when dried out at the end of the experiment, was only 'about two ounces' less in weight than the original weight of 200 pounds, Helmont concluded that 164 pounds of 'Wood, Barks, and Roots' must have been formed out of water alone.[2] Helmont did not know (or suspect) that the air itself might supply some of the weight of the tree, a discovery made by Hales, who repeated Helmont's experiment with the added precision of weighing the water added to the plant and measuring the plant's rate of 'perspiration' (Hales, 1969, Ch. 1, expts. 1–5). The original of this experiment had been proposed by Cusanus, but there is no certainty as to whether or not he may have actually performed it.

I have purposely chosen these first examples from the life (or biological) sciences, since it is usually supposed that in the Scientific Revolution, numerical reasoning was the prerogative of the physical sciences. One of the most famous uses of numerical reasoning in the Scientific Revolution occurs in Harvey's analysis of the movement of the blood. A central argument in Harvey's demonstration of the circulation is quantitative, based on an estimate of the capacity of the human heart; the left ventricle, he finds, when full may contain 'either 2, or 3, or 1½ oz.; I have found in a dead man above 4 oz.' Knowing that 'the heart in one half hour makes above a thousand pulses, yea in some, and at some times, two, three or four thousand', simple calculation shows how much blood the heart discharges into the arteries in a half hour—at least 83 pounds 4 ounces, 'which is a greater quantity than is found in the whole body'. He made similar calculations for a dog and a sheep. These numbers showed 'that more blood is continually transmitted through the heart, than either the food which we receive can furnish, or is possible in the veins'.[3] Here we may see how numerical calculation provided an argument in support of theory: an excellent example of how numbers entered theoretical discussions in the new science.

Despite the force of the foregoing examples, however, it remains true that the major use of numerical reasoning in the science of the seventeenth century occurred in the exact physical sciences: optics, statics, kinematics and dynamics, astronomy, and parts of chemistry.[4] Numerical relations of a special kind tended to become all the more prominent in seventeenth-century exact science because at that time the laws of science were not yet written in equations. Thus we today would write Galileo's laws of uniformly accelerated motion as $v = At$, and $S = \frac{1}{2} At^2$, but he expressed the essence of naturally accelerated motion (free fall, for example, or motion along an inclined plane) in language that sounds much more like number theory than like algebra: 'the spaces run through in equal times by a moveable descending from rest maintain among themselves the same rule [*rationem*] as do the odd numbers following upon unity'.[5] Galileo's rule, that these first differences (or 'the progression of spaces') accord with the odd numbers, led him to another form of his rule, that the 'spaces run through in any times whatever' by a uniformly accelerated body starting from rest 'are to each other in the doubled ratio of the times [or, as the square of the times]' in which such spaces are traversed. This form of his rule, expressed in the language of ratios, comes closer to our own algebraic expression.[6] Thus while speeds increase with time according to the natural numbers, total distances or spaces traversed increase (depending on the chosen measure) according to the odd numbers or the squares[7] of the natural numbers.[8]

In the exact science of the seventeenth century, considerations of shape, or of geometry, are to be found alongside rules of numbers. In a famous statement about the mathematics of nature, Galileo said:

> Philosophy [i.e., natural philosophy, or science] is written in that vast book which stands forever open before our eyes, I mean the universe; but it cannot be read until we have learnt the language and become familiar with the characters in which it is written. It is written in mathematical language, and the letters are triangles, circles and other geometrical figures, without which means it is humanly impossible to comprehend a single word.[9]

This is not the philosophy of Newton, where mathematics suggests at once a set of equations or proportions (which may be verbal), infinite series, and the taking of limits.[10] In fact, the above-quoted statement almost sounds like Kepler, rather than Galileo. It was Kepler who found in numerical geometry a reason why the Copernican system is to be preferred to the Ptolemaic. In one of these systems—the Ptolemaic—there are seven 'planets' or wanderers (sun and moon; Mercury and Venus; and Mars, Jupiter, and Saturn), but in the other there are only six planets (Mercury and Venus; the earth; and Mars, Jupiter, and Saturn). Suppose that each planet is associated with a giant spherical shell in which it moves (or which contains its orbit). Then there would be five spaces between each pair of successive spheres. Kepler knew of Euclid's proof that there are only five regular solids that can be constructed by simple geometrical rules (cube, tetrahedron, dodecahedron, icosahedron, octahedron). By choosing them in the above order, Kepler found that they would just fit into the spaces between the spheres of the planetary orbits, the only error of any consequence occurring in the case of Jupiter. Hence number and geometry showed that there must be six planets, as in the Copernican system, and not seven, as in the Ptolemaic.[11]

Rheticus, Copernicus's first and only disciple, had proposed a purely numerical argument for the Copernican system. In the sun-centered universe there are six planets, he said, and 6 is the first 'perfect' number (that is, it is the sum of its divisors, $6 = 1 + 2 + 3$).[12] Kepler, however, rejected the perfect-number argument of Rheticus, preferring to base his advocacy of the Copernican system on the five perfect solids. He said:

> I undertake to prove that God, in creating the universe and regulating the order of the cosmos, had in view the five regular bodies of geometry as known since the days of Pythagoras and Plato, and that He has fixed, according to those dimensions, the number of heavens, their proportions, and the relations of their movements.[13]

It is not without interest, accordingly, that when Kepler heard that Galileo had discovered some new "planets" by using a telescope, he was greatly concerned lest his own argument should fall to the ground (cf. Kepler, 1965, p. 10). How happy he was, he recorded, when the "planets" discovered by Galileo turned out to be secondary and not primary "planets", that is, satellites of planets.

Two reactions to Galileo's discovery of four new "planets" may show us that the use of numbers in the exact sciences in the seventeenth century was very different from what we might otherwise have imagined. Francesco Sizi, in opposition to Galileo, argued that there must be seven and only seven "planets"; hence Galileo's discovery was illusory. His assertion about the number seven was based on its occurrence in a number of physical and physiological situations, among them the number of openings in the head (two ears, two eyes, two nostrils, one mouth).[14] Kepler, who applauded Galileo, proposed that he look next for the satellites of Mars and of Saturn, since the numerical sequence of the satellites (one for the earth and four for Jupiter) seemed to demand two for Mars and eight (or possibly six) for Saturn: 1, 2, 4, 8.[15] This type of numerical reasoning had deleterious effects on astronomy in the case of at least one major scientist: Christiaan Huygens. For when Huygens had discovered a satellite of Saturn, he did not bother to look for any further ones. He was convinced, as he boldly declared in the preface to his *Systema Saturnium* (1659), that there could be no others (Huygens, 1888–1950, vol. 15, pp. 212sq). With his discovery of a new satellite, he said, the system of the universe was complete and symmetrical: one and the same "perfect" number 6 in the primary planets and in the secondary "planets" (or planetary satellites). Since his telescope could resolve the ring of Saturn and solve the mystery of its strange and inconstant shape, it could have revealed more satellites had Huygens not concluded that God had created the universe in two sets of planetary bodies, six to a set, according to the principle of "perfect" numbers.[16] Such examples all illustrate some varieties of the association of numbers with actual observations. That we today would not accept such arguments is probably less significant than the fact that those who did included some major founders of our modern science, among them Kepler and Huygens, and Cassini.[17]

1.4   *Mathematics in the new science (2): exact laws of nature and the hierarchy of causes*

In addition to the search for special numbers (odd, prime, perfect, the number of regular solids), which did not always lead to useful results, the scientists of the seventeenth century—like scientists ever since—also sought exact relations between the numbers obtained from measurement, experiment, and observation. An example is Kepler's third (or "harmonic") law. In the Copernican system, each of the planets has a speed that seems related to its distance from the sun: the farther from the sun, the slower the speed. Both Galileo and Kepler were convinced that the speeds and distances could not be arbitrary; there must be some exact relation between these two quantities; for God, in creating the universe, must have had a plan, a law. The Keplerian scheme of the five regular solids imbedded in a nest of spheres showed an aspect of mathematical "necessity" in the distribution of the planets in space, but it did not include the data on their speeds. Thus it only partly satisfied Kepler's goal as a Copernican, expressed by him as follows: 'There were three things in particular, namely, the number, distances and motions of the heavenly bodies, as to which I [Kepler] searched zealously for reasons why they were as they were and not otherwise.'[1]

In the *Mysterium cosmographicum* (1596), in which he had used the five regular solids to show why there were five and only five planets spaced as in the Copernican system, Kepler had also tried to find 'the proportions of the motions [of the planets] to the orbits'. The orbital speed of a planet depends upon its average distance from the sun (and hence the circumference of the orbit) and its sidereal period of revolution, both values given by Copernicus in his *De revolutionibus* (1543) with a reasonably high degree of accuracy. Kepler decided that the 'anima motrix' which acts on the planets loses its strength as the distance from the sun gets greater. But rather than assuming that this force diminishes as the square of the distance (which would mean that it spreads out uniformly in all directions, as light does), Kepler thought it more likely that this force would diminish in proportion to the circle or orbit over which it spreads, directly as the increase of the distance rather than as the square of the increase of the distance. The distance from the sun, according to Kepler, 'acts twice to increase the period' of a planet; for it acts once in slowing down the planet's motion, according to the law by which the force that moves the planet weakens in proportion as the distance increases, and again because the total path along which the planet has to move to complete a revolution increases as the distance from the sun increases. Or, 'conversely half the increase of the period is proportional to the increase of distance'.[2] This relationship comes near to the truth, Kepler observes, but he sought in vain for more than two decades for an accurate relation between the average distance (*a*) of the planets and their periods (*T*). Eventually it occurred to him to use higher powers of *a* and *T*, and on 15 May 1618 he found that the 'periodic times of any two planets are in the sesquialteral [3/2] ratio of their mean distances', that is, the ratio of the squares of the periods is the same as the ratio of the cubes of their average distances, a relationship which we express as $a^3/T^2 = $ const. and call Kepler's third law.[3] It should be noted that Kepler's discovery apparently resulted from a purely numerical exercise and insofar differed from his discovery of the area law and of the law of elliptic orbits, both of which were presented originally (and may have been discovered) in association with a definite causal concept of solar force and a principle of force and motion.[4]

Galileo's approach to this problem was based on a kinematical law rather than on purely numerical considerations: the principle of naturally accelerated motion, which he had discovered in his studies of freely falling bodies.[5] He thought so well of his solution to the cosmic problem that he introduced it into both his *Dialogo* (1632) or the *Dialogue Concerning the Two Chief World Systems*, and his *Discorsi* (1638), or the *Two New Sciences* (Galileo, 1953, pp. 29sq; 1890–1909, vol. 4, pp. 53sq; also 1974, pp. 232–234; 1890–1909, vol. 8, pp. 283sq). He attributed the basic idea to Plato, but there is nothing even remotely resembling it in any of the Platonic works, nor has this idea been found in any known Neoplatonic composition or commentary, ancient, medieval, or modern.[6] Galileo said that there was a point out in space from which God had let fall all of the planets. When each planet arrived at its proper orbit, it would have attained its proper orbital speed and would have needed only to be turned in its path to accord with the known values of planetary distances and speeds. Galileo did not specify where that point is located, and (as an analysis by Newton revealed) the point would in fact have to be infinitely far away.[7] Galileo, furthermore, did not understand that such a descent toward the sun would require a constantly changing acceleration, which in dynamics would correspond to a constantly changing solar–planetary force that varies inversely as the square of the distance. In this example we may see that Galileo could have had no conception of a solar gravitating force. His discussion does not contain the slightest hint of a relation between force and acceleration that might be said to have contained a germ of Newton's second law of motion.[8]

Galileo was primarily successful in applying mathematics to such areas as statics and kinematics, in both of which there is no need to take account of physical causes, such as quantifiable forces. As he himself says, in his *Two New Sciences*:

> The present does not seem to me to be an opportune
> time to enter into the investigation of the cause of the
> acceleration of natural motion, concerning which various
> philosophers have produced various opinions. . . . For
> the present, it suffices . . . that we . . . investigate and
> demonstrate some attributes of a motion so accelerated
> (whatever be the cause of its acceleration) that the
> momenta of its speed go increasing . . . in that simple
> ratio with which the continuation of time increases. . .
> [Galileo, 1974, pp. 158sq; 1890–1909, vol. 8, p. 202].

In part, but only in part, his procedure resembles that of the late medieval kinematicists. Like them, he defines uniform motion and then proceeds to uniformly accelerated motion. Almost at once, he reveals the mean-speed law: In uniformly accelerated motion during a time t the distance traveled is the same as if there had been uniform motion at the mean value of the changing speeds during that same time (Galileo, 1974, p. 165; 1890–1909, vol. 8, p. 208). Since the motion is uniform, the mean value is one half of the sum of the initial and final speeds. If we may somewhat anachronistically translate Galileo's verbal statements of ratios into their equivalent equations, we may show that he has proved that $s = \bar{v}t$ where $\bar{v} = (v_1 + v_2)/2$. Since $v_2 = v_1 + At$, it would follow at once that $s = v_1 t + \frac{1}{2}At^2$. In the special case of motion starting from rest, $v_1 = 0$ and $s = \frac{1}{2}At^2$.

Thus far, except for the final result (in which the relation $s = [(v_1 + v_2)/2]t$ leads to $s = vt + \frac{1}{2}At^2$), Galileo could be proceeding much like his fourteenth-century predecessors.[9] But there are significant differences of such consequence that we may easily discern in Galileo's *Two New Sciences* the beginnings of our own science of motion, whereas this feature is lacking in the medieval treatises. The major difference is that the writers of the fourteenth century were not concerned with the physics of motion, with nature as revealed by experiment and observation. Thus they constructed a generalized "latitude of forms", a mathematico-logical analysis of any quality that can be quantified, of which motion (in the sense of "local" motion, from one place to another) is but one example, along with such other quantifiable qualities as love, virtue, grace, whiteness, hotness, and so on. Even in the case of motion, they were dealing with Aristotelian "motion", defined in very general terms as the transition from actuality to potentiality. For two centuries, there is no record of any scholar ever applying the principles of uniform and accelerated motion to actual motions as observed on earth or in the heavens. Prior to Galileo, only Domingo de Soto made such an application, and he appears as a *lusus naturae* of no real importance (see §1.1, n. 13).

How different it is with Galileo! He based his very definitions on nature herself. His aim was not to study motion in the abstract, but the observed motions of bodies. The true test of his mathematical laws (as $s = \frac{1}{2}At^2$) was not their logical consistency but their conformity with the results of actual experimental tests. Thus much is said in the public record.[10] But now we know additionally, thanks to the studies of Galileo's manuscripts by Stillman Drake, that Galileo was making experiments not only to relate his discovered laws to the world of nature, but also as part of the discovery process itself.

Galileo's laws of the uniform and uniformly accelerated motion of physical bodies were derived by mathematics from sound definitions, guided to some degree by experiment, but without consideration as to the nature of gravity or the cause of motion. The concept of a physical cause did enter his analysis of projectile motion, however, but only to the limited extent of establishing that the horizontal component of the motion is not accelerated while the vertical component is. Galileo recognized that there is a force of gravity producing a downward acceleration, but that in the horizontal direction the only force that can affect the projectile's motion is the resistance of the air, which is slight (Galileo, 1974, pp. 224–227; 1890–1909, vol. 8, pp. 275–278). But he did not analyze the cause of acceleration to any degree further than being aware that acceleration requires a cause in the form of some kind of downward force. That is, he did not explore the possibility that the gravitational accelerating force may be caused by the pull of the earth on a body, or by something pushing the body toward the earth; nor whether such a cause or force is external or internal to a body; nor whether the range of this force is limited and, if so, to what distance (as far as the moon, for example); nor whether this force is constant all over the earth's surface; nor whether gravity may vary with distance from the earth's center. Galileo eschewed the search for causes, describing most causes assigned to gravity as 'fantasies' which could be 'examined and resolved' with 'little gain'. He said he would be satisfied 'if it shall be found that the events that . . . shall have been demonstrated are verified in the motion of naturally falling and accelerated heavy bodies' (Galileo, 1974, p. 159; 1890–1909, vol. 8, p. 202). In this point of view, as Stillman Drake has wisely remarked, Galileo was going against the main tradition of physics, which had been conceived as 'the study of natural motion (or more correctly, of change) in terms of its causes'. Drake would thus see 'Galileo's mature refusal to enter into debates over physical causes' as epitomizing 'his basic challenge to Aristotelian physics' (Galileo, 1974, editorial introduction, pp. xxvi–xxvii). As we shall see below, however, there is a middle ground between a study of physical or even metaphysical causes and the mathematical elucidation of their actions and prop-

erties. The recognition of this hierarchy and the exploration of
the properties of gravity as a cause of phenomena (without any
overt commitment to the cause of gravity) was a great advance over
the physics of Galileo and may be considered a main feature of the
Newtonian revolution in science (see Ch. 3).

Thus in the exact sciences of the seventeenth century we may
observe a hierarchy of mathematical laws. First, there are mathematical laws deduced from certain assumptions and definitions,
and which lead to experimentally testable results. If, as in Galileo's
case, the assumptions and definitions are consonant with nature,
then the results should be verifiable by experience. When Galileo
sets forth, as a postulate, that the speed acquired in naturally accelerated motion is the same along all planes of the same heights,
whatever their inclination, he declares that the 'absolute truth' of
this postulate 'will be later established for us by our seeing that
other conclusions, built on this hypothesis, do indeed correspond
with and exactly conform to experiment'. This reads like a classic
statement of the hypothetico-deductive method; but it is to be
observed that it is devoid of any reference to the physical nature of
the cause of the acceleration. Such a level of discourse is not essentially different in its results from another seventeenth-century way
of finding mathematical laws of nature without going into causes:
by the direct analysis of the data of experiment and observation.
We have seen this to have been in all probability Kepler's procedure in finding his third (or "harmonic") law of planetary motion.
Other examples are Boyle's law of gases and Snel's law of refraction
(see Mach, 1926, pp. 32–36; Sabra, 1967; Hoppe, 1926, pp. 33sq).

The second level of the hierarchy is to go beyond the mathematical description to some sort of causes. Boyle's law, for example, is
a mathematical statement of proportionality between two variables, each of which is a physical entity related to an observable or
measurable quantity. Thus the volume (V) of the confined gas is
measured by the mercury level according to some volumetric scale,
and the pressure of the confined gas is determined by the difference
between two mercury levels (h) plus the height of the mercury
column in a barometer (h_1). Boyle's experiments showed that the
product of V and $h + h_1$ is a constant. The sum $h + h_1$ is a height
(in inches) of a mercury column equivalent to a total pressure exerted on and by the confined gas; what is measured directly in this
case is not the pressure but a quantity (height of mercury) which

itself is a measure of (and so can stand for) pressure. But nothing is said concerning the cause of pressure in a confined gas, nor of the reason why this pressure should increase as the gas is confined into a smaller volume, a phenomenon known to Boyle before he undertook the experiments and which he called the "spring" of the air. Now the second level of hierarchy is to explore the cause of this "spring". Boyle suggested two physical models that might serve to explain this phenomenon. One is to think of each particle being itself compressible, in the manner of a coiled spring or a piece of wood, so that the air would be 'a heap of little bodies, lying one upon another, as may be resembled to a fleece of wool'. Another is to conceive that the particles are in constant agitation, in which case 'their elastical power is not made to depend upon their shape or structure, but upon the vehement agitation'. Boyle himself, on this occasion, did not choose to decide between these explanations or to propose any other (see Cohen, 1956, p. 103; Boyle, 1772, vol. 1, pp. 11sq). But the example does show that in the exact or quantitative sciences of the seventeenth century, there was a carefully observed distinction between a purely mathematical statement of a law and a causal mechanism for explaining such a law, that is, between such a law as a mathematical description of phenomena and the mathematical and physical exploration of its cause.

In some cases, the exploration of the cause did not require such a mechanical model, or explanation of cause, as the two mentioned by Boyle. For example, the parabolic path of projectiles is a mathematical statement of a phenomenon, with the qualifications arising from air resistance. But the mathematical conditions of a parabola are themselves suggestive of causes: for—again with the qualifications arising from air resistance—they state that there is uniform motion in the horizontal component and accelerated motion in the downward component. Since gravity acts downward and has no influence in the horizontal component, the very mathematics of the situation may lead an inquirer toward the physical causes of uniform and accelerated motion in the parabolic path of projectiles. Similarly, Newton's exploration of the physical nature and cause of universal gravity was guided by the mathematical properties of this force: that it varies inversely as the square of the distance, is proportional to the masses of the gravitating bodies and not their surfaces, extends to vast distances, is null within a uniform spherical shell, acts on a particle outside of a uniform spherical shell (or a body made up of a series of uniform spherical shells) as if the mass of that shell (or body made up of shells) were concentrated at its geometric center, has a value proportional to the distance from the center within a uniform sphere, and so on.

Such mathematical specifications of causes are different from physical explanations of the origin and mode of action of causes. This leads us to a recognition of the hierarchy of causes which it is important to keep in mind in understanding the specific qualities of the Newtonian revolution in science. For instance, Kepler found that planets move in ellipses with the sun at one focus, and that a line drawn from the sun to a planet sweeps out equal areas in equal times. Both of these laws encompass actual observations within a mathematical framework. The area law enabled Kepler to account for (or to explain) the nonuniformity of the orbital motion of planets, the speed being least at aphelion and greatest at perihelion. This is on the level of a mathematical explanation of the nonuniform motion of planets. Kepler, however, had gone well beyond such a mathematical explanation, since he had assigned a physical cause for this variation by supposing a celestial magnetic force; but he was never successful in linking this particular force mathematically to the elliptical orbits and the area law, or in finding an independent phenomenological or empirical demonstration that the sun does exert this kind of magnetic force on the planets (see Koyré, 1973, pt. 2, sect. 2, ch. 6; Aiton, 1969; Wilson, 1968).

Newton proceeded in a different manner. He did not begin with a discussion of what kind of force might act on planets. Rather he asked what are the mathematical properties of a force–whatever might be its causes or its mode of action, or whatever kind of force it might be–that can produce the law of areas. He showed that, for a body with an initial component of inertial motion, a necessary and sufficient condition for the area law is that the said force be centripetal, directed continually toward the point about which the areas are reckoned. Thus a mathematically descriptive law of motion was shown by mathematics to be equivalent to a set of causal conditions of forces and motions. Parenthetically it may be observed that the situation of a necessary and sufficient condition is rather unusual; most frequently it is the case that a force or other "cause" is but a sufficient condition for a given effect, and in fact only one of a number of such possible sufficient conditions. In the *Principia* the conditions of central forces and equal areas in equal times lead to considerations of elliptical orbits, which were shown by Newton to be a consequence of the central force varying inversely as the square of the distance (see Ch. 5).

Newton's mathematical argument does not, of course, show that in the orbital motion of planets or of planetary satellites these bodies are acted on by physical forces; Newton only shows that within the conceptual framework of forces and the law of inertia, the forces acting on planets and satellites must be directed toward a center and must as well vary inversely as the square of the distance. But in the hierarchy of causal explanation, Newton's result does finally direct us to seek out the possible physical properties and mode of action of such a centrally directed inverse-square force.[11] What is important in the Newtonian mode of analysis is that there is no need to specify at this first stage of analysis what kind of force this is, nor how it acts. Yet Newton's aim was ultimately to go on by a different mode of analysis from the mathematical to the physical properties of causes (or forces) and so he was primarily concerned with 'verae causae', causes—as he said—that are 'both true and sufficient to explain the phenomena'.[12]

This hierarchy of mathematical and physical causes may be seen also in Newton's analysis of Boyle's law, that in a confined gas (or "elastic fluid", as it was then called) the pressure is inversely proportional to the volume. We have seen that Boyle himself suggested two alternative physical explanations of the spring of the air in relation to his law, but declined to declare himself in favor of either of them. In the *Principia* (as we shall see in §3.3) Newton showed that, on the supposition that there is a special kind of force of mutual repulsion between the particles composing such an "elastic fluid", Boyle's law is both a necessary and sufficient condition that this force vary inversely as the distance. Again there is a hierarchy of mathematical and physical analyses of cause. In this second Newtonian example, it is more obvious that the physical conditions assumed as the cause of the law are themselves open to question. Newton himself concluded his discussion of this topic (*Principia*, bk. two, prop. 23) by observing that it is 'a physical question' as to 'whether elastic fluids [i.e., compressible gases] do really consist of particles so repelling one another'. He himself had been concerned only with the mathematical demonstration, so he said, in order that natural philosophers (or physical scientists) might discuss the question whether gases may be composed of particles endowed with such forces. With regard to the hierarchy of mathematical and physical causes, there is of course no real formal difference between the Newtonian analysis of Kepler's laws and of Boyle's law. In the case of Kepler's laws, however, Newton could take the law of inertia for granted, as an accepted truth of the new science, so that there would have to be some cause for the planets to depart from a rectilinear path and to trace out an elliptical orbit. If this cause is a force, then it must be directed toward a point (the sun, in the case of the planets), since otherwise there can be no area law. But in the case of compressible gases or elastic fluids, the situation is somewhat different. In the first place, in Newton's mind there was no doubt whatsoever that such 'elastic fluids do really consist of particles', since he was a firm believer in the corpuscular philosophy; but it is to be observed that there were many scientists at that time who, like the followers of Descartes, believed in neither atoms nor the void. But even if the particulate structure of gases could be taken for granted, there was the additional property attributed to such particles by Newton, that they be endowed with forces which enable them to repel one another. Many of those who believed in the "mechanical philosophy" and accepted the doctrine of particularity of matter would not necessarily go along with Newton's attribution of forces to such particles, whether atoms, molecules, or other forms of corpuscles. Furthermore, as Newton makes plain in the scholium which follows his proposal of an explanatory physical model for Boyle's law, 'All these things are to be understood of particles whose centrifugal forces terminate in those particles that are next to them, and are diffused not much further.' Accordingly, there is a great and wide gulf between the supposition of a set of mathematical conditions from which Newton derives Boyle's law and the assertion that this is a physical description of the reality of nature. As will be explained in Ch. 3, it is precisely Newton's ability to separate problems into their mathematical and physical aspects that enabled Newton to achieve such spectacular results in the *Principia*. And it is the possibility of working out the mathematical consequences of assumptions that are related to possible physical conditions, without having to discuss the physical reality of these conditions at the earliest stages, that marks the Newtonian style.

The goal of creating an exact physical science based on mathematics was hardly new in the seventeenth century. O. Neugebauer has reminded us that Ptolemy, writing in the second century A.D., had declared this very aim in the original title of his great treatise on astronomy, which we know as the *Almagest*, but which he called 'Mathematical Composition' (or 'Compilation') (Neugebauer, 1946, p. 20; cf. Neugebauer, 1948, pp. 1014–1016). But there was a fundamental difference between the old and the new mathematical physical science, which may be illustrated by an aspect of planetary theory and the theory of the moon.

It is well known that in the *Almagest* Ptolemy was concerned to produce or develop geometric models that would serve for the computation of the latitudes and longitudes of the seven "planetary bodies" (the five planets plus sun and moon) and hence would yield such special information as times of eclipses, stationary points, conjunctions and oppositions. These were quite obviously mathematical models, and were not intended to partake of physical reality. Thus there was no assumption that the true motion of these planetary bodies in the heavens necessarily is along epicycles moving around on deferents and controlled by equal angular motion about an equant. In particular, Ptolemy was perfectly aware that his order of the planets (in terms of increasing distance from the earth: moon, Mercury, Venus, sun, Mars, Jupiter, Saturn) was somewhat arbitrary for the five "planets", since their distances cannot be determined by parallaxes. In fact, Ptolemy admits that some astronomers would place Mercury and Venus beyond the sun, while others would have Mercury on one side of the sun and Venus on the other.[13] Again, in the theory of the moon, Ptolemy introduced a "crank" mechanism, which would increase the 'apparent diameter of the epicycle' so as to make the model agree with positional observations. As a result Ptolemy was able to make an accurate representation of the moon's motion in longitude, but only at the expense of introducing a fictitious variation in the distance of the moon from the earth, according to which 'the apparent diameter of the moon itself should reach almost twice its mean value, which is very definitely not the case' (Neugebauer, 1957, p. 195). This departure from reality was one of the telling points of criticism raised by Copernicus in his *De revolutionibus* (1543). Descartes also proposed hypothetical models that, according to his own system, had to be fictitious.

Newton believed that he had proved that gravity, the cause of terrestrial weight and the force producing the downward acceleration of freely falling bodies, extends as far as the moon and is the cause of the moon's motion. He gave a series of arguments that it is this same force that keeps the planets in their orbits around the sun and the satellites in their orbits about their respective planets. He showed, furthermore, how this force of gravity can account for the tides in the seas and the irregularities (as well as the regularities) in the moon's motion. He set forth the goal of explaining the moon's motion in a new way—not by celestial geometry and models which (like Ptolemy's) obviously cannot possibly correspond to reality, but rather by 'true causes' ('verae causae') whose properties could be developed mathematically. Thus Newtonian theory would reduce the features of the moon's motion to two sources: the interactions of the earth and the moon, and the perturbing effects of the sun. It is to be remarked that this procedure does not depend on the origin, nature, or physical cause of the gravitating force but only on certain mathematically elucidated properties, as that this force varies inversely as the square of the distance, that it is null within a spherical shell (or within a homogeneous sphere or a sphere made up of homogeneous shells), that the action of a sphere on an external particle is the same as if all the mass of the sphere were concentrated at its geometric center, that within a solid sphere the force on a particle is as the distance from the center, and so on. Such investigations did not depend on whether a planet is pushed or pulled toward the center, whether gravitation arises from an aether of varying density or a shower of aether particles or is even a simple action-at-a-distance. For Newton these latter questions were far from irrelevant to a complete understanding of the system of the world, and we know that he devoted considerable energy to them. Furthermore, the mathematical analysis had revealed some of the basic properties of the force and thus made precise the analysis into its cause. But in Newton's hierarchy of causes, the elucidation of the properties of universal gravity was distinct from—that is, on a different level from—the search for the cause of gravity. He thus put forth the radical point of view in the concluding General Scholium to the *Principia*: It is enough ('satis est') that gravity exists and that it acts according to the laws he had mathematically demonstrated, and that this force of gravity suffices 'to explain all the motions of the heavenly bodies and of our sea' (see n. 12 *supra*). How revolutionary this proposal was can be seen in the number of scientists and philosophers who refused to accept it and who rejected the *Principia* together with its conclusions because they did not approve of the concept of "attraction".

1.5   *Causal mathematical science in the Scientific Revolution*

In the last section, an outline was given of a hierarchy in the mathematical science of nature. On the lowest or most primitive level, this phrase may mean no more than mere quantification or calculation. Numerical data may provide arguments to test or to buttress essentially nonmathematical theories such as Harvey's. On a simple level, primarily in the realms of physics and astronomy, mathematics came to signify not only the measurements of positions and apparent (observed) angular speeds, and the rather straightforward application of plane and spherical trigonometry to the solutions of problems of the celestial sphere, but also the increasing quantification of qualities ranging from temperature to speeds. The ideal was to express general laws of nature as mathematical relations between observable physical quantities, notably in relation to the science of motion: kinematics and then dynamics. Such laws expressed number-relations or geometrical properties, and they were formalized in ratios or proportions, algebraic equations (or their equivalents in words), together with geometric properties and trigonometric relations, and eventually the infinitesimal calculus and other forms of higher mathematics, notably infinite series.

Since such mathematical laws use physically observable quantities (volume, weight, position, angle, distance, time, impact, and so on), they can to a large degree be tested by further observations and direct experiment, which may limit the range in which they hold: examples are Boyle's, Snel's, and Hooke's law, and the forms of Kepler's law of refraction.[1] Or, the test may be the verification or nonverification of a prediction (as the occurrence or nonoccurrence of a lunar or solar eclipse or a particular planetary configuration), or the accurate retrodiction of past observations. Obviously, some kind of numerical data must provide the basis for applying or testing such general or specific mathematical laws or relations. In all of this, there is and there need be no concern for physical causes. Galilean science is a foremost example of the successful application of mathematics to physical events on this level. Cause enters in the argument only to the extent of an awareness that air resistance may cause a slowing down of an otherwise uniform rectilinear motion (or component of motion), and that weight may cause an acceleration downward. Thus, for Galileo, motion could continue uniformly in a straight line only if there were no air resistance and if there were an extended horizontal plane to support the mobile, on which it could move without friction.[2]

We have seen, however, that in the seventeenth century there were found to be significant quantitative laws that cannot be directly tested, such as the law of uniform motion for falling bodies that speeds acquired are as the times elapsed ($v_1{:}v_2 = t_1{:}t_2$). Galileo, as we saw, could do no more than confirm another law of falling bodies, that the distances are in the squared ratio of the times ($s_1{:}s_2 = (t_1{:}t_2)^2$); and then, since the distance law is a consequence of the speed law, he supposed that the truth (verified by experiment) of the distance law guaranteed the truth of the speed law. In our modern language, the testability of $s \propto t^2$ is the way to confirm $v \propto t$. This is a classic and simple example of what has generally come to be called the hypothetico-deductive method. Galileo tested the distance–time relation for the accelerated motion upon an inclined plane, for various angles of inclination, and showed that s does maintain a constant proportion to t^2. Since this relation was an inference (or deduction) from an assumption (or hypothesis) that v is proportional to t, the hypothetico-deductive method assumes that experimental confirmation of the deduced result $s_1{:}s_2 = t_1{}^2{:}t_2{}^2$ guarantees the validity of the hypothesis $v_1{:}v_2 = t_1{:}t_2$ from which the relation of s to t^2 had been deduced (see §1.4). As Ernst Mach (1960, p. 161) put it, in his celebrated *Science of Mechanics*, 'The inference from Galileo's assumption was thus confirmed by experiment, and with it the assumption itself.' The limitations to this mode of confirmation are twofold. One is philosophical: are there any ways to be sure that only $v \propto t$ implies $s \propto t^2$? That is, granted that $v \propto t$ is a sufficient condition for $s \propto t^2$, is it also a necessary condition?[3] The second is historical as well as philosophical. That is, a scientist may make an error in logic or mathematics: This is illustrated by the fact that at one stage in his career Galileo believed that the verifiable relation $s_1{:}s_2 = (t_1{:}t_2)^2$ follows from the speeds being proportional to distances ($v_1{:}v_2 = s_1{:}s_2$) rather than the relation of speeds to times ($v_1{:}v_2 = t_1{:}t_2$) (see Galileo, 1974, pp. 159sq; 1890–1909, vol. 8, p. 203).

The Galilean science of motion embodies only one part of the revolution in the exact sciences in the seventeenth century. For, in addition to the production of exact or mathematical laws, systems, and general constructs that may or may not be like models that conform to the direct experience of nature (experiment and observation), there arose the ideal of finding the true physical causes of such laws, systems, constructs, and models, in a hierarchy of causes that began with the mathematical elucidation of the properties of forces causing motions and only then proceeded to the analysis of the nature and cause of such forces.[4] The degree to which this goal was first achieved in Newton's *Principia* set the seal on an accomplished Scientific Revolution and was in and of itself revolutionary. Lest my readers should suppose that this is an anachronistic judgment of the twentieth century superimposed upon the events of the past, let me anticipate here one aspect of the next chapter, by indicating that this was an unequivocal judgment in the Age of Newton. Clairaut, Newton's immediate intellectual successor in celestial mechanics, declared unambiguously in 1747, 'The renowned treatise of Mathematical principles of Natural Philosophy [of Isaac Newton] inaugurated a great revolution in physics', a sentiment echoed by Lagrange and others (Clairaut, 1749; see §2.2).

The program for this revolution in physical science was first clearly set forth in astronomy, in the declared goal to put aside all noncausal and nonphysical computing schemes and to discover how the sun, moon, and planets actually move in relation to the physical ("true") causes of their motions. This aspect of the revolution found its first major spokesman in Kepler, whose *Astronomia nova* (1609), or *Commentary on the Motion of Mars*, was also described by him as a 'physica coelestis', a celestial physics (see Caspar, 1959, pp. 129sqq; Koyré, 1973, pp. 166sqq, 185sqq). What made this work 'new' was that it was not merely an *Astronomia nova*, but an *Astronomia nova* αἰτιολογητος, a 'new astronomy based on causes'; and this was the sense in which Kepler declared it to be a 'celestial physics'.[5] That is, Kepler was not content with the limited goal of previous astronomers (including Ptolemy, Copernicus, and Tycho Brahe) of choosing a fitting center of motion and then determining planetary motions by judicious combinations of circular motions that would 'save the phenomena' (cf. Duhem, 1969). He wanted to derive planetary motions from their causes, from the forces that are the causes of the motions. He thus rejected one of the basic aspects of Copernican astronomy: that planetary orbits be computed with respect to an empty point in space corresponding to the center of the earth's orbit, rather than the sun itself. Kepler reasoned that forces originate in bodies, not in points in space; hence the motion of the planets must be reckoned in relation to the center of planetary force, the central body, the sun. As a result, Kepler attempted a dynamical rather than a kinematical astronomy, based on laws of force and motion rather than on applied geometry and arithmetic (see Koyré, 1973; Cohen, 1975a; Beer & Beer, 1975, sect. 10). Certain of Kepler's fellow astronomers disapproved of his thus introducing into astronomy a set of physical causes and hypotheses; it were better (said his former teacher, Michael Maestlin) to stick to the traditional geometry and arithmetic (letter to Kepler, 21 Sept. 1616; Kepler, 1937–, vol. 17, p. 187). Of course, it was easier to effect this radical change in Kepler's day than it would have been earlier, since Tycho Brahe had effectively demonstrated that comets move through the solar system. As Tycho himself put it: Had there ever existed crystalline spheres to which the planets were attached, they were now shattered and existed no longer. Hence, for anyone who went along with Tycho's conclusions, there was need for a wholly new scheme for explaining how the planets can possibly move in their observed curved paths.[6]

And so we are not surprised to find that Descartes also sought for causal explanations of the celestial motions, and so did certain other astronomers of the early seventeenth century, such as Bullialdus and Borelli.[7] But others were content to confine their attention more nearly to the phenomenological level of prediction and observation, without exhibiting any concern as to physical causes, or the possible reality (or lack of reality) of geometric computational schemes. From this point of view, one of the most astonishing aspects of Galileo's *Dialogue concerning the Two Chief World Systems* is the absence of any celestial physics. Galileo, in fact, seems never to have concerned himself with any speculations on the possible forces that might be acting in the operation of a Copernican system.[8] In this sense, Galileo was not at all a pioneer in celestial mechanics, as Kepler and Descartes were, however much his personal contributions to the science of motion influenced the course of development of theoretical dynamics at large. But he was con-

cerned with the truth and reality of the Copernican system, and
he even advanced an explanation of the tides that seemed to him
to require that the earth rotate about its axis while revolving
around the sun.

The enormous advance in the exact physical sciences in the seventeenth century may be gauged by the gap between both Galileo's
kinematics and Kepler's faulty and unsuccessful dynamics[9] on the
one hand and Newton's goal of a mathematical dynamics congruent with the phenomenological kinematical laws and disclosing
their physical cause on the other. Kepler, who in so many of his
precepts resembles Newton, nevertheless represents a wholly different level of scientific belief and procedure. Kepler starts out
from the causes, whereas Newton concludes in causes. Kepler accepts a kind of celestial attraction, based on an analogy with terrestrial magnetism, and seeks its consequences; Newton arrives at
his concept of universal gravitation only after the logic of the study
of forces and motions leads him in that direction (see Ch. 5). Newton's philosophy directs him from effects to causes, and from par-
ticulars to generalities. But Kepler believed it best to proceed in
the reverse direction. 'I have no hesitation', he wrote, 'in asserting
that everything that Copernicus has demonstrated *a posteriori* and
on the basis of observations interpreted geometrically, may be
demonstrated *a priori* without any subtlety or logic'.[10]

Like Galileo's laws of falling bodies, Kepler's laws were shown
by Newton to be true only in rather limited circumstances which
Newton then actually specified. Newton sought to determine new
forms of these laws that would be more universally true. As we
shall see in Ch. 3, the revolutionary power of Newton's method
came from his ability to combine new modes of mathematical
analysis with the study of physical causes, controlled constantly by
the rigors of experiment and observation. But an essential ingredient was Newton's clear recognition of the hierarchy of causes and
his ability to separate the mathematical laws from the physical
properties of forces as causes. In this process he did not produce
mere mathematical constructs or abstractions that were devoid of
any content of reality other than "saving the phenomena", but he
did create what he conceived to be purely mathematical counterparts of simplified and idealized physical situations that could later
be brought into relation with the conditions of reality as revealed
by experiment and observation. It was this aspect of Newtonian

science, in my opinion, that produced so outstanding a result that his *Principia* was conceived to have been or to have inaugurated the epoch of a revolution in science, or at least to have brought to a level of revolutionary fruition the goals of creating a mathematical science of nature that had been expressed, however imperfectly, by Galileo and by Kepler.

2

Revolution in science and the Newtonian
revolution as historical concepts

2.1   *The concept of revolution*

Many historians of science believe that the concept of revolution in science is of fairly recent origin, but I have found that
during some three centuries there has been a more or less unbrok-
en tradition (though by no means shared by all scientists) of viewing scientific change as a sequence of revolutions. In the eighteenth
century, when this tradition appears to have taken its first rise, the
word "revolution" continued to be used, as in the past, as a technical term in mathematics and astronomy; but it also gained currency in a general sense in two very distinct meanings, both of
which are found in writings about scientific change and in historical accounts of political events. Of these, one which came into general usage during the eighteenth century denotes a breach of continuity or a secular (i.e., noncyclical) change of real magnitude,
usually accompanied—at least in political events—by violence. The
other is the older sense, used in relation both to the history of
science and the history of political events, signifying a cyclical phenomenon, a continuous sequence of ebb and flow, a kind of circulation and return, or a repetition. After 1789, the new meaning
came to predominate, and ever since, "revolution" has commonly
implied a radical change and a departure from traditional or accepted modes of thought, belief, action, social behavior, or political
or social organization.[1] Thus in early modern times there occurred
a double transformation of "revolution" and the concept for which
it is the name. First, a scientific term, taken from astronomy and
geometry, came to be applied to a general range of social, political,
economic, and intellectual or cultural activities; second, in this
usage the term gained a new meaning that was radically different
from, if not diametrically opposite to, the original and strict ety-

mological sense of "revolution" (*révolution, rivoluzione*), which is derived from the medieval Latin *revolutio*, a rolling back or a return, usually with an implied sense of revolving in time.[2]

During the eighteenth century, the point of view emerged that scientific change is characterized by an analogue of the revolutions that alter the forms of society and the political affairs of the state (see Cohen, 1976a): now visualized as a series of secular discontinuities of such magnitude as to constitute definite breaks with the past. A revolution no longer implied a cyclical continuous process, or an ebb and flow, or a return to some better or purer antecedent state. The notion of revolution in science, in the new sense of a single dramatic change producing something new, has been part of science's historiography ever since the opening years of the eighteenth century, and has been constantly influenced by the development of concepts and theories of political and social (and cultural) revolution.

One possible link between the original cyclical meaning and today's common usage of "revolution"—for a "complete change of affairs" or a "reversal of conditions", an overthrowing (usually accompanied by violence) of established government or society and institutions—lies in the close association of a cyclical "turning-over" and a secular "overturning". Today, the associated verb used to denote cyclical phenomena is "revolve"; whereas the verb "revolt" implies an uprising against the political state or social order. Both "revolve" and "revolt" come from the same verb: *revolvere, revolutus*. In the eighteenth century, prior to 1789, both of these distinct and very different senses of "revolution" are apt to occur together (even in the same work) in discussions of history and politics as well as the course of development in literature, the arts, and the sciences. Accordingly, it is not always a simple task to discover whether a given eighteenth-century author may have had in mind a cyclical return (an ebb and flow) or a secular change of a significant magnitude (often, but not necessarily, accompanied by violence). This ambiguity was particularly a feature of the years between the English revolutions of the seventeenth century and the American and French revolutions: the era of the Newtonian revolution in science and of the emergence of the concept of revolution as a mode of scientific change.

There is one term, however, whose usage generally enables the modern (i.e., post-1789) reader to distinguish between the two senses of "revolution", that is, the word "epoch". Thus there is no ambiguity whatsoever in Clairaut's blunt assertion in 1747, coupling "epoch" and the "revolution" produced by Newton's *Principia*.[3] Here "epoch" is not used in the current meaning of an era or an age (the primary sense in American English), but rather signifies an event that inaugurates a new age or that is the initial or major occurrence of or in a revolution: the beginning of a new era, as in "epoch-making". Often, in the late seventeenth and in the eighteenth century, this word appears in its Late Latin form as *epocha*, in historical and political writings and in scientific works (see §2.2).

The Glorious Revolution apparently was of chief importance in the development of the concept of revolution between 1688 and 1789, as it gradually became more and more evident that there *had* been a revolution in England, possibly the first true revolution in the modern era. In Samuel Johnson's *Dictionary of the English Language* (1755), this revolution appears in the third definition of this term: 'Change in the state of a government or country. It is used among us . . . for the change produced by the admission of king William and queen Mary'. The Glorious Revolution may not seem as revolutionary to us—with our cognizance of such cataclysms as the French, Russian, and Chinese revolutions—as it did to the men and women of the eighteenth century, for whom it was the first true revolution, in a modern sense. But to such different men as Joseph Priestley and David Hume, it was indeed a revolution, and a rather glorious one at that (see Cohen, 1976*a*, esp. p. 263, n. 17).

The word "revolution" had already achieved its new (noncyclical) sense by the time of the great *Encyclopédie*, where it signifies a secular political change of great magnitude and is even applied specifically to advances in the sciences. Under the entry "Révolution", first place is assigned to political revolutions, the cyclic scientific phenomena appearing only toward the end:

> RÉVOLUTION means, when used as a political term,
> a major change which occurs in the government of a state.
> This *word* comes from the Latin *revolvere*, to roll.
> There are no states which have not undergone a revolution of one sort or another. The Abbé Vertot has given us several excellent histories of revolutions in various countries . . .

There follows a long note (by *D.J.* = Chevalier de Jaucourt) on British history, with the observation that 'the English have applied this term especially to the revolution of 1688, in which the Prince of Orange . . . gained the throne'.[4] Thus the *Encyclopédie* displays its modernity in the priority given to the new political sense of the word "revolution" at the expense of the classical and cyclical original sense, as found in geometry and astronomy. Even more notable in relation to the concept of revolution is the fact that in their separate contributions both Diderot and d'Alembert wrote of revolutions in the sciences.

2.2   *The introduction of the concept of revolution to describe scientific progress*

Although it might seem that the development of science from the time of Copernicus and Vesalius to the end of the seventeenth century would have been then described in terms of radical changes, if not revolutions, this does not seem to have been the case at all. It is not merely that there are no specific references to "revolutions" in the sciences earlier than 1700, but rather that those who were writing about the sciences, even when referring to the "newness" of the experimental sciences then being forged, did not generally consider that their science had made so complete a break with tradition as to constitute what we today would think of as a revolution. Many scientists, of course, were aware that they were doing something new. For example, Galileo's last great treatise is called *Two New Sciences,* and William Gilbert declared that his book was intended only for those "new" philosophers who sought truth in nature herself and not in books. But even such pioneers often tended to assume that they were going back to the science of antiquity (by-passing the medievals) and not overturning everything in the sense that we would suppose a revolutionary would do. It was not until the end of the seventeenth century that the general concept of social and political revolution (as we would understand these terms in a full post-1789 sense) came into being. Before the time of Newton, there had been revolts and dynastic changes, but not revolutions of the sort that completely alter the fabric of social and economic, or even political, life. It is not surprising, then, that despite extensive research, I have not been able to find any reference to revolutions in science or in intellectual realms earlier than 1700.

In the early stages of my research, one source that held promise of possible usage of "revolution" was the literature concerning the Battle of the Books (the Quarrel between the Ancients and the Moderns). One would have guessed that in the sciences the great superiority of the moderns was so obvious that it would have implied an order-of-magnitude break with the past. But a close examination of the main writers disclosed that apparently they never used the term "revolution" to denote a sudden change in the sciences but rather tended to invoke "improvement" of knowledge; although two of the protagonists (Fontenelle and Swift) did write of "revolutions" in other contexts and one of them (Fontenelle) applied this very word to the development of mathematics. Nor did I find any explicit reference to a revolution in the sciences (in today's sense) in Thomas Sprat's defense of the Royal Society of 1667 (see Cohen, 1977e).

An unambiguous reference to a revolution as a radical change occurs in Fontenelle's preface to his *Éléments de la géométrie de l'infini* (1727). The context of Fontenelle's discussion is the newly discovered (or invented) infinitesimal calculus (*le calcul de l'infini*) of Newton and Leibniz, and the several ways in which 'Bernoulli, the Marquis de l'Hôpital, Varignon, all the great mathematicians [géomètres]' carried the subject forward 'with giant steps'. Fontenelle says that the new mathematics introduced 'a previously unhoped-for level of simplicity; and thus was inaugurated an almost total revolution in mathematics [*géométrie*]'.[1] The conjunction of words "époque" and "révolution" leaves no doubt that Fontenelle had in mind a change of such an order of magnitude as to alter completely the state of mathematics. And he went on at once to emphasize that this revolution was progressive or beneficial to mathematical science, although not unaccompanied by several problems.

The revolution to which Fontenelle referred was the discovery or invention of the calculus, which he attributed to Newton as the first discoverer and to Leibniz as an independent co-discoverer (though first in publication).[2] Another early eighteenth-century reference to Isaac Newton and a revolution in science is found in Clairaut's statement of 1747, quoted earlier, that Newton's *Principia* had marked 'l'époque d'une grande révolution dans la Physique' (Clairaut, 1749, p. 329). The fact that these earliest references to a revolution in science occur in relation to the infinitesimal calculus and to the *Principia* is worthy of notice, since it was Newton's achievement in pure mathematics coupled with his analysis of the system of the world on the basis of gravitational dynamics that actually set the seal on the Scientific Revolution and caused scientists and philosophers to recognize that a revolution had in fact taken place. In this sense, Newton's *Principia* of 1687 would appear to have played the same role in the recognition that a scientific revolution may occur that the Glorious Revolution of 1688 apparently did for political revolution.

The *Encyclopédie* of Diderot and d'Alembert has many references to revolutions in science, a concept of radical scientific change introduced at the very start of this collective work, in d'Alembert's *Discours préliminaire* (published in 1751). This contains a thumbnail sketch of the rise of modern science or, rather, of a philosophy associated with modern science. The aim of the essay was to sketch out a methodological and philosophical analysis of all knowledge (including science, which occupies a central place in his scheme) and not to portray the sciences themselves. D'Alembert begins his historical presentation with 'le Chancelier Bacon', who occupies a grandfatherly position, and then moves on to a brief résumé of Descartes's radical innovations. He calls particular attention to the great 'révolte' of Descartes, who had shown 'intelligent minds how to throw off the yoke of scholasticism, of opinion, of authority . . .'. D'Alembert has in mind a clear image of the action of political revolutionary forces, and he portrays Descartes 'as a leader of conspirators who, before anyone else, had the courage to rise against a despotic and arbitrary power and who, in preparing a resounding revolution, laid the foundations of a more just and happier government, which he himself was not able to see established'. Descartes's role in thus preparing the revolution (or his 'révolte') was 'a service to philosophy perhaps more difficult to perform than all those contributed thereafter by his illustrious successors'. Although d'Alembert does not say so specifically, he then implies that the revolution prepared by Descartes was achieved by Newton. For d'Alembert not only proceeds at once to spell out at length the accomplishments of Newton in general physics, celestial mechanics, and optics, in terms of high praise; he specifically says that when Newton 'appeared at last', he 'gave philosophy a form which apparently it is to keep' (d'Alembert, 1963, pp. 80–84).

The concept of revolution in science also appears explicitly in

the article written by d'Alembert for the *Encyclopédie*, entitled
'Expérimental'. Here, as in the *Discours préliminaire*, d'Alembert
includes a brief history of the subject. First of all, d'Alembert observes that Bacon and Descartes had introduced 'l'esprit de la physique expérimentale'; then the Accademia del Cimento, Boyle,
Mariotte, and others took up the work. Then,

> Newton appeared, and was the first to show what his
> predecessors had only glimpsed, the art of introducing
> Mathematics [*géométrie*] into Physics and of creating—
> by uniting experiment and calculation [*expérience et
> calcul*]–an exact, profound, brilliant, and new science.
> At least as great for his experiments in optics as for his
> system of the world, Newton opened on all sides an im-
> mense and certain pathway. England took up his views;
> the Royal Society considered them as their own from the
> beginning. The academies in France adopted them
> more slowly and with more difficulty. . . . The light at last
> prevailed; the generation hostile to these men had died
> out in the academies and universities. . . . A new genera-
> tion has arisen; for once a revolution has been started,
> it is almost always the case that the revolution is brought
> to fruition in the next generation.

In this notable passage, d'Alembert not only has expressed a philosophy of historical development in science according to generations; he has also centered the great revolution in science on the
work of Isaac Newton.[3]

By the time of the publication of the *Encyclopédie*, "revolution"
had gained currency (at least in French) in its new meaning of a
change of great magnitude that is not necessarily cyclical at all.
During the second half of the eighteenth century, this concept, and
the word to express it, were notably applied to realms of the mind,
and in particular to writings about science. Various authors, however, dated the revolutions at different times, according to their
subject. Thus in 1764, Joseph Jérôme Le Français de Lalande [La
Lande] saw a revolution in astronomy in the era after Hevelius,
when 'all the nations vied with one another for the glory of discovery and improvement. The Academy of Sciences at Paris and
the Royal Society of London played the greatest parts in this revolution; the number of illustrious men and celebrated astronomers
which they have produced is enormous . . .' (Lalande, 1764, vol.

1, p. 131). But Lalande did not use the word "revolution" for
Copernicus's revolt against the authority of Ptolemy, nor for the
radical novelties discovered or introduced by a Galileo or a Kepler;
he apparently reserved the designation of "revolution" for the
process of discovery and improvement that he conceived to have
been part and parcel of the establishment and elaboration of the
subject of astronomy in more recent times.[4]

The writings of Jean Sylvain Bailly, published in the decade before the French Revolution, show the degree to which the concept
of revolution in the sciences had achieved the form in which, with
variations, it continued well established during the nineteenth
century. In his *Histoire de l'astronomie moderne*, Bailly introduced revolutions of several sorts and magnitudes. These range in
scope all the way from revolutionary innovations in the design and
use of telescopes to the elaboration of the Copernican system of the
world and the Newtonian natural philosophy. They include revolutions of the past and of the recent present, and even forecasts of
revolutions to come. Bailly conceived that revolutions in science
on a grand scale, such as the establishment of a new system of the
world (the Copernican) or a new natural philosophy (the Newtonian), required two stages. One was the revolt that destroyed the
accepted system in science (Aristotelian physics, epicyclic planetary
orbits, etc.); the other was the introduction of something new to
take its place. Descartes and Galileo were usually seen as having
accomplished the first stage only, by which Bailly meant that they
had not introduced a satisfactory replacement, which awaited the
genius of Isaac Newton. Bailly did not refer to a Galilean or a
Cartesian "revolution", even though he considered Descartes's concept of explaining all natural phenomena on a mechanical basis an
outstanding feat of intellect, however vitiated by his poor–if not
useless–system of vortices as an example thereof.

Bailly does not use the actual expression "Copernican Revolution", but he leaves no doubt of his belief that one of the major
revolutions in science was inaugurated (if not, however, accomplished) by Copernicus. Copernicus, according to Bailly, was responsible for the introduction of the true system of the world, just
as Hipparchus was to be credited with the true system of astronomy. Bailly said that a radical step had to be taken at the time of
Copernicus: It was necessary for man to forget the apparent motions that can actually be seen, in order to be able to believe in

those motions that cannot be known directly through the senses. Copernicus thus fulfilled the two necessary functions that, according to Bailly's implied standards, made his work qualify as a revolution. He undermined the authority of the old or accepted system and he set up a better one in its place (see Cohen, 1977*a*, 1977*c*).

> Walther and Regiomontanus, in Germany, constructed [astronomical] instruments and renewed the practise of observing. In each of its new homes, the science [of astronomy] was subjected to a new examination and the transmitted knowledge was subjected to the test. At that time [*époque*], however, a great revolution occurred which changed everything. The genius of Europe revealed itself through Copernicus.

While declaring that 'Copernicus had taken a great step toward the truth', Bailly (1785, vol. 3, pp. 320sq) pointed out that the 'destruction of the system of Ptolemy was an indispensable precondition, and this first revolution [of Copernicus] had to precede all the others'.

By the 1780s, there is no difficulty in finding French authors who refer explicitly to one or another revolution in the sciences. But the case of Condorcet may especially attract our attention since he is said to have been an originator of the term "révolutionnaire". The concept of a revolution in science (and the use of "revolution" to express it) occurs frequently in the *éloges* of deceased academicians which it was Condorcet's duty to write and read, in his capacity of *secrétaire perpétuel*.[5]

The major work of Condorcet in which the term and the concept of revolution figure most prominently is his *Sketch for a Historical Picture of the Progress of the Human Mind*, first published in 1795. Condorcet's primary example of a revolution came from chemistry, rather than physics or astronomy or the life-sciences: a natural result of the fact that he had actually been witness to the recent Chemical Revolution.[6] This revolution had been invented by Lavoisier in a double sense, for he both gave the Chemical Revolution its name and was its chief architect. Lavoisier referred to his own work in terms of "revolution" in at least three manuscripts: two letters and an entry in his laboratory register. The publication of the latter by Marcelin Berthelot in 1890, in a book on Lavoisier entitled *La révolution chimique: Lavoisier*, fixed the name "Chemical Revolution" on the historical record. Lavoisier's own first statement is notable. In writing out his plans and hopes for research, he could not help but be conscious of their ultimate significance. 'The importance of the subject has engaged me to take up anew', he wrote in 1773, 'all this work which has seemed to me made to be the occasion of a revolution in physics and in chemistry'.[7] The most remarkable aspect of this note is that Lavoisier referred to his own work in explicit terms of revolution.

At the century's end, the concept of revolutions in science had become firmly established. The first overall review of the intellectual accomplishments of the eighteenth century—Samuel Miller's *Brief Retrospect*, published in 1803—stated in its title that it contained 'A Sketch of the Revolutions and Improvements in Science, Arts, and Literature, during that Period'. His work was in some ways more a compilation than an original essay, as he himself admitted; accordingly, he would have encountered the concept of revolution in science and in the arts in the course of his readings (including works in French, which are prominent among his footnotes and references).

In his 'Recapitulation' at the end of the second volume, Miller turned to 'the revolutions and progress of science', observing that the 'last age was remarkably distinguished by REVOLUTIONS IN SCIENCE':

> Theories were more numerous than in any former period, their systems more diversified, and revolutions followed each other in more rapid succession. In almost every department of science, changes of fashion, or doctrine, and of authority, have trodden so closely on the heels of each other, that merely to remember and enumerate them would be an arduous task.

Miller set himself the problem of accounting for this 'frequency and rapidity of scientific revolutions'. His solution is a most modern one, since he saw the primary cause to be the emergence of what we now call a "scientific community". He pointed in particular to the 'extraordinary diffusion of knowledge'; the 'swarms of inquirers and experimenters every where'; and, above all, 'the unprecedented degree of intercourse which men of science enjoyed', the consequence of which was 'the thorough and speedy investigation which every new theory was accustomed to receive', resulting in 'the successive erection and demolition of more ingenious and splendid fabrics than ever previously'. Thus 'the scientific world [was kept] more than ever awake and busy' by a 'rapid succession of discoveries, hypotheses, theories and systems'. With an insight that shows how far Miller surpassed the bounds of a mere compiler, he concluded his 'Recapitulation' by observing: 'The eighteenth century was pre-eminently THE AGE OF LITERARY AND SCIENTIFIC INTERCOURSE' (Miller, 1803, vol. 2, pp. 413, 438).

2.3   *The Newtonian revolution in the sciences*

There is an obvious temptation to aggrandize the possible significance of the conjunction of references to the scientific work of Isaac Newton with some of the earliest occurrences of the word "revolution" in the context of tremendous changes occurring in the sciences. For in part this may seem to be no more than an accidental feature of history: that Newton's *Principia* was published within a year of the Glorious Revolution.

Once the concept of revolution emerged as a characteristic of social and political thought and action, the consequence was inevitable that it should be applied to other areas of human activity— if, and only if, they showed signs of revolutionary change in this new sense. We have seen that the first such area was in the exact sciences, specifically the development of the calculus and that part of physics encompassed by dynamics and celestial mechanics. As the eighteenth century wore on, the designation of revolution was extended to the achievements of Descartes and Copernicus, and then to various other scientific events. It is thus a historical fact that Newton's *Principia* in particular, and the Newtonian achievement in science in general, came to be the first recognized scientific[1] revolution. And it is further to be noted that, generally speaking, this revolutionary feature was not at once conceived to have been characteristic of Galilean, Keplerian, or Huygenian science. In the following chapters, I shall indicate the grounds of validity for such a judgment.

In short, rather than attempt to define what constitutes a scientific revolution, and then see whether the definition fits the Newtonian achievement, I have chosen instead to go back to the historical scientific record. There I have found that the new concept of revolution was applied to Newtonian science just as soon as that concept began to gain general currency after the Glorious Revolution. I have taken the writings of the scientists of the age of Newton as my guides in defining the features of the Newtonian revolution in science and I have also used them to confirm the views I have been developing over three decades of study of Newtonian science and its immediate background.

The identification of a Newtonian revolution in science does not mean that Newton was alone in effecting the revolutionary change that was associated with his name. The roots of the revolution go back to at least the sixteenth century, and there were crucial innovations in the early seventeenth century—associated primarily with Galileo, Kepler, and Descartes—that were essential. Nor could the Newtonian revolution have succeeded without the contributions of Wallis, Wren, Huygens, Hooke, and others. Hence there arises a question as to whether Newton merely brought to fruition what others had begun, that is, carried the subject further and was more successful than they had been. Or, did his achievement actually constitute something so new that it was in and of itself revolutionary, and not merely revolutionary in the magnitude or scope and profundity of Newton's successful application of a science already to some degree established? This and similar questions, hinging on the study of the fine structure of scientific change rather than its gross features, shall occupy much of the discussion in the following pages.

I believe that there can be no doubt whatever that the scientists in the age of Newton who wrote of a Newtonian revolution in the sciences had in mind specifically the revolution wrought by the *Principia*. One of the main aims of the present work is to try to make precise exactly what that revolution was. The *Principia* was not revolutionary in its goal of applying mathematics to the study of natural philosophy, since this had been done in the writings of Galileo, Kepler, and more recently Huygens. Even in Greek times, Archimedes and Ptolemy had introduced mathematics into the study of problems of the external world, and Copernicus's book *On the Revolutions of the Celestial Spheres* (1543) had indicated its mathematical quality by the placement on the title page of the statement supposed to have been written over the entrance to Plato's Academy, that no one ignorant of geometry should enter here. In fact, Copernicus made this point even more explicitly in his introduction by saying, 'Mathematics is for mathematicians'. In particular, the two areas explored by Newton in the *Principia*, rational mechanics and the motion of the heavenly bodies, had been those most subject to mathematical analysis in antiquity, the Middle Ages, the Renaissance, and in his own century. Nor was it particularly revolutionary to have produced a physics based upon causes, since the title of Kepler's *New Astronomy* of 1604 had specifically stated that it was 'new' because it was a 'physics based upon causes'. Nor do I believe that the revolutionary aspect of the *Principia* was found in the subject matter with which Newton dealt, although almost every subject to be found in his treatise was presented in a somewhat new way. Even his success in using the same forces (or causes) to explain terrestrial and celestial events had been presaged by Kepler and others.

Nor was the revolutionary aspect of the *Principia* to be found wholly in the introduction of the Newtonian force of universal gravity, since many of Newton's contemporaries and successors objected most strongly to Newton's having introduced this particular concept, a grasping force which can extend itself over many hundreds of millions of miles, in the way that the sun affects the motion of a comet at its aphelion. Of course, the magnificence of the *Principia* was not so much the individual novelties it contained, but the collective effect of there having been so many of them. Similarly, we may see in retrospect that the greatness of the *Principia* did not lie entirely in the fact that it solved so many problems as that it showed new ways by which traditional problems, as well as new ones, might be solved. This collective effect was that of a "blockbuster", producing an advance in the state of the exact sciences of several orders of magnitude, a series of genuine quantum jumps. But looking back at the *Principia* from our vantage point of three centuries later, it seems to me (as I believe it seemed to certain of the Newtonians in the Age of Newton) that the greatest thing about Newton's magnificent book was not the individual successes or new methods and concepts that were disclosed, nor even the totality of the Newtonian innovations, but rather the Newtonian style that made them possible. As I see it, the most revolutionary aspect of Newton's *Principia* was the elaboration of an incredibly successful method of dealing mathematically with the realities of the external world as revealed by experiment and observation and codified by reason. This is what I have called the Newtonian style, the style adopted in Newton's *Principia* toward the development of mathematical principles that could be applied in a significantly fruitful way to natural philosophy.

3

The Newtonian revolution and the
Newtonian style

3.1 Some basic features of Newtonian exact science: mathematics and the disciplined creative imagination

An outstanding feature of Newton's scientific thought is the close interplay of mathematics and physical science. It is, no doubt, a mark of his extraordinary genius that he could exercise such skill in imagining and designing experiments, and in performing such experiments and drawing from them their theoretical significance. He also displayed a fertile imagination in speculating about the nature of matter (including its structure, the forces that might hold it together, and the causes of the interactions between varieties of matter). In the present context, my primary concern is with mathematics in relation to the physical sciences of dynamics and celestial mechanics, and not with these other aspects of Newton's scientific endeavors. We shall see that although Newton expressed the pious wish that optics might become a fully developed branch of mathematical science in the Newtonian style, this subject never achieved that state during his lifetime (see §3.11); hence Newton's optical researches are not given major consideration here.

The 'principles of natural philosophy' that Isaac Newton displayed and elaborated in his *Principia* are 'mathematical principles'. His exploration of the properties of various motions under given conditions of forces is based on mathematics and not on experiment and induction. What is not so well known is that his essays in pure mathematics (analytic geometry and calculus) often tend to be couched in the language and principles of the physics of motion. This interweaving of dynamics and pure mathematics is also a characteristic feature of the science of the *Principia*. We shall see that Newton shows himself to be a mathematical empiricist to the extent that he believed that both basic postulates and the final results of mathematical analysis based on those postulates could be consonant with the real or external world as revealed by experiment and critical or precise observation.[1] But his goal was attained by a kind of thinking that he declared explicitly to be on the plane of mathematical rather than physical discourse and that corresponds to what we would call today the exploration of the consequences of a mathematical construct or a mathematical system abstracted from, yet analogous to, nature.

Newton's achievement in the *Principia* was, in my opinion, due to his extraordinary ability to mathematicize empirical or physical science. Mathematics at once served to discipline his creative imagination and thereby to sharpen or focus its productivity and to endow that creative imagination with singular new powers. For example, it was the extension of Newton's intellectual powers by mathematics and not merely some kind of physical or philosophical insight that enabled him to find the meaning of each of Kepler's laws and to show the relationship between the area law and the law of inertia.[2] The power of mathematics may also be seen in Newton's analysis of the attraction of a homogeneous sphere (or a homogeneous spherical shell, and hence of a sphere composed of such shells). Newton proves that if the force varies either as the distance directly or as the square of the distance inversely, then the gravitational action of the sphere will be the same as if the whole mass of the sphere were to be concentrated at the geometric center. The two conditions (as Newton points out in the scholium to prop. 78, bk. one) are the two principal cases in nature. The inverse-square law applies to the gravitational action on the surface or at a point outside the sphere (the force inside having been proved to be null). The direct-distance law applies to the action on a particle within a solid sphere. It might have been supposed that in any solid body, the centripetal force (as Newton says) of the whole body would 'observe the same law of increase or decrease in the recess from the center as the forces of the particles themselves do', but for Newton this is a result that must be attained by mathematics. Mathematics shows this to be the case for the above two conditions, a fact which Newton observes 'is very remarkable'.[3]

As to the discipline imposed by mathematics on the free exercise of the critical imagination, a single example may suffice. During the seventeenth century, there were current two different laws of

planetary speed. One was the area law and the other a law according to which a planet has a speed inversely proportional to its distance from the sun. Both laws had been discovered by Kepler, who
had abandoned the speed-distance law by the time he had found
the elliptical orbits.[4] But as late as 1680, as may be seen in a famous
letter written to Newton, Hooke believed that these two laws of
planetary speed could both be valid, and that both were derivable
from an inverse-square centripetal force.[5] In the *Principia*, Newton proved that the true law, consistent with the area law, is that
the speed of a planet is inversely proportional not to a planet's distance from the sun, but to the distance from the sun to a tangent
line drawn through the planet. As may be seen in Figure 3.1, the
difference between the direct distance and the tangential distance
becomes smaller and smaller as the planet approaches either perihelion or aphelion, and at either of these apsides that difference
vanishes altogether. Hooke seems to have had neither the mathematical ability nor the mathematical insight to see that the two
speed laws he had put forth in his letter to Newton could not be
true simultaneously; he was apparently endowed with no mathematical censor to tell him right from wrong in a nonelementary
problem. The stern discipline of Newton's superior mathematical
powers ruled out the false speed-distance law, which is inconsistent
with the area law for this kind of orbit.[6]

The main thrust of the present discussion is the way in which
Newton's mathematical thought was especially suitable for the
analysis of physical problems and the construction and alteration
of models and imaginative constructs and systems, but it must be

Figure 3.1. If the planet is at the point p_1, the speed is inversely
proportional to the distance Sp_2, not Sp_1, but at perihelion (P) and
aphelion (A), these two "distances" become identical. Of course, Sp_2
is drawn from S perpendicular to the tangent.

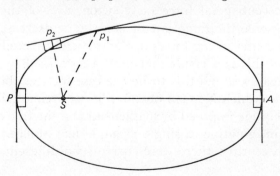

kept in mind that some of Newton's basic mathematical concepts were themselves derived from physical situations. Since Newton tended to think in terms of curves that are drawn or traced out by moving points, his primary independent variable was time. In fact, his discussion of time in his purely mathematical treatises so closely resembles the presentation of time in the *Principia* (under the heading of 'absolute, true, and mathematical time'[7]) that it would be difficult to tell them apart, out of context.

There is an obvious pitfall in making too much of the language (images as well as metaphors) of physics in Newton's mathematics. For, when in his October 1666 tract on fluxions (or calculus), he writes 'To resolve Problems by Motion',[8] he is actually concerned with pure mathematics, even though the language may suggest physics; but this would have been true of all writers on "locus" problems since Greek times, who would trace out a curve or a line by a moving point or construct a solid by revolving a plane figure about an axis (see Whiteside, 1961*b*; also Newton, 1967–, vol. 1, p. 369, n. 2). As we shall see in the following chapters, Newton's success in analyzing the physics of motion depended to a large degree on his ability to reduce complex physical situations to a mathematical simplicity, in effect by studying the mathematical properties of an analogue of the reality that he eventually wished to understand. Thus we shall see him exploring by mathematics the motion of a mass point in a central force field as a first step toward understanding the significance of Kepler's area law as a general rule and not in relation to any specific orbital system. Newton was quite aware of the difference between the mathematical properties of such simplified analogous constructs and the physical properties expressed in mathematical relations or rules or principles of the physical world as revealed by experiments and observations; but later readers and some scholars today have tended to blur Newton's usually clear distinctions. In formalizing and developing his mathematical principles of natural philosophy, Newton made use of his own new mathematics, although this fact is apt to be masked from the reader by the general absence of a formal algorithm for the calculus in the *Principia*. This new mathematics appears in his early papers both in a purely algebraic or symbolic presentation, much as in a present-day treatise on analysis (although with different symbols), and in a discussion of motion from a mathematical point of view. The latter is of interest to us here, because what is at issue is not merely a vague kinematical tracing out of the conditions of a locus, but rather the elaboration for the purposes of pure mathematics of the geometry of curves based on principles of motion that are also used in physical kinematics.[9] What is perhaps even more significant than the close conceptual fit of Newton's pure mathematics to solutions of physical problems is that, while there is a mode of thought common to both his mathematics and physics, there is in the *Principia* a continual awareness of the fundamental difference between mathematical principles and natural philosophy expressed through mathematical principles.

By this I mean that for purely mathematical purposes–that is, in a mathematical context and not with the aim of elucidating problems of physics–Newton uses principles of motion that read just as if they were physical principles being applied to physical local motion (or locomotion), including the resolution and composition of vector speeds and the concept of inertial or uniform motion.[10] Care must be exercised, according to D. T. Whiteside's warning, lest we assume too hastily that Newton was using physical principles in pure mathematics. Rather he was constructing a mathematical system that was analogous to (but not identical with) a physical system. That is, his mathematical "time" is not the physical time of experience; and it is the same with respect to mathematical "speed", and so on. Nevertheless, he did use the same language in both his writings on the physics of motion and his development of mathematics by a mathematics of motion. I believe, although of course there can be no proof, that there is a close bond between Newton's tendency to think about pure mathematics in terms that are the same as those arising in the physics of motion and his insight and skill in using pure mathematics to solve problems in physical motion. Yet one should not make too much of such a linkage, which would have been operative only on the subconscious level, since Isaac Barrow (for one) had also written pure mathematics in the language of motion, which may have been Newton's direct source of inspiration.

The use of the principles of motion in attempting to solve problems in pure mathematics may be seen clearly in a simple example in the *Waste Book*, a paper of 8 November 1665 entitled 'How to Draw Tangents to Mechanicall Lines' (Fig. 3.2). It begins:

> *Lemma.* If one body move from *a* to *b* in the same time
> in which another moves from *a* to *c*, & a 3rd body move from

a with motion compounded of these two, it shall (completing the parallelogram) move to *d* in the same time.
For those motions would severally carry it the one from
a to *c*, the other from *c* to *d*, &c [Newton, 1967–, vol. 1,
p. 377].

This leads him to the following principle: 'In the description of
any Mechanicall line what ever, there may bee found two such
motions which compound or make up the motion of the point de-
scribeing it, whose motion being by them found by the Lemma,
its determinacion shall bee in a tangent to the mechanicall line.'
He has started out from the parallelogram rule for velocities in
physical motions (entered earlier in the *Waste Book* in January
1665), and has then gone on to generalize this rule, in what D. T.
Whiteside has called 'a complex of early notes on motion and force',
in which we find him writing: 'Two bodys being uniformely moved
in the same plaine, their center of motion will describe a streight
line. . . . They doe the same in divers plaines' (Newton, 1967–, vol.
1, p. 377, n. 2; see also vol. 4, pp. 270–273). Then, using these principles of motion in a purely mathematical sense, he addresses himself to 'Mechanicall' curves, and goes further to consider motions
that generate a 'helix' (an Archimedean spiral), a curtate 'Tro-
choides' (a general cycloid), and a quadratrix. Applications are
then made to the ellipse, the hyperbola, and the parabola: 'The
tangents of Geometricall lines may be found by their descriptions
after the same manner' (Newton, 1967–, vol. 1, p. 380, scholium).

Newton's program of solving problems in geometry by the application of principles of uniform (inertial) motion and of the
parallelogram law for combining vector motions suggests an intimate bond in his mind between his explorations of physics and
mathematics (starting from Cartesian principles and even examples and ending up with the method of fluxions applied to 'limit-
motions' in relation to general problems of tangents and curva-

Figure 3.2

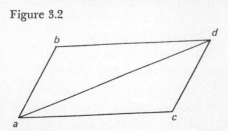

ture) (Newton, 1967–, vol. 1, pp. 369sqq). Futhermore, he was not only using the parallelogram law and principle of uniform motion at about the same time in a mathematical sense in tracts on pure mathematics and in a physical sense in tracts on the physics of motion; these tracts are also physically similar, having been written out on pages originating from one and the same notebook, the *Waste Book*.[11] Furthermore, I believe that in Newton's purely mathematical discussions of the motions of a point in an ellipse, or other conic section, one can see the early preparation of his creative mind to deal with planetary and cometary motions in ellipses and parabolas some twenty years later in the *Principia*: a transition, however, that would have been far from obvious at the time, invoking a shift from a conceptual framework of geometry to conditions of physical dynamics.

The reader who has never studied Newton's mathematical writings can have no idea of the quasi-physical imagery of motion in his presentation of the method of fluxions. For example, in his *Treatise of the Method of Fluxions and Infinite Series*, he observes that 'all the difficulties' may be 'reduced to these two Problems only, which I shall propose, concerning a space described by local Motion, any how accelerated or retarded':

> I. The length of the Space described being continually (that is, at all times) given; to find the velocity of the motion at any time proposed.
>
> II. The velocity of the motion being continually given; to find the length of the Space described at any time proposed.[12]

As an example Newton proposes the equation $x^2 = y$, where 'y designates the length of the space described in any time which is measured by a second space x as it increases with uniform speed'. Then, $2\dot{x}x$ (the first fluxion, or derivative with respect to time, of x^2) will 'designate the speed [$\dot{y}$] with which the space y at the same moment of time proceeds to be described'. In other words, Newton will 'consider quantities as though they were generated by continuous increase in the manner of a space which a moving object describes in its course'. Newton eventually used the letters $a,b,c,$ ... for constants in equations (quantities that 'are to be looked on as known and determined') and v,x,y,z for variables ('quantities which I consider as just perceptibly but indefinitely growing from others'). The latter are called 'fluents' (*fluentes*) and 'the speeds

CAMROSE LUTHERAN COLLEGE
LIBRARY

with which they each flow and are increased by their generating motion' are called 'fluxions' (*fluxiones*); thus, he says, 'for the speed of the quantity v I shall put $\dot{v}$, ...'.[13] Here is the language of the physics of motion used by Newton in displaying the mathematics of motion in pure analysis.

An example from the article on fluxions in John Harris's *Lexicon technicum* (1704) illustrates the Newtonian solution of problems by motion. It is proposed to demonstrate that the fluxion of xy is $x\dot{y} + \dot{x}y$. First, suppose '$xy =$ to any Rectangle made or encreased by a Perpetual Motion in Fluxions of either of Sides x, or y, along the other'. The fluxions of the sides are $\dot{x}$ and $\dot{y}$, by which 'we understand the *Velocity* with which either Side moves to form the Rectangle'. The proof then proceeds in a familiar manner.[14]

It is to be observed that Newton refers (in his *Treatise on Fluxions*) not to abstract motion in an Aristotelian general sense but to 'local motion' or 'locomotion', a transition in time from one point in space to another. This must, however, be a motion in geometrical space, rather than in physical space, for which the late scholastic physico-mathematicians had developed concepts and laws of uniform and accelerated ("difform") motion. As D. T. Whiteside has pointed out, Newton was to some degree following Isaac Barrow, whose *Lectiones geometricae* (1670) were published a few weeks before Newton began to write out the portion of the *Treatise on fluxions* in which concepts of motion are introduced. Barrow, according to Whiteside, 'deals at some length with "local" motion of "crescent" and "decrescent" magnitudes and their "flux" in time' (Newton, 1967–, vol. 3, p. 71, n. 80). I do not wish to enter here into the problem of Newton's indebtedness to Barrow for concepts of the fluxional calculus, so much as to indicate that both Newton and Barrow were using concepts and principles originating in the study of physical motion in a mathematical sense that is essentially removed from its physical sources.

If Newton thus conceived of fluxions and limits in terms of a mathematical local motion, it may not be surprising that he should have developed a powerful tool for analyzing local motion in a physical sense by means of mathematics that used the method of limits, as in the *Principia*. Many years later, in about 1714, in a draft of his anonymous book review of the *Commercium epistolicum* (London, 1712), in which the Newtonian priority in the discovery of the calculus is asserted, Newton made it clear once again that mathematical concepts similar to those used in the physics of motion were basic to his own version of the calculus[15]. 'I consider time', he wrote, 'as flowing or increasing by continual flux & other quantities as increasing continually in time, & from the fluxion of time I give the name of fluxions to the velocitys with which all other quantities increase'. His method was to 'expose time by any quantity flowing uniformly' and, in a manner reminiscent of Galileo, he said that his 'Method is derived immediately from Nature her self' (Newton, 1967–, vol. 3, p. 17).

In an introduction to an English version of Newton's *Methodus fluxionum*, the translator (John Colson) explained at length the connections between the physics of motion (rational mechanics) and Newton's mathematics:

> The chief Principle, upon which the Method of Fluxions is here built, is this very simple one, taken from the Rational Mechanicks; which is, That Mathematical Quantity, particularly Extension, may be conceived as generated by continued local Motion; and that all Quantities whatever, at least by analogy and accommodation, may be conceived as generated after a like manner. Consequently there must be comparative Velocities of increase and decrease, during such generations, whose Relations are fixt and determinable, and may therefore (problematically) be proposed to be found [Newton, 1736, p. xi].

We shall see below (§4.4) that it is only by taking cognizance of the mathematical nature of time in Newtonian mathematical physics that we may understand one of the important aspects of the *Principia*: the relation between motion under the individual action of discrete forces and motion produced by a continuously acting force.

This intimate connection between pure mathematics and the physics of motion is, I believe, a characteristic feature of Newton's *Principia*, wherein certain aspects of natural philosophy are reduced to mathematical principles, then developed as mathematical exercises, and finally reapplied to physical problems. The main subject of the *Principia* is terrestrial and celestial dynamics: the physics of motion,[16] or the motion of bodies under the action of various kinds of forces and different conditions of restraint and resistance, and the mathematical method is fluxional[17] and uses vanishing infinitesimals; it is a characteristic feature to apply the limit process to geometric conditions and to proportions (or equations) arising from or representing those conditions. Hence the quasi-physical nature of Newton's mathematics was eminently suited to the solution of the problems to which he addressed himself in his *Principia*. But while this intermingling of a pure mathematics derived from or related to motion and the physical problems of motion may have led Newton to achieve unheard-of results of astonishing fecundity, this very aspect of his work has caused great confusion among his commentators and interpreters ever since (see §3.6). In particular, they have not always known when Newton was speaking on the level of mathematics and when on the level of physics. Or, they have perhaps assumed this to be a distinction without a difference and have not bothered to ascertain whether Newton as a mathematician was—in the *Principia*—everywhere intending to be understood as a physicist.[18] It will be seen below that a major aspect of Newton's method in the *Principia* (and possibly in other aspects of his work in the exact sciences) was his intuitive separation of these two levels of discourse and then, on the proper occasion, using his mathematical results to illuminate the physical problem. The blurring of Newton's distinctions, which has led to continual misunderstandings concerning Newton's method and his intentions, probably derives from a reading of certain scholia and introductory sections of the *Principia* out of the context of the mathematical physics in which they are imbedded and which they were intended to illuminate.[19]

3.2 *Mathematics and physical reality in Newton's exact science*
One of the clearest statements Newton ever made of his own position was in reply to a criticism made by Leibniz. The details of this criticism would take us far afield, and we need only take notice here that Newton held that what his critic 'saith about Philosophy is foreign to the Question & therefore I shall be very short upon it'. Newton's grounds for this statement of disagreement with Leibniz concerning the fundaments of 'Philosophy' (i.e., natural philosophy) were threefold. First, 'He [Leibniz] denys conclusions without telling the fault of the premisses.' Second, 'His arguments against me are founded upon metaphysical & precarious hypotheses & therefore do not affect me: for I meddle only with experimental Philosophy.' Third, 'He changes the significa-

tion of the words Miracles & Occult qualities that he may use them in railing at universal gravity . . .' In writing the last sentence, he had at first used the words 'railing at me', which shows the degree to which in his own mind he had equated himself with the conceptual fruit of his intellect.[1]

As Newton said again and again, there was a fundamental difference in philosophy between himself and Leibniz. To deny universal gravitation could be legitimate in Newton's philosophy only by going back to the arguments given by Newton, and to the premises of those arguments: a combination of empirical findings, mathematical developments, and sound logic. It was not enough merely to say that a concept of universal gravitation is not acceptable in philosophy. And so to understand the fundaments of Newtonian exact science (i.e., the exact science of the *Principia*), and the principle features of the Newtonian revolution in science, it is necessary to see what in fact were the stages by which Newton got to universal gravitation. In so doing, we shall see why Newton held that there is a profound difference between 'metaphysical & precarious hypotheses' and 'experimental Philosophy'. Finally, Newton was particularly concerned about 'Miracles & Occult qualities'. He stoutly denied the relevance of 'Miracles' to his natural philosophy in the sense of a suspension of the ordinary laws of nature, and he equally denied that he had reintroduced into science the 'Occult qualities' of late Aristotelian-scholastic philosophy. Gravitation itself was not 'Occult', but its cause was, in the degree that it was still hidden from us.[2]

Newton's spectacular achievement in producing a unified explanation of the events in the heavens and on our earth, and in showing how such diverse phenomena as the ebb and flow of the tides and the irregularities in the moon's motion might be derived from a single principle of universal gravity, drew attention to his mode of procedure—a special blend of imaginative reasoning plus the use of mathematical techniques applied to empirical data—which I have called the Newtonian style. Its essential feature is to start out (phase one) with a set of assumed physical entities and physical conditions that are simpler than those of nature, and which can be transferred from the world of physical nature to the domain of mathematics. An example would be to reduce the problems of planetary motion to a one-body system,[3] a single body moving in a central force field; then to consider a mass point rather than a physical body, and to suppose it to move in a mathematical space in mathematical time. In this construct Newton has not only simplified and idealized a system found in nature, but he has imaginatively conceived a system in mathematics that is the parallel or analogue of the natural system. To the degree that the physical conditions of the system become mathematical rules or propositions, their consequences may be deduced by the application of mathematical techniques.

Because the mathematical system (to use an expression of Newton's in another context) duplicates the idealized physical system, the rules or proportions derived mathematically in one may be transferred back to the other and then compared and contrasted with the data of experiment and observation (and with experiential laws, rules, and proportions drawn from those data). This is phase two. For instance, the condition of a mass point moving with an initial component of inertial motion in a central force field is shown (*Principia*, props. 1 and 2, bk. one) to be a necessary and sufficient condition for the law of areas, which had been found to be a phenomenologically verifiable relation in the external astronomical world.

The comparison and contrast with the reality of experiential nature (that is, with the laws, rules, and systems based upon observations and experiments) usually require a modification of the original phase one. This leads to further deductions and yet a new comparison and contrast with nature, a new phase two. In this way there is an alternation of phases one and two leading to systems of greater and greater complexity and to an increased *vraisemblance* of nature. That is, Newton successively adds further entities, concepts, or conditions to the imaginatively constructed system, so as to make either its mathematically deduced consequences or the set conditions conform more exactly to the world of experience. In the example under discussion, the first of these additional steps is to introduce Kepler's other laws of planetary motion. The third law, applied to uniform circular motion in combination with the Newtonian (Huygenian) rule for centripetal (centrifugal) force yields the inverse-square law of force. An elliptical orbit is then shown to require an inverse-square force, as does a parabolic or hyperbolic orbit.

In the next stage of complexity or generality, Newton adds to the system a second body or mass point, since (as Newton says at the start of sect. 11, bk. one, of the *Principia*) attractions are not made toward a spatial point but rather 'towards bodies', in which case the actions of each of the bodies on the other are always equal in magnitude though oppositely directed. Yet additional conditions include the introduction of bodies with finite sizes and defined shapes, and of a system of more than two interacting bodies. (There is also the question of whether bodies move through mediums according to some specified law of resistance.)

For Newton there is a final stage in this process: when the system and its conditions no longer represent merely nature simplified and idealized or an imaginative mathematical construct, but seem to conform to (or at least to duplicate) his realities of the external world. Then it becomes possible, as in bk. three of the *Principia*, to apply the aggregate of mathematical principles to natural philosophy, to elaborate the Newtonian system of the world. This is the final phase three of the Newtonian style, the crown of all, to display the variety of natural phenomena that can be attributed to the action of universal gravity. It is only after this stage, not earlier, that Newton himself would have to yield to the demand for investigation into the nature, cause, or mode of operation of such forces as he had used in accounting for the motions of terrestrial bodies, the planets, their moons, our moon, the comets, the tides, and various other phenomena. This additional inquiry, a kind of sequel to phase three, went beyond the requirements of the Newtonian style, however, at least insofar as the *Principia* is concerned. Even in the general scholium with which the later editions of the *Principia* conclude, Newton insisted that his gravitational dynamics and his system of the world could be accepted even though he had said nothing about the cause of gravity. But he did then express his personal conviction that gravity 'really exists'.

It is a feature of the Newtonian style that mathematics and not a series of experiments leads to the most profound knowledge of the universe and its workings. Of course, the data of experiment and observation are used in determining the initial conditions of the inquiry, the features that yield the mathematical principles that are applied to natural philosophy, and Newton was also aware that the success of the eventual natural philosophy (or of the system of the world) must rest ultimately on the accuracy or validity of the empirical data of which it was constructed. Furthermore, the test of the end result was necessarily the degree and extent of the

ability to predict and to retrodict observed phenomena or phenomenologically determined "rules" (such as Kepler's laws). Even
so, on some significant occasions Newton seems to have given priority to the exactness of mathematical system rather than the
coarseness of empirical law. In the case of Kepler's laws, the reason
is that they prove in Newton's analysis to be exact only of a very
limited situation and to be no more than phenomenologically
"true" (that is, "true" only within certain conventionally acceptable limits of observational accuracy) of the real world as revealed
by experience. Hence, even the Newtonian system of the world,
when said by Newton (in the final editions of the *Principia*) to be
based on "phenomena", is in fact based also to some degree on
truths of mathematical systems or idealizations of nature that are
seen to be approximate to but not identical equivalents of the conditions of the external world.

The simplified physical system (and its mathematical analogue
that Newton develops at the beginning of the *Principia*) exhibits
all three of Kepler's laws and in fact serves to explain these laws by
showing the physical significance of each one separately. In short,
this system or construct is not a figment of the free imagination, or
a purely arbitrary or hypothetical fiction created by the mind,[4] but
is rather closely related to the real Copernico-Keplerian world that
is made known to us by phenomena and by laws that are phenomenologically based. In his first flush of victory, just before writing
the *Principia*, when he had completed his analysis of this system,
Newton himself thought that it was more than an imagined construct. We shall see below that in a first version he expressed his
belief that he had now explained exactly and fully how nature
works in the operation of the solar system. But not for long. It was
almost at once plain that the construct he had been studying did
not accord with the real world. And so, bit by bit, it was endowed
with more and more features that would bring it into closer harmony with the world of reality. In the course of these transformations of his construct, Newton was led by degrees to the concept of
a mutual gravitating force, a concept all the more conspicuous by
its absence from the first considerations. As a result, it is possible
to assign a precise limiting date to the first step toward that great
concept: no earlier than December 1684.[5]

The advantages of the Newtonian method, as I have outlined it,
are manifold. First of all, by making the construct simple at the

start, Newton escapes the complexities of studying nature herself. He starts out with an idealized version of nature, in which certain descriptive laws of observed positions and speeds—Kepler's planetary laws—hold exactly. Then, on the basis of the laws and principles that underlie these descriptive laws, Newton proceeds to new constructs and to more general underlying laws and principles, and eventually gets to the law of universal gravity in a new system in which the original three planetary laws as stated by Kepler are—strictly speaking—false (see §5.8).

Is Newton's ultimate system still only an imagined construct? Or is it now so congruent with reality that its laws and principles are the laws and principles of the universe? Newton does not tell us what he believes on this score, but we may guess at his position. His first construct, for which Kepler's laws are valid, proved to be a one-body system: essentially a single mass-particle moving under the action of a force directed toward a fixed center.[6] Next he extended and modified the results found true of the one-body system so that they would apply to a two-body system, in which each of the two bodies acts mutually on the other with the same inverse-square force that acts on the single body in the first one. Then there are many bodies, each acting on all the others mutually with an inverse-square force; and, finally, the bodies have physical dimensions and determinate shapes, and are no longer mere mass points or particles. The force is shown to duplicate gravity and to be mutually acting, and is then found to be a universal force proportional to the product of the masses. In this way, Newton extended his construct from one to two mass points and then to many, and from particles or mass points to physical bodies. Since there are no more bodies to be added, I believe he would have argued that the system was complete. Further physical complexities or mathematical conditions might be imagined and put in: for instance, gross bodies with negative masses; or bodies that would interact with other bodies, but that might have negative gravitational forces (repulsions) as well as positive ones (attractions), just as in electrical and magnetic phenomena. But in terms of accumulated observations over many centuries, any such condition would then have been ruled out as highly improbable, if not downright impossible.[7] Of course, since Newton was unable to give a general solution to the problem of three mutually gravitating bodies, there might have been unforeseen complications in a many-body system. But we may guess that these speculative conditions would not have carried much weight with him. He had found the system of the world.

Furthermore, the final system would certainly have seemed to have transcended the status of being merely an imagined construct in the degree to which its results agreed with many different kinds of observations. The Newtonian theory could explain not only why all bodies fall with the same acceleration at a given place on earth, but also the observed fact that the acceleration varies in a certain definite manner with latitude (as shown by the concomitant variation in the period of a freely swinging simple pendulum) and other factors. The theory of gravity could also explain the tides[8] and many features of the motion of the moon, and could even predict the oblate shape of the earth from the known facts of precession. The variety and exactness of the verifiable predictions and retrodictions of experience gave every reason to believe that the Newtonian system of the world, displayed in bk. three of his *Principia*, and developed and extended by others, was indeed the true system of the world. And so it was conceived for over two hundred years.[9]

A general limiting condition on that system was not found until Einstein's theory of relativity.[10] During that long period a major unresolved failure to predict or retrodict the experiential data of our universe arose in the case of the advance of the perihelion of Mercury, produced by a very slow rotation of the planetary orbit. This failure of the developed Newtonian system is small, requiring a century and a half to accumulate a discrepancy of as much as a single degree of arc. This is a measure of the tremendous increase in precision of positional astronomy since Newton's day. Yet this unaccounted-for or anomalous advance per century is also a large quantity, in that it 'alters the calculated position of Mercury at transit by more than the planet's diameter—an amount impossible to ignore' (Russell, Dugan, & Stewart, 1926, p. 306).

Because the final system achieved by Newton was found to work so well, it no longer had to be treated as an imagined construct. According to a declaration by Newton, universal gravity 'really exists' and it served to account for a wide range of phenomena, on a scale and to a degree never before achieved in the exact sciences. In this sense, Newton was justified in believing that he had elucidated the system of the world and had not merely an imagined construct that satisfied the needs of computing, or that was contrived to 'save the phenomena' (see Duhem, 1969). As he himself put it in the general scholium, there are three conditions for gravity that are enough, that suffice in natural (or experimental) philosophy. It is enough ('Satis est'), first of all, 'that gravity really exists'; second, that gravity 'acts according to the laws that we have set forth'; third, that gravity 'is sufficient to explain all the motions of the heavenly bodies and of our sea'. For Newton there then arose two wholly different sets of questions. The first were technical: to work out, as he saw it, the "details" of gravitational celestial mechanics and thus to get better results for such problems as the motion of the moon. This range of activity can be described as completing the *Principia* on the "operative" level.[11] The second set of questions were another type altogether: to explain gravity and its mode of action or to assign 'a cause to gravity'. His critics, however, proceeded in just the opposite manner, starting out with the vexing problem of how such a force as the proposed Newtonian universal gravity can possibly exist and act according to the Newtonian laws, and not accepting the formal results of the *Principia* so long as they did not find its conceptual basis to be satisfactory. These critics, in other words, were not willing to go along with the procedural mode of the Newtonian style.

3.3 Newton's use of imagined systems and mathematical constructs in the Principia

Newton's approach to celestial dynamics was based on the concepts of the "mechanical philosophy" (matter and motion) plus the very significant additional concept of force. In Newtonian science, "force" could act in two major different modes: instantaneously or continually. One is the action on a body when struck by or when encountering a second body that alters its state of rest or motion on impact. This is the familiar situation observed in examples of tennis balls being struck by racquets, of billiard balls either hitting other billiard balls or receiving a blow from the cue, or of billiard balls altering their motion on reflection by the cushion. In these examples the cause of motion (the blow or impact) is plainly discernible at the moment of contact. We shall see below that this situation is embodied in the primary statement of the second law of motion in the *Principia*: that the change in what Newton calls the 'quantity of motion' (measured by the product of mass and velocity) is proportional to the impulsive force. The second mode of action of forces is illustrated in the motion of planets. There is no visible blow, no impact; and yet there is a constantly changing state of motion. Hence at every instant there must be a force in action, according to the first law or principle of inertia. Since this is a continuously acting force, as manifested by a continuous change in the 'quantity of motion', there is a different law for this kind of force: the change in the Newtonian 'quantity of motion' (or momentum) in any given time is proportional to the force or, more strictly, the rate of change in the 'quantity of motion' is proportional to the force.[1]

In Newton's day, the primary controversy about the science of the *Principia* was not about technical matters, such as whether the law of inertia implies that every curvilinear motion (as in the case of planets) requires a constant force, nor even that the planetary force must diminish as the square of the distance.[2] Disagreement with, and even rejection of, the Newtonian system was based on a genuine concern as to whether one body could really and truly "attract" another body over vast distances of many hundreds of millions of miles. Even before the *Principia* had been published, Huygens voiced his uneasiness about this very question in a letter to Fatio de Duillier (11 July 1687), saying that he hoped Newton would not 'give us suppositions like that of attraction' (Huygens, 1888–1950, vol. 9, p. 190; cf. Koyré, 1965, p. 116).

The first criticisms directed at Newtonian physics on the Continent–by Huygens, Leibniz, Fontenelle, and an anonymous reviewer in the *Journal des Sçavans*–all hinge on a point of metaphysics and not really on the subject matter of physics, or of rational mechanics (dynamics) and celestial mechanics: Can one admit into the domain of science anything other than matter and motion? In particular, can one make one's peace with attraction, a grasping force that causes bodies to act mutually on one another "at a distance" that may be as great as hundreds of millions of miles? Later Newtonians would point to the phenomena of electricity and magnetism, in arguing for the existence of a universal force of attraction.[3] Not that gravity was *in esse* analogous to electricity and magnetism, since gravity differs from them in its universality and in its not ever manifesting itself in repulsion. But electrical and magnetic forces are "real", as may be seen in their evident power to act on bodies at a distance; and so, it was argued, if forces of attraction do exist in nature, why not universal gravitation? In the *Principia*, however, no such argument is made.

Bks. one and two of the *Principia* are, as Newton states unambiguously, primarily mathematical and not physical. That is, they contain 'principles of philosophy that are not, however, philosophical but strictly mathematical'. They are, nevertheless, 'principles . . . on which natural philosophy can be based': 'laws and conditions of motions and of forces, which especially relate to philosophy'. Knowing that these 'principles' might seem 'barren' to his readers, Newton 'illustrated them with some philosophical scholia, treating topics that are general and seem to be the principal foundations of philosophy, such as the density and resistance of bodies, spaces void of bodies, and the motion of light and sounds' (*Principia*, bk. three, introduction).

Bk. one opens with a sect. 1 on the theory of limits[4] that is purely mathematical.[5] Sect. 2 is concerned with centripetal forces acting on point-masses, and sect. 3 with motions in conic sections. While these too are purely mathematical,[6] they obviously are being directed toward uses in natural philosophy, since they deal with such physical (or astronomical) topics as the area law, motion in elliptic orbits, and so on. In prop. 3, dealing with a body moving in orbit about another body which is itself in motion, Newton (in the second and third editions) calls these bodies respectively L and T; as a result, the reader cannot help but think at once of Luna moving in orbit about a moving Terra. And in prop. 4, on centripetal forces in uniform circular motion, Newton (in a philosophical scholium) remarks that corol. 6–'If the periodic times are as the $3/2$ power of the radii . . . centripetal forces will be inversely as the squares of the radii; and conversely'–applies to the motion of the 'celestial bodies', as Wren, Hooke, and Halley have severally observed. And, accordingly, he will 'treat more at large of those things which relate to centripetal forces decreasing as the squares of the distances from centres'. This declaration by no means implies that he will forbear to study questions that have no immediate application in physics. For, although his goal is plainly prop. 11, on motion in an ellipse with a force directed to one of the foci, he first studies bodies revolving in the circumference of a circle with a 'centripetal force tending toward any given point' ('tending toward a given point in the circumference': first edition) and then bodies revolving in an ellipse with a 'centripetal force tending toward the center of the ellipse'. Curiously enough, there is no 'philosophical' scholium accompanying Newton's demonstration that orbital motion in an ellipse implies a centripetal force directed to a focus that varies inversely as the square of the distance.[7] For this the reader must await bk. three, 'On the System of the World'.

A little later on, in sect. 6, which is basically concerned with mathematical aspects of "Kepler's problem", Newton observes of his solution that 'the description of this curve is difficult' and accordingly he proposes 'a solution by approximation [which] will be preferable'.[8] Then, at the end of some 80 percent of sect. 6, Newton says that he has indicated how to obtain 'a general analytical resolution of the problem' of finding 'the place of a body moving in a given elliptic trajectory at any assigned time'. This has been primarily of mathematical rather than practical interest. The solutions are not of any real use in making astronomical calculations, and so (in the second and third editions) he proposes yet another 'particular calculus that . . . is better fitted for astronomical purposes', now actually reckoning the error that arises from its use in the orbital motion of Mars (where 'the error will scarcely exceed one second'). Obviously, then, the major thrust of this section is purely mathematical. That the goal of the next section is also mathematical appears from the fact that, after considering centripetal forces varying as the distance along with those that vary inversely as 'the square of the distance of places from the centre',[9] there is a concluding generalization concerning the motion of bodies on the supposition of 'a centripetal force of any kind'. In the succeeding sect. 8 (where Newton assumes the ability to perform integration, 'granting the quadratures of curvilinear orbits'[10]) there occurs prop. 41, the crown of all: 'Supposing a centripetal force of any kind, . . . to find the trajectories [i.e., the curves] in which bodies will move, and the time of their motions in the trajectories so found'. This is a mathematical result of great generality and power in which there is no restriction placed on the force, which can be as any (unspecified) function of the distance whatsoever.[11]

That the first nine sections of bk. one are primarily mathematical is evident from the general lack of what Newton has designated as "philosophical" scholiums; although plainly Newton is developing propositions concerning the motion of bodies under a variety of conditions of force or constraint that may eventually be of service to problems in the physics of motion. In other words, he is dealing with the mathematics of limited or arbitrary conditions; or, he is exploring the mathematical properties of artificial situations or imagined constructs and not studying nature as revealed in all her complexities by experiment and observation. On this point there can be no doubt whatever.[12]

For example, in sect. 8 he says that he has 'considered the motions of bodies in immovable orbits' and in sect. 9 goes on to a more complex system in which he 'adds' further propositions 'concerning their motions in orbits which revolve round the centres of forces'. Thus, as Newton approaches a condition like that of the real world, he adds complexities which declare that the previous set of conditions were so simplified as, by definition, to constitute a mathematical construct rather than nature itself.[13]

On this basis there should be no real ambiguity about the much misunderstood introduction to sect. 11, 'The motion of bodies that by centripetal forces tend toward one another'. Newton starts out:

> Hitherto I have set forth the motions of bodies attracted
> toward an unmoving center, such as, however, hardly
> exists in the real world. For attractions customarily are
> directed toward bodies, and—by the third law—the actions
> of attracting and attracted bodies are always mutual
> and equal; so that if there are two bodies, neither the
> attracting nor the attracted body can be at rest, but both
> (by the fourth corollary of the laws) revolve about a
> common center, as if by a mutual attraction; and if there
> are more than two bodies that either are all attracted by a
> single body and all attract it, or all mutually attract one
> another, these bodies must move among themselves
> in such a way that the common center of gravity either
> is at rest or moves uniformly straight forward.

Observe Newton's opening statement, that he has been thus far considering a wholly artificial situation, 'quale tamen vix extat in rerum natura'. But what of bodies being 'attracted'—and the statement that 'attractions customarily are directed toward bodies'? Does this imply that Newton has turned to physical nature where he believes there are attractions and forces acting at a distance?

Let us allow him to answer for himself. First of all, in the remainder of the above quoted paragraph, he says:

For this reason I now go on to explain the motion of bodies that mutually attract one another, considering centripetal forces as attractions, although perhaps–if we speak in the language of physics–they might more truly be called impulses. For we are here concerned with mathematics; and therefore, putting aside any debates concerning phys- ics, we are using familiar language so as to be more easily understood by mathematical readers.[14]

What does this actually mean? First of all, Newton wants us to know that he is 'concerned with mathematics' and has accordingly put 'aside any debates concerning physics'. He is now introducing the term 'attraction' for what he has up until now been calling 'centripetal force'. And he is merely 'using familiar language' in order that he may be 'more easily understood by mathematical readers'. In the prior sects. 1–10, he had been considering the problem of a single center of force, so that the expression "centrip- etal force" had been appropriate. But now there is to be not only a whole set of centripetal forces, but each is to be one of a pair, mutual and opposite. In short, in a system of n bodies, there are not only n centers of force, but $n(n-1)$ forces directed toward these n centers. That is, each body would be the source of $n-1$ forces, directed toward $n-1$ centers. These centers to which the $n-1$ forces are directed are no longer mere centers of force, but are other bodies. Under these new conditions, when there would be $n(n-1)$ forces directed toward n bodies rather than toward a single point or center, it would be confusing to continue to use the expression "centripetal forces". In this context, Newton is using the word "attraction" merely as a generalization of "cen- tripetal force" to the case of more than a single force-center. New- ton hoped that his use of the common word "attraction" would be understood by his 'mathematical readers', and he meant what he said. That is, he was exploring the consequences of his mathemati- cal construct, by deriving mathematical properties of an interactive system of two or more bodies, the properties of "mutually-acting centripetally-directed forces", which for convenience he called "at- tractions".[15] "Attraction" is, however, a loaded word, and the phys- ical implications of it are as evident to us as they were to Newton's contemporaries.[16] Yet it would be difficult to find another word that would be so expressive of the situation on the mathematical level of discourse, that would so readily convey the properties of the imagined "mutually-acting centripetally-directed forces". Since Newton was concentrating on the mathematical properties, he believed that he did not need to validate the use of "attraction", and says so. In the foregoing quotation, he even admits that if he had been using 'the language of physics' these forces 'might more truly be called impulses'.[17]

The logical structure of the *Principia* leads the reader to go back from this statement to def. 8, where Newton discusses centripetal force. With respect to 'attractions and impulses', he says, 'I use words such as "attraction", "impulse", or words for any sort of propensity toward a center interchangeably and indiscriminately, considering these forces not from a physical but only from a mathematical point of view'. Clearly, in the language I have been using, Newton is conceiving of an imagined system, a mathematical construct. And he then especially enjoins the reader '. . . not to imagine that by those words I anywhere take upon me to define the kind, or the manner of any action, the cause or the physical reason thereof, or that I attribute forces, in a true and physical sense, to certain centres (which are only mathematical points); when at any time I happen to speak of centres as attracting, or as being endowed with attractive powers'.[18]

This point is emphasized again in the scholium with which sect. 11 concludes. Once more he insists that

> I use the word *attraction* here in a general sense for any
> endeavor whatever of bodies to approach one another,
> whether that endeavor occurs by the action of the bodies
> either tending toward one another or agitating one
> another by means of spirits emitted, or whether it arises
> from the action of aether or of air or of any medium
> whatsoever—whether corporeal or incorporeal—in any way
> impelling toward one another the bodies floating therein.
> I use the word *impulse* in the same general sense,
> considering in this treatise not the species of forces and
> their physical qualities but their quantities and mathe-
> matical proportions, as I have explained in the definitions.[19]

By using such plain language, Newton hoped that he might not be misunderstood. But the fact of the matter is that critics from that day to this have misunderstood him. Such critics have tended to assume that either Newton did not mean what he said, or did not say what he meant.

The misunderstanding of Newton's avowed intention probably arises from the fact that the *Principia* is usually read in bits and not in its entirety *a capite ad calcem*. In bk. three there is a transition from mathematical systems to the realities of the system of the world. The results he has obtained from considerations of a mathematical system or imagined construct[20] fit the conditions of the astronomical and terrestrial world, and hence Newton can conclude that his mathematical construct with its unspecified "attraction" is the analogue of the world of reality and seems to represent and even be the real world. Then, and only then, does the question arise as to what can possibly "cause" such an "attraction". Newton believes he has shown that this attraction or universal gravity is merely the same force–i.e., gravity–which is operative when bodies fall down to the earth or are heavy with respect to the earth. That is, strictly speaking, what Newton has shown is that there exist "forces" which (to use his own expression) "duplicate" the action of forces at a distance. If challenged to prove the physical existence of such forces, he can retreat to the kind of positivist position that is consistent with the Newtonian style, as in the statement in a draft preface (1717) to the *Principia*, in which he says 'Causam gravitatis ex phaenomenis nondum didici' ('I have not as yet learned the cause of gravity from phenomena'). The power of the Newtonian method of thinking in terms of mathematical constructs, however, is that it enables him to develop the properties of a two- and three-body gravitating system independently of the question as to whether bodies can or do attract one another. Or, it enables him to postpone this question until a later time.

In short, had Newton suppressed bk. three, 'On the System of the World', as had once been his intention, bks. one and two would perhaps have been read in the context that, as Newton continually reminds his readers, is the only context in which they were conceived. I use the word "perhaps", however, because most readers would probably have disregarded Newton's warnings and would have interpreted the mathematical development of the laws of mathematical constructs and imagined systems as if they were the simple and straightforward elaboration of the laws of the physical universe, as indeed they often–but not always–turn out to be. Since Newton's eventual goal was to produce 'mathematical principles' of 'natural philosophy' and not 'mathematical principles' of arbitrary systems, his insistence on the separate levels of mathematical and physical discourse may have seemed no more than a distinction without a difference.

The confusion made by readers can be seen in a rather striking example from bk. two of the *Principia*, in which it is all too easy to forget (or ignore) that Newton is dealing with a mathematical system (or, in this case, an explanatory model) and to believe that he is discussing nature. In prop. 23, Newton analyzes Boyle's law: that in a confined gas (or "elastic fluid", as it was then called) the pressure is inversely proportional to the volume. In the *Principia*, Newton explores the consequences of supposing that there is a force of mutual repulsion between the particles composing such an "elastic fluid"; that is, he sets up a model of a compressible gas 'composed of particles that repel one another'. He then proves that Boyle's law is both a necessary and sufficient condition that this force must vary inversely as the distance between the centers of particles.

As is his custom, Newton sets forth a hierarchy of mathematical and physical analysis of cause. In this proposition, as he tries to make it absolutely clear to his readers, he is only dealing with a model, an explanatory model based on inter-particulate force, with the consequences of an assumed mathematical condition of force, not with the secondary question of physical reality. And he says so explicitly in a 'philosophical' scholium: 'But whether elastic fluids do really consist of particles so repelling each other, is a physical question.' Newton had only '. . . demonstrated mathematically the property of fluids consisting of particles of this kind, that hence philosophers [i.e., natural philosophers or physical scientists] may have occasion to discuss that question.' What may, however, have been a source of confusion to his readers is the fact that he has proved that the supposed law of force is both a necessary and a sufficient condition for Boyle's law. In other words, this is not a simple instance of hypothetico-deductive reasoning. Newton does not merely prove that if there is a force of repulsion $f \propto 1/r$, then Boyle's law must follow, and accordingly that the truth of Boyle's law may serve as assurance that there actually is a force of repulsion which accords with the law $f \propto 1/r$. Rather he has shown that, under the very special conditions that he posits, it can be proved additionally that the law $f \propto 1/r$ follows from Boyle's law. It is easy, therefore, to understand how John Dalton, on encountering this proposition, merely assumed that, because of the obvious truth of

Boyle's law, Newton had proved that gases are composed of mutually repelling particles.[21] He did not take Newton's *caveat* at its face value.

This example has been discussed earlier (in §1.4) in relation to the hierarchy of mathematical and physical causes, where a comparison was made between this explanatory model and the system Newton had imagined in which Kepler's laws are valid. There can be little doubt that Newton conceived of the two in quite different ways. The simple system for Kepler's laws could be easily modified and by degrees led to more complex systems and eventually to what Newton conceived to be the system of the world. The only conceptual problem was to admit a universal force of attraction. But in the model for Boyle's law, there was not only the problem of corpuscular forces of repulsion, but the forces would have to possess the additional property of terminating at or near neighboring corpuscles, even though the law of force $(f \propto 1/r)$ gives us no hint that this should be so. Newton knew that his simple system for Kepler's laws was a construct that does not correspond to reality; accordingly, he introduced more complex conditions that brought it into conformity with the real world as revealed by experiment and observation. He then found out how Kepler's laws must accordingly be modified. But for the Boyle's law model, even the law of force was not expressed clearly and mathematically without the addition of arbitrary boundary conditions to limit the range of the force's action. No *caveat* was needed for the simple mathematical or imagined construct in which Kepler's laws are true; it would be obvious to anyone that a one-body system cannot correspond to the world of nature. Nor was a *caveat* needed for the final complex system, because Newton believed that it was no longer a mathematical construct but corresponded to nature.

In regard to the Boyle's law model, however, there are a number of aspects that must be considered. First, there is the corpuscular composition of matter, to which Newton was firmly committed, and which he probably did not conceive to need any justification, although not all scientists at that time were corpuscularians or atomists. But the proposal that the particles of matter have forces associated with them was then quite radical and as yet to be tested. Furthermore, even if such corpuscular forces were to be admitted, there was the additional question of their short range. And, finally, there was the problem as to whether such a static model would correspond to nature (or even could be stable). If all these things were to be admitted, then Newton's proof would show that the forces must vary inversely as the distance. As opposed to what Newton considered the proven reality of universal gravity, here was a model that was based on many assumptions and hence required a public *caveat*. Whatever Newton believed in his heart of hearts about the forces of repulsion between the particles of elastic fluids, he needed a public *caveat* since he had no way of convincing any doubter as to the possible physical correctness of his model.[22] The model for Boyle's law was but a small part of a general theory of matter and of the reactions of matter (i.e., chemistry) based on the forces arising from and acting between the particles of matter. This Newtonian theory never attained either the mathematical level of the *Principia* or the physical completeness of the Newtonian system of the world. Possibly, therefore, this static model of a compressible gas remained no more than a mere explanatory model, even for Newton himself.

I have used the word "model" in discussing the system devised by Newton in order to explain Boyle's law. I have done so because it is a conceptual structure postulated to account for a certain range of evidence. In this it is to some degree similar to the kinetic-molecular model postulated to account for the gas laws in a later age. One could probably equally justify the appellation "model" for the aether of varying degrees of density which Newton put forth in order to explain the action of universal gravity. But in order to qualify fully as models in the sense that is accepted today, both of these explanatory proposals of Newton's would have to be shown to be metaphors rather than possible literal descriptions. That question would, however, take us too far afield. It can be noted, however, that what Newton uses in what I have called phase one and phase two of the Newtonian style are not models in this sense. They are imagined constructs, conceived often (but not necessarily) as mathematical analogues of nature simplified and idealized. But in the sequel to phase three, after the mathematical principles established in phases one and two have been applied to natural philosophy, Newton would appear to have used models (or to have used what are very like models) in order to explain the mode of action or of transmission of the force of universal gravity or the spring of the air, respectively according to the law of the inverse square or Boyle's law (see n. 13).

3.4   *Gravitation and attraction: Huygens's reaction to*
        the Principia

Newton had no doubt that universal gravity exists: i.e.,
that there is a force that draws together every two bodies in the
universe and that is quantitatively determined by the product of
their masses and (inversely) by the square of the distance between
them. He denied that gravity could be an essential property of
matter, and yet it was found in all matter. And, in fact, he devoted
a great deal of intellectual energy to the attempt to find a cause of
this force. In the *Principia*, however, this question was put aside.
Let us recall the famous words of the concluding general scholium:
'. . . I have not been able to deduce from phenomena the reason
[that is, the physical reason or cause] of these properties of gravity
and I do not feign [I don't frame] hypotheses'.[1] His critics would
agree that Newton had found the moon to move as if the same
causa gravitatis that causes weight or heaviness on earth acts to
cause the motion of the moon, and they could even go along with
Newton in assigning the same cause to the motion of the planets
around the sun, and of satellites around planets. Where they balked
was at conceiving this centripetal force to be "caused" by a force
of attraction.[2]

    Thus Huygens wrote:

> I have nothing against the *Vis Centripeta*, as M. Newton
> calls it, by which he makes planets gravitate toward the sun
> and the moon toward the earth, but am in agreement
> [with him] without [feeling] any difficulty: because
> not only is it known by experience that there is in nature
> such a manner of attraction or impulsion, but also that
> it is explainable by the laws of motion as one has seen
> in what I have written *supra* concerning gravity. Indeed,
> nothing prohibits that the cause of this *Vis Centripeta*
> toward the sun be similar to that which makes the bodies
> that we call heavy descend toward the earth. It was
> long ago that it was imagined that the spherical figure
> of the sun could be produced by the same [cause] that,
> according to me, produced the sphericity of the earth.[3]

In other words, Huygens knew quite well that there must be a
cause of some sort acting on the planets, since otherwise they
would move in straight lines according to the principle of inertia.
And the same is true of our moon in its motion around the earth.

And this may well be the same cause that makes terrestrial bodies be heavy and to descend in free fall toward the earth. This cause, for Huygens, could be the physical action of a set of vortices in a Cartesian or neo-Cartesian manner.

Huygens next explains how Newton made a great advance in knowledge which it had never occurred to him himself to make:

> I had not extended the action of gravity to such great distances as those between the sun and the planets, or between the moon and the earth; this because the vortices of M. Descartes which formerly seemed to me rather likely, and that I still had in mind, hindered me from doing so. I did not think, either, about this regular diminishing of gravity, namely, that it was in reciprocal proportion to the squares of the distances from the centers: which is a new and remarkable property of gravity, which it is, indeed, worth while to investigate [Huygens, 1888–1950, vol. 21, p. 472].

And then Huygens makes a rather surprising admission:

> But seeing now by the demonstrations of M. Newton that, supposing such a gravity toward the sun, and that it diminishes according to the said proportion, it counterbalances so well the centrifugal forces of the planets and produces precisely the effect of the elliptical motion that Kepler had guessed and proved by observation, I cannot doubt the truth either of these hypotheses concerning gravity or of the System of M. Newton, in so far as it is based upon it . . . [ibid.].

But planetary motion apart, attraction was something else. As Huygens put it:

> Concerning the Cause of the tides given by M. Newton, I am by no means satisfied [by it], nor by all the other Theories that he builds upon his principle of Attraction, which to me seems absurd, as I have already mentioned in the addition to the *Discourse on Gravity*. And I have often wondered how he could have given himself all the trouble of making such a number of investigations and difficult calculations that have no other foundation than this very principle.[4]

Furthermore,

> I do not agree with a Principle . . . according to which

all the small parts that we can imagine in two or several different bodies mutually attract each other or tend to approach each other.

That is something I would not be able to admit because I believe that I see clearly that the cause of such an attraction is not explainable by any of the principles of mechanics, or of the rules of motion. Nor am I convinced of the necessity of the mutual attraction of whole bodies, since I have shown that, even if there were no earth, bodies would not cease to tend toward a center by that which we call gravity.[5]

On reading the *Principia*, Huygens was forced to admit that 'Vortices [have been] destroyed by Newton',[6] but he eventually substituted for the destroyed Cartesian vortices a new kind of vortex so that the effects of gravity might still be explained by "matter" and "motion", according to the fundaments of the mechanical philosophy.[7]

Huygens, disturbed by the intrusion of the concept of attraction, failed to discern that this term appears primarily toward the end of bk. one of the *Principia*, where Newton is still concerned with mathematics rather than with physics, with what I have chosen to call here a mathematical construct, and not with physical reality. This was a distinction that Huygens himself was not able to make, or was not willing to make (hence he could not understand how Newton could have taken the trouble to concern himself with an investigation of the implications of a principle so absurd as attraction). A consequence was that Huygens was effectively denied the possibility of having discovered the inverse-square law. As Koyré (1965, p. 116) wisely remarked, 'Huygens paid a tremendous price for his fidelity to the Cartesian rationalism *à outrance*.' There can be no plainer testimony to the inhibiting effect of stringent philosophical presuppositions upon the creative force of a scientist of the first rank.

Huygens accepted the Newtonian demonstration concerning the cause (or force) operative in planetary motion and its identification with terrestrial gravity. For him Newton had made two 'suppositions' or 'hypotheses': that there is 'such a gravity toward the sun' and 'that it diminishes according to the said proportion [of the square of the distance]'. And he could not 'doubt the truth of either of these hypotheses concerning gravity or the truth of the system of M. Newton in so far as it is based upon it'. In short, Huygens was willing to go along with Newton's simplest construct, found at the beginning of bk. one of the *Principia*, because the gravity posited by Newton 'counterbalances so well the centrifugal forces of the planets and produces precisely the effect of the [Keplerian] elliptical motion'. Apart from the fact that Huygens had not really got the message of the *Principia*, and still thought of an interplay or counterbalancing of centripetal and centrifugal forces rather than the action of a centripetal force on a body with inertial motion, his statement is of interest to us because it shows his willingness to accept the one-body mathematical construct or the system of simplified and idealized version of nature of which it was the analogue. He did not balk at the concept of a center of force, since this could be harmonized with some kind of vortex concept. But Huygens did not and could not give credence to Newton's two-body construct or system, much less the many-body construct or system which came to have all the properties of the system of the world, because then there would have to be a force toward each of the bodies. This cannot be accommodated in a theory of vortices, in which the central body plays no physical role whatever. And this mutual action of two bodies on one another suggests attraction, for Huygens a concept that was absolutely taboo.[8]

We have seen how strongly Huygens inveighed against attraction, even to the point of asking himself how Newton could have spent so many tedious hours investigating and calculating the effects of a supposed principle of attraction, 'which to me seems absurd'. And in this connection Huygens referred specifically to the Newtonian theory of the tides. But Newton never uses the word "attraction" in relation to the tides. In fact, he tried to keep to the distinction he had made between mathematical models and physical reality, between the levels of discourse of bks. one and two of the *Principia* and bk. three ('On the System of the World'). Gravity and gravitation are physical concepts appropriate to bk. three, but we have seen that "attraction" was said by Newton to have been introduced in a mathematical and not a physical sense, and so was appropriate only in bks. one and two.

The way in which Newton kept the distinction he had made between "attraction" and "gravity" (or "gravitation") is shown by the *Index verborum*[9] to the *Principia*, which records a little more than three hundred occurrences of the noun *attractio* or of the verb *attrahere* in all forms of accidence. Of these, more than 90 percent appear in bks. one and two; there are only a mere 18 occurrences in bk. three, of which nine refer to electric or magnetic attractions. Of the others, two appear in unimportant parts of the discussion of comets, four are concentrated in the proof of a single proposition (prop. 28) and not particularly prominent.[10] Thus the reader, truly on the alert to see how Newton uses 'attract' or 'attraction' in bk. three, would be drawn to three examples (two only in the first edition). The first of these is to be found in corol. 1 to prop. 5, where in discussing gravity (and the gravitation of Jupiter 'toward all its satellites, Saturn toward its satellites, and the earth . . . toward the moon, and the sun toward all the primary planets'), Newton observes as a general principle that 'every attraction is mutual, by the third law of motion'. And in corol. 3, he says: 'Jupiter and Saturn, near conjunction, by attracting each other, sensibly perturb their mutual motions'.[11] And midway through the long proof of prop. 6, there is a reference to an 'inequality of attraction' that would perturb the motions of Jupiter's satellites. Elsewhere in bk. three, Newton rather uses *gravitas* and *gravitatio* (and not *attractio*) and *gravitare* (not *attrahere*). The *Index verborum* shows also that in bk. one of the *Principia*, there are only two occurrences of any form of the verb *gravitare*, both in examples to illustrate the definitions. The nouns *gravitas* or *gravitatio* are not found anywhere in bks. one and two. The words "gravity" and "gravitate" belong to the discourse of terrestrial and celestial physics and are appropriate to bk. three, but they have no place in the mathematical elaboration of the properties of imagined constructs in bk. one.

3.5   *Newton's path from imagined systems or constructs and mathematical principles to natural philosophy: the system of the world*

Newton's use of mathematical systems or constructs in relation to natural philosophy is nowhere better illustrated than in bk. two of the *Principia*. Each of the first three sections, in fact, explores the consequences of a different mathematical construct. In sect. 1, the motion of bodies is resisted 'in proportion to the velocity'; in sect. 2, the resistance is as 'the squares of the velocities'; in sect. 3, the resistance is 'partly in proportion to velocity and partly as the square of that proportion'. Obviously, all three cannot conform to the same physical reality, nor can all three be applied simultaneously to mathematicize the same physical phenomena. Newton himself emphasizes that his procedure is making use of imagined mathematical constructs when, in a concluding scholium to sect. 1, he informs his reader that the condition of resistance being proportional to the velocity 'is more a mathematical hypothesis than a physical one'.

Again, at the end of bk. two, he brings an introductory hypothesis into sect. 9, on the basis of which he proves that the theory of Cartesian vortices is inconsistent with Kepler's third law:

> *Hypothesis*
>
> The resistance arising from the want of lubricity in the parts of a fluid is, *caeteris paribus*, proportional to the velocity with which the parts of the fluid are separated from one another.

In a scholium further on in sect. 9, however, he says that this 'hypothesis' had been introduced merely 'for the sake of demonstration'. He has, in other words, proposed a construct or imagined system that can be no more than an approximation to physical reality, and has then used it in order to prove that the Cartesian theory of vortices is contradictory to the physical reality of Kepler's laws. The proof is not definitive, however, because it is based on such a construct, which is not a direct analogue of experiential reality. He then points out that if you do not use this construct but choose one that is more closely analogous to reality, the contradiction is even greater. The most likely conditions of reality give the greatest contradiction of all, the final confutation of the Cartesian theory. Thus, in relation to the construct (or "hypothesis") that he has introduced, Newton says that 'it is in truth probable that the resistance is in a less ratio than that of the velocity'; in this case the discrepancy between the vortex theory and Kepler's third law would prove to be even greater. In this example, be it noted, the destruction of the vortex theory of Descartes does not require an exact law of nature; a construct suffices, as long as it is known whether the difference between the true law and the one being supposed in the construct introduces a positive or a negative correction factor.

For physics, in the Newtonian style of the *Principia*, a major question must be how to advance from mathematical systems or constructs to physical reality, from 'mathematical principles' of such systems or constructs to 'mathematical principles of natural philosophy'. In the conclusion to the scholium at the end of sect. 11, bk. one, Newton gives explicit rules for proceeding from mathematics to physics, from imagined systems or constructs to natural philosophy:

> In mathematics, the task is to investigate the quantities of forces and those ratios that follow from any conditions that may be supposed. Then, moving on to physics, these ratios must be compared with the phenomena, so that it may be found out what conditions of forces apply to each kind of attracting bodies. And then, finally, it will be possible to dispute more safely about the physical species, physical causes, and physical proportions of these forces.

Each of the sentences in this paragraph corresponds to one of the three successive phases of the Newtonian method in the *Principia*. These are discussed more fully in §3.7, but at this point we may emphasize again that the power of the method derives from the fact that in phase one there is a complete freedom from any constraints of physical or even experiential actuality or of considerations of what may be permissible according to the "themata" or canons of acceptability that are imposed by the metascientific standards of the age. It is in phase two that comparisons are made between constructs and physical reality as revealed by experiment and observation and calculations based on real data. And it is only in a sequel to phase three, after the mathematical principles (established in phases one and two) have been applied to natural philosophy, that such questions as physical cause or the nature of a force need arise.

Let us see how Newton exemplifies these precepts in bk. three of the *Principia*. Newton's procedure is somewhat different from what might be imagined. He does not begin bk. three (on the System of the World) with Kepler's three laws as given by observation, and then apply the theorems from bk. one about the area law and elliptical orbits. Rather, he first proposes a set of 'Rules for the Study of Natural Philosophy' and then a set of 'Phenomena'.[1] The first two phenomena state the area law and harmonic law (but not the law of elliptical orbits) for the satellites of Jupiter and of Saturn.[2] The harmonic law is confirmed by measurements which are tabulated; the area law follows from the near-circularity of the orbits of Jupiter's satellites and their uniform motion. Phen. 3 shows observational evidence favoring the Copernican position that the orbits of Mercury, Venus, Mars, Jupiter, and Saturn 'encircle the sun'; phen. 4 establishes the harmonic law for these five planets and the earth. Phen. 5 and phen. 6 state the area law for the planets and for our moon.[3]

Since Newton omits the elliptical orbits from these phenomena, he cannot use directly the construct that led in bk. one to the law of the inverse square. But, for Jupiter's satellites the area law (plus props. 2 and 3, bk. one) shows that the forces by which these satellites 'are continually drawn away from their rectilinear orbits' are 'directed to the centre of Jupiter'. In the second part of prop. 1, bk. three, Newton uses corol. 6, prop. 4, bk. one, which deals with uniform motion in circular orbits, to show that the forces on these satellites are 'inversely as the squares of the distances of their places from that centre'.[4] The same is said to be true of the satellites of Saturn. The circular model of prop. 4, bk. one, is accurate enough in the case of the satellite systems of Jupiter and Saturn to serve in natural philosophy.

It is from phenomena again (Kepler's third law and the observed law of areas) that Newton sees how to apply his simplest construct, considered in terms of a one-body system with circular orbits (prop. 2 and prop. 4, bk. one), to show that there is a force—continually drawing the planets 'away from their rectilinear orbits' and maintaining them 'in their respective orbits'—that is directed to the sun and varies inversely as the square of the distance (prop. 2, bk. three.) Newton then proves this same prop. 2 in a different way by using a more advanced construct (introduced in prop. 45, corol. 1, bk. one); he now demonstrates from the observed 'fact that the aphelia are at rest' that the force must vary as the inverse square of the distance.[5]

Thus Newton begins by using the simplest possible construct, that of a one-body system and a circular orbit, for the satellites of Jupiter (and of Saturn in the second and third editions), and also for the planets. But he then goes to a more complex construct for the planets. This is taken from sect. 9 of bk. one, in which Newton shifts from considerations of bodies in immovable orbits to the motion of bodies in movable orbits, a topic which leads to 'the motion of the apsides'. He investigates here the difference between the forces producing uniform area–motion in an orbit at rest and in an orbit that is revolving about the center of force. For the case of nearly circular orbits, Newton explores the motion of the apsides (prop. 45, sect. 9, bk. one). This construct is a curious one, because Newton has not yet introduced the two-body system (which he will not present until sect. 11), nor the perturbations that arise if there are three or more interacting bodies. And yet the construct used in prop. 45 is that of a single body moving around a center of force, while being acted on by a second force (from an as yet unspecified source). This procedure in introducing movable orbits this early in bk. one may show how he was proceeding by orders of mathematical complexity in developing his constructs, rather than by making closer and closer approximations to physical nature.

In the construct in sect. 9, Newton ends up by considering (prop. 45, corol. 2) a double set of conditions: (1) that a body under the action of a centripetal force that is inversely as the square of the altitude 'revolves in an ellipse having a focus in the center of force', and (2) that there is some other 'extraneous force' (outside, or foreign, force) that is to be 'added to or taken away from this centripetal force'. Under these conditions, corol. 2 to prop. 45, bk. one, says that the 'motion of the apsides that will arise from that extraneous force can be found out . . . and conversely'. This 'extraneous force' is understood to be instantaneously directed through the primary center of force.

In the preceding corol. 1, Newton shows how to compute the magnitude of this 'extraneous force' from the motion of the apsides, one result being that it is only when the centripetal force is exactly as the inverse square of the distance that the apsides can be at rest.[6] For the planets, the apsides are at rest and thus the force must be as the inverse square (prop. 2, bk. three). The apsides of the moon are not truly at rest; nevertheless, the motion of the moon's apogee is 'very slow' and hence (prop. 3, bk. three) 'can be ignored'. In fact, for the observed motion of 'three degrees and three minutes forward' in every revolution, Newton says that the force would be as the minus $2\frac{2}{243}$ power of the 'distance of the moon from the center of the earth'.[7] This analysis of the opening of bk. three (both the nature of the evidence for the area law and the use of constructs from bk. one) indicates that Newton can hardly be said to have founded his system of the world on a simple phenomenological basis.

Having shown that there is an inverse-square force of the earth acting on the moon, and directed toward the earth's center, Newton proceeds (prop. 4, bk. three) to identify this force with terrestrial gravity. Essentially, the proof is as follows. He has shown that the earth's force on the moon is as the inverse square of the distance. Hence, at the earth's surface this force would be 60x60 times greater than it is out at the moon's orbit. Accordingly, the earth's force would (by the second law of motion) cause an object to fall on earth in one second through a distance 60x60 times greater than at the moon's orbit. This computation is confirmed by pendulum experiments, which yield the acceleration of gravity at the earth's surface. Accordingly, 'that force by which the moon is kept in its orbit, in descending from the moon's orbit to the surface of the earth, comes out equal to the force of gravity with us [here on earth]'.[8] The first stage of this proof is based on a simple construct, 'the hypothesis that the earth is at rest'. Then, Newton advances to a more complex construct (based on prop. 60, bk. one), in which 'the earth and moon move around the sun and in the meanwhile also revolve around their common centre of gravity'; this construct obviously corresponds more closely to reality.[9] By means of a procedural precept enunciated at the beginning of bk. three ('Hypothesis II' in the first edition, 'Regula Philosophandi II' in the later editions)—that 'the causes assigned to natural effects of the same kind must be, so far as possible, the same'—he argues (prop. 5, bk. three) that the force exerted by Jupiter on its satellites and by the sun on the planets must also be gravity, which is the force—whatever it is—that is the cause of weight on earth. Jupiter, Saturn, and the earth show by the motion of their satellites (or satellite) that they are centers to which a force (now identified as gravity) is directed. Since all planets 'are bodies of the same kind', there must be a force of the same kind toward planets without satellites; i.e., gravity occurs toward 'all planets universally', toward 'Venus, Mercury, and the rest' (prop. 5, corol. 1, bk. three). Furthermore, by Newton's own third law of motion, 'every attraction is mutual'; hence each planet 'will gravitate toward all its satellites . . . and the sun toward all the primary planets'. Then he concludes (prop. 5, corol. 2, bk. three) that the 'gravity that is directed toward every planet is inversely as the square of the distance of places from its centre'.

In the first edition, Newton proceeds at once to prop. 6, that 'all bodies gravitate toward each of the planets' and that at one and the same distance from the center of any planet the weights (or gravities) of all bodies are as their masses.[10] And this will lead to prop. 7, that all bodies universally gravitate toward one another, with a force proportional to the product of their masses. This, incidentally, is as close as Newton ever gets to a full and explicit statement of the law of universal gravitation.[11] While he does not here mention the factor of the distance between bodies, he has already shown at large that this force varies inversely as the square of the distance; and in the following prop. 8 he shows how to reckon the effective distance for the gravitational action of either homogeneous spherical bodies or bodies composed of concentric homogeneous shells.[12]

In the second edition of the *Principia*, Newton evidently felt that his procedure should be clarified and his case strengthened. After the two corollaries to prop. 5, therefore, he introduced a new corol. 3, in which he points out that if all planets are centers toward which a gravitating force is directed and if all planets are affected by the gravitating force, it follows that all 'planets are heavy toward one another'; that is, this result follows from the preceding corols. 1 and 2. Newton was most likely aware of the criticism that had been directed against his introduction of the concept of universal gravity, considering it to be an 'attraction', and of the further criticism that he had not been dealing with physics; he decided to stress the fact that here in bk. three he was indeed concerned with physics and phenomena and not just mathematics, with natural philosophy and not merely imagined constructs or even models.[13] And so, in the new corol. 3 Newton points to the phenomenological evidence of the universal gravity of planets and satellites: 'Jupiter and Saturn near conjunction, by attracting each other, sensibly perturb their mutual motions, the sun perturbs the lunar motions, and the sun and moon perturb our sea, as will be explained in what follows'.[14] Then, summarizing his procedure in a new scholium, Newton says: 'Hitherto we have called centripetal that force by which celestial bodies are kept in their orbits. It is now established that this force is gravity, and hereafter we shall call it gravity. For the cause of that centripetal force by which the moon is kept in its orbit ought to be extended to all the planets, by rules 1, 2, and 4.'[15] These 'rules' 1 and 2, to which Newton refers, had been 'hypotheses' 1 and 2 in the first edition. The point of hypothesis 2/rule 2 is that the same causes be assigned to effects of the same kind; hypothesis 1/rule 1 posits that no 'more causes of natural things' are to be admitted in natural philosophy 'than are both true and sufficient to explain their phenomena'.

The succeeding props. 9–12 introduce the 'force of gravity' within the body of planets, the stability of the solar system, the immovability of 'the common centre of gravity of the earth, sun, and all the planets',[16] and the motion of the sun being such as never to 'recede far from the common centre of gravity of all the planets'. Then in prop. 13, Newton introduces (for the first time in bk. three) the elliptical orbits of planets–a topic that must have seemed conspicuous in its absence from the phenomena at the beginning of bk. three and from the first twelve propositions. By this point Newton has established that there is an inverse-square force of gravity acting between sun and planets, so that he can use the properties of such a force that he established mathematically in bk. one. He says that he has thus far (in bk. three) discussed planetary motions 'from the phenomena'. Now, however, 'that we have found the principles of motions, we deduce the celestial motions from these principles a priori'. The planetary 'orbits would be elliptical, having the sun in their common focus, and they would describe areas proportional to the times' *if* 'the sun were at rest and the remaining planets did not act upon one another'. In other words, the Kepler-law solar system is not an exact representation of the world of nature; it is especially poor for Saturn (because of the perturbation caused by Jupiter) and for the earth, since 'the orbit of the earth is sensibly perturbed by the moon' (from the discussion of prop. 13, bk. three). It is, in fact, the common center of gravity of the earth–moon system that traverses an elliptic orbit about the sun at a focus, and it is by a radius-vector from the sun to this center of gravity that equal areas are described in equal times (from the discussion of prop. 13, bk. three). (Other departures from the simple system in which Kepler's laws are true are discussed in Ch. 5.)

Whoever reads bk. three carefully will be impressed by the continual display of the difference between physics or observation and the exactness of mathematics applied to the final complex system. For example, we have seen that the mutual actions of the planets are mentioned in prop. 13 on elliptical orbits: 'The mutual actions of the planets upon one another, however, are very small' and so 'can be ignored' except for Jupiter's action on Saturn; these mutual actions 'perturb the motions of the planets in ellipses about the mobile sun less (by prop. 66, bk. one) than if those motions were performed about the sun at rest'. In prop. 14 the motions of the aphelia 'are ignored . . . because of their smallness'. In prop. 21 there is predicted that there must be a nutation of the earth's axis, but it 'must be very small and either scarcely or not at all perceptible'.

Beginning with prop. 25, Newton explores the moon's motion and its inequalities. Here it is necessary to introduce simplifying assumptions, or to consider a series of constructs rather than the full reality. Thus, in prop. 26 Newton says: 'To make the computation easier, let us suppose that the orbit of the moon is circular, and let us ignore all inequalities with the sole exception of the one under discussion.' In prop. 28 he makes this assumption: 'Since the figure of the moon's orbit is unknown, in its place let us assume an ellipse . . . and let the earth be placed in its center . . .' In prop. 29 he concludes that he has thus far 'investigated the variation in a non-eccentric orbit, in which of course the moon in its octants is always at its mean distance from the earth'. Following prop. 34, he says, 'These things are so on the hypothesis that the moon revolves uniformly in a circular orbit'. In an elliptical orbit, the mean motion of the nodes 'will be diminished in the ratio of the minor axis to the major axis' and the 'variation of the inclination also will be diminished in the same ratio'. In prop. 35 he impatiently declares that he has supposed a certain angle to increase uniformly, for 'there is no time to consider all the minute details of inequalities'. In his rules for determining the moon's motion, given in the scholium following prop. 35 (appearing for the first time in the *Principia* in the second edition),[17] the careful reader will not fail to observe that the antepenultimate paragraph begins with an 'approximation' to make 'easier' the 'computation of this motion [which] is difficult'.[18]

It should be abundantly clear from such examples that even in the system of the world, particularly in the theory of the moon's motion, Newton had to use idealized systems or simplified constructs and that he introduced simplifying assumptions with respect to effects that were mathematically provable but small enough in magnitude to be ignored in a system of the world that was true only within the limits of observation. By the principles of Newtonian celestial dynamics, neither the pure elliptical orbits, nor the simple law of areas, nor the simple harmonic law could be accurate descriptions of the solar system, if that is a system of gravitationally interacting real bodies. We may well understand, therefore, why in the first edition of the *Principia* these were included among the "hypotheses" at the beginning of bk. three: they were the planetary hypotheses of the system of the world. At that time, any system of the world (the Ptolemaic, the Copernican, the Tychonian) was known as a "hypothesis"; accordingly, Newton quite legitimately could refer to the basic rules of such a system as "hypotheses". When Newton later changed the designation of these "planetary hypotheses" so that they became "phenomena" (not, however, "laws"), as I have said, he was probably indicating that these statements about the motions of primary and secondary planets are not true in the sense that mathematical laws are true, but are "true" only to within certain limits of accuracy of observation. Or, they are "physically" exact even if they are not "mathematically" exact. This distinction between the two types of exactness is introduced by Newton in prop. 48, bk. two, of the *Principia* in relation to a proportion that is not exact ('Accurata quidem non est haec proportio'). Nevertheless, unless certain contractions and expansions of a given elastic fluid are not too great, he says that this proportion 'will not be wrong so far as the senses can perceive and so it can be considered as physically exact' ('. . . non errabit sensibiliter, ideoque pro physice accurata haberi potest').

To what conclusion are we led? That mathematics is exact and that nature is not.[19] The fine mathematical structure shown by Newton's analysis led to complexities and difficulties which even Newton himself could not fully resolve, and so he was forced to make approximations. Or, to put it in another way, in dealing with the physical system of the world, it was possible to ignore certain aspects of the system which were shown by mathematical analysis but which were (Newton hoped) of so small a magnitude that they could be ignored within the limits of observation, even with the best telescopes of the time. I believe it is important to keep this distinction in mind, because one is otherwise apt to suppose that there was for Newton an exact correspondence between the mathematical constructs or imagined systems and the physical reality, whereas in fact bk. three ('On the System of the World') is itself replete with reasoning employing mathematical constructs or imagined systems or results derived from such constructs and systems.

Supplement to 3.5. Newton's first version of his System of the World and his "mathematical way" in fact and fiction

It will be helpful in understanding Newton's path from 'mathematical principles' to 'natural philosophy' if we examine his treatment of universal gravity in an early version of bk. three of the *Principia*, known today as the *System of the World*.[1] This is the work to which Newton refers in the beginning of bk. three, when he says: 'I composed an earlier version of book 3 in popular form, so that it might be more widely read . . . I have [here] translated the substance of the earlier version into propositions in a mathematical style [*more mathematico*], so that they may be read only by those who have first mastered the principles.' In the present context, we may be concerned not only with the meaning of the phrase '*more mathematico*', but with the treatment of gravity.

Whoever makes a comparison of these two texts cannot help being struck by the fact that the expression "universal gravity" does not occur *expressis verbis* anywhere in the *System of the World*, which is quite different from what we have seen to be the case in bk. three of the *Principia*. In sect. 2 of the *System of the World* ('The principle of circular motion in free spaces'), Newton briefly reviews some theories as to how planets stay in their orbits, including the vortices of Kepler and Descartes, and 'some other principle whether of impulse or attraction'. In the first draft this is followed by a statement: 'That some force is required is most certain from the first law of motion. By us that force, lest we determine its type hypothetically, is called by the general name "centripetal".' Then, after an intermediate version, Newton decided to say:

> From the first law of motion it is most certain that some
> force is required. We have undertaken to find out its
> quantity and properties and to investigate mathemati-
> cally its effects in the moving of bodies; consequently, lest
> we determine its type hypothetically, we have called by
> the general name "centripetal" the force that tends
> toward some center, or even (taking the name from
> the center [toward which the force tends]) "circumsolar"
> the force that tends toward the sun, "circumterrestrial"
> the one that tends toward the earth, "circumjovial" the
> one that tends toward Jupiter, and so on.[2]

There can be no doubt that the adverb "mathematically" modifies the verb "investigate";[3] Newton is merely saying that he is using the methods of mathematics to explore the effects of the force on the moving of bodies. But, when he rewrote the *System of the World*, turning it into bk. three of the *Principia*, he introduced much more mathematics, producing an account of the system of the world that required a deeper acquaintance with the mathematical principles developed in bk. one than had been the case for the earlier version. At the same time he changed the external form from a series of prose paragraphs to numbered propositions, corollaries, and scholia, plus lemmas and problems, in the manner of the preceding bks. one and two.[4] Thus Newton tells us that in the first *System of the World*, he had 'undertaken to find out the quantity and properties' of the force that keeps planets in their orbits and 'to investigate mathematically its effects in the moving of bodies'; and that in bk. three of the *Principia* he had translated 'the substance of the earlier version into propositions in a mathematical style'.

If I have insisted, in what may seem to the reader to be an unnecessarily pedantic manner, on what Newton has actually said, the reason is that there has been a quite different explanation given of Newton's use of the word "mathematical" in the *System of the World*. The latter is incorporated in an English phrase: Newton's "mathematical way".[5] This derives from the English version of the *System of the World* first published in 1728, where Newton is made to say

> ... from the laws of motion, it is most certain that these effects must proceed from the action of some force or other.
>
> But our purpose is only to trace out the quantity and properties of this force from the phenomena, and to apply what we discover in some simple cases as principles, by which, in a mathematical way, we may estimate the effects thereof in more involved cases; for it would be endless and impossible to bring every particular to direct and immediate observation.
>
> We said, *in a mathematical way*, to avoid all questions about the nature or quality of this force, which we would not be understood to determine by any hypothesis; and therefore call it by the general name of a centripetal force, as it is a force which is directed towards some centre; and as it regards more particularly a body in that centre, we

call it circumsolar, circumterrestrial, circumjovial, and so in respect of other central bodies.[6]

These last two paragraphs sound very "Newtonian", but they have no basis of authenticity in Newton's own MS copy of the text of this work, which served as the basis of the printed Latin text (U.L.C. MS Add. 3990), nor are they to be found in the transcript made by an amanuensis at Newton's direction and which was deposited in the university library (U.L.C. MS Dd. 4.18). There are a number of manuscript copies of this work in existence, and these two paragraphs appear in none of them. Despite these facts, writers on Newton's method continue to discuss 'Newton's "mathematical way" ' as if this were an authentic Newtonism,[7] which it certainly cannot be said to be. The mathematical method of the *Principia* embodied the use of mathematical systems or constructs and the application of mathematical techniques from geometry and algebra, the mathematical theory of proportions, the application of infinite series, and above all the method of limits. To have Newton's "mathematical way" be anything less in the language of today is to make a travesty of his magnificent achievement.

As to Newton's name for the force active in celestial motions, we may observe that in the *System of the World* he does indeed write about planets being kept 'in definite orbits by centripetal forces' (sect. 3) and uses the words "gravitates" and "gravity" exclusively in relation to bodies on or near the surface of the earth.[8] He writes about 'centripetal forces [that] tend towards the bodies of the sun, the earth, and the planets' (sect. 5), and shows that these 'centripetal forces decrease as the square of the distances from the planets' centres' (sect. 6), that 'the circumsolar force decreases . . . as the square of the distance from the sun' (sect. 9). Even the moon test is presented as proof that 'the circumterrestrial force decreases as the square of the distance from the earth' (sects. 10, 11), not as a proof that the 'circumterrestrial force' acting on the moon is none other than gravity. Of course, when he comes to the application of the third law to show that satellites exert a force on planets, he no longer uses the word "centripetal" but introduces "attraction", presumably in the sense of sect. 11 of bk. one, in which he had said explicitly that he did not have in mind a particular physical significance. But he does not use "attraction" consistently.[9] Thus he writes (sect. 22) of 'the forces of small bodies', but discusses these forces in the following text in terms of the words "attract" and "mutual attraction"; and (in sect. 21), he writes that the 'sun attracts Jupiter and the other planets, Jupiter attracts its satellites . . .' In the *System of the World* (sects. 23, 24), Newton shows that 'forces proportional to the quantity of matter . . . tend towards all terrestrial bodies' and that 'these same forces tend toward the celestial bodies'. But he also discusses (sect. 23) 'the attractive forces of all terrestrial bodies' and he introduces (sect. 24) 'the attraction of all the planets' toward any given planet along with 'the circumsolar force' and 'the circumjovial force'.

Thus, in the *System of the World*, Newton's transition from the systems or constructs of bk. one to the world of physical reality did not go quite so far as in the *Principia*. He establishes a universal force and shows that the same force acts on planetary satellites, planets, and terrestrial bodies, but he uses what he conceives to be the neutral term "attraction"[10] (along with "circumsolar", "circumterrestrial", "circumjovial", and "centripetal" force) and he does not ever write of universal gravity as a force or of gravitation as such. Only after 1685, when he recast the *System of the World* into bk. three, did he apparently decide that the universal force should be given the concreteness of positive identification with the terrestrial force of gravity, so as to become the universal gravity for which the *Principia* is famous.

3.6 *Mathematical systems or constructs and the review of the Principia in the Journal des Sçavans*

Newton's use of mathematical systems and constructs in a physical setting could easily lead to a total misunderstanding in the mind of a hostile critic. One such, a strict Cartesian, who may have been Pierre Silvain Régis,[1] expressed his views in the *Journal des Sçavans* (2 August 1688) as follows:

> The work of M. Newton is a mechanics, the most perfect that one could imagine, as it is not possible to make demonstrations more precise or more exact than those he gives in the first two books on lightness, on springiness, on the resistance of fluid bodies, and on the attractive and repulsive forces that are the principal basis of Physics. But one has to confess that one cannot regard these demonstrations otherwise than as only mechanical; indeed, the author recognizes himself at the bottom of page 4 and the

top of page 5 that he has not considered their Principles as a Physicist, but as a mere Mathematician [*Géomètre*].[2] Although the tone is unmistakably pejorative, there can be no doubt that the reviewer may well have understood the character of bks. one and two. The reference (the end of page 4 and the beginning of page 5) is to a statement of Newton's toward the end of the discussion of def. 8: 'has vires non physice sed Mathematice tantum considerando'. It is to be noted that the reviewer calls what Newton has done a "mechanics", where we would talk of an imagined system or a mathematical construct, even a kind of mathematical model or a hypothetical situation or condition. Furthermore, even in his reference to the sentence just quoted, the reviewer transforms Newton's *mathematice* into *geometrice*, and changes Newton's 'forces' into 'principles'.[3]

The reviewer's reference to the bottom of page 4 and the top of page 5 is especially interesting because Newton is there taking pains to differentiate the mathematical system or construct from physical reality. It is there that he says: 'I use the words *attraction, impulse,* or *propensity* of any sort whatever toward a center for one another without distinction and indiscriminately, considering such forces not physically but merely mathematically [*non physice sed Mathematice tantum*].' He is not, in bks. one and two, 'defining a species or mode of action or a physical cause or reason', and he specifically admonishes the reader to 'beware of thinking that by words of this kind' he has done so. Furthermore, he is definitely not 'attributing forces truly and physically to centers (which are mathematical points)' when he 'happens to say that centers attract or that centers have forces'.[4]

The reviewer, however, failed to notice the careful distinction Newton had made between the 'mathematical principles' of bk. one (and bk. two) and their application to 'natural philosophy' in bk. three. The reviewer actually supposed bk. three itself to be only mathematical and hypothetical, thus at best displaying a "mechanics" and not a physics or a natural philosophy: 'He [Newton] confesses the same thing at the beginning of the third book, where he endeavors nevertheless to explain the System of the World. But it is [done] only by hypotheses that are, most of them, arbitrary, and that, consequently, can serve as foundation only to a treatise of pure mechanics.' In fact, bk. three *does* begin with a set of 'hypotheses', at least in the first edition. That is, Newton states the phenomenological basis of his physics, his natural philosophy, as a series of 'hypotheses', along with two methodological precepts, and an un-provable statement about the 'center of the system of the world'.[5] Thus, when Newton invokes the results of observation to show the degree to which aspects of the system conform to *phenomena*, he does so (in the first edition) by referring to a particular *hypothesis*. For example, in the proof of prop. 1—that the satellites of Jupiter are drawn toward the center of Jupiter by a continual force that varies inversely as the square of the distance—Newton says that this appears from 'Hypoth. V. & Prop. II. vel. III. Lib. I.' and 'Hypoth. V. & Corol. 6. Prop. IV. ejusdem Libri'.[6] The reviewer may have been guilty of a little willful or purposeful misunderstanding, but Newton had given him a possible ground for belief that bk. three rested on "hypotheses", since this was literally true.

The reviewer also took Newton to task because he '. . . bases the explanation of the inequality of the tides on the principle that all the planets gravitate reciprocally toward each other. . . . But this supposition is arbitrary as it has not been proved; the demonstra-tion that depends on it can therefore only be mechanics.' And he then concluded: 'In order to make an *opus* as perfect as possible, M. Newton has only to give us a Physics as exact as his Mechanics. He will give it when he substitutes true motions for those that he has supposed.'[7] The subsequent alteration of the designation of 'Hypotheses' to 'Regulae Philosophandi' and 'Phaenomena' may have been Newton's direct response to this criticism.[8] For thus he could make it clear that bk. three presented a phenomenologically based physics or natural philosophy and not a purely hypothetical or imagined system or a merely mathematical construct.

Another review, some thirty years later, also anti-Newtonian, took a somewhat different tack, not making a distinction between a "mécanique" and a "physique" but between the point of view of a geometer and a physicist. The review opens as follows: 'The reputation of this work is constant among geometers, who admire the force and profundity of the genius of the author; and it is chal-lenged by physicists, who for the most part have not known how to reconcile themselves to a natural attraction [misprinted as *atten-tion*], that he alleges to exist between all bodies' (*Mémoires pour l'histoire des sciences & des beaux arts* [Trévoux, February 1718], vol. 67, pp. 466–475). The reviewer observes correctly that the first two books of the *Principia* are characterized by the exercise of New-

ton's 'esprit Géométrique' and that it is only in bk. three that Newton moves on to physics. He summarizes Newton's views quite correctly at this point. In physics, or natural philosophy, we are not at
liberty to imagine any hypothesis we may wish. He delineates the
Newtonian point of view in these words: '[Newton] says that in
order to arrive at the knowledge of the true system of the world,
one must not put one's faith in one's imagination, but must consult nature; that fictions, however ingenious they may be, are none-
theless fictions, whereas experience leads to reality' (ibid., p. 470).
He then shows how Newton gives evidence to support the idea of
gravity, describing in detail Newton's argument that the moon
falls constantly to the earth with a force varying as the inverse
square of the distance. In the present context, it is less significant
that the reviewer carps at a technical point in Newton's argument
than that he clearly sees the distinction between the mathematical
character of bks. one and two and the reality factor that is introduced into bk. three in a shift from mathematical systems or constructs to the 'true system of the world'.

3.7   *Newton's three-phase procedure in action: Newton's
constructs compared to Descartes's models and to those in
use today*

Newton's procedure in the *Principia*, which I have designated the Newtonian style, is displayed in an alternation of two
phases or stages of investigation. In the first, the consequences of
an imaginative construct are determined by applying mathematical techniques to the initial conditions concerning mathematical
entities in a mathematical domain. In the second phase the physical counterpart of the initial conditions or of the consequences are
compared or contrasted with observations of nature or with ex-
perientially based laws and rules. This usually gives rise to some
alteration of the conditions of the initial construct, producing a
new phase one, followed by a new phase two, and so on. Such a
mathematical construct is usually founded on a simplified and
idealized natural system, of which it is the mathematicization and
the analogue. The succession of phases one and two may eventually
generate a system which seems to embody all the complexities of
nature.

There is a great temptation to think of such constructs or mathematical systems and their sets of initial conditions as some sort of

"hypotheses"; but in doing so there is a real danger to our understanding of Newton's procedure. In this connection, something must be said about the expression "hypothesis". This word is simply the Greek for "supposition", or for a presupposition in an argument. In Latin texts of the seventeenth century, it is used somewhat interchangeably with *suppositio*, a late (i.e., nonclassical) noun. Thus Descartes, writing in French, will use *une supposition*, which may appear in a Latin version as either *suppositio* or *hypothesis*. In 1672, Newton objected strongly when Father Pardies called the Newton theory of light a "hypothesis"; the reason was that Newton believed he had not merely supposed his conclusions but had derived them from (and proved them by) experiments. At the time of writing the *Principia*, the word "hypothesis" did not as yet have for Newton the extreme pejorative sense of the later slogan, 'Hypotheses non fingo'. Not only do plainly marked "hypotheses" appear in the beginning of bk. three (1687) and in sect. 9 of bk. two, but many mathematical deductions contain the expression "per hypothesin", referring to the conditional clause of the proposition being proved. (Later on, in 1729, these would be translated by Motte not as "by hypothesis" but as "by the supposition".) By the 1690s, however, Newton began to take a hard line on hypotheses. He became troubled by those who concocted a new ad hoc hypothesis for every phenomenon, so that (as he put the matter) there would be as many hypotheses as phenomena, which would hardly advance true science. In the succeeding years, he began to use the word "hypothesis" sparingly in his own writings, often in relation to an unprovable or possibly unproved proposition (as in the two "hypotheses" in bk. three in the second and third editions). But he did label as "hypotheses" those theories of rivals and opponents that he wished to dismiss out of hand.

We must, therefore, be very cautious about the word "hypothesis" in reading Newton's early writings or in discussing his methodology. Each of the constructs proposed by Newton in the *Principia* has a set of initial conditions or assumptions; these could properly be called "suppositions" and translated into Latin as *hypotheses*. But these constructs are not "hypothetical" in a general sense because they are not proposed as purely imaginary systems to account for physical nature or to explain particular phenomena. Many of Newton's constructs are only mathematicizations of simplified or ideal natural conditions, or may be based on either generalizations of such conditions or imagined variations of such conditions. They would not be acceptable if they were found to contradict the laws of experiment or the results of observation, in the sense that Descartes's vortices lead to a contradiction of Kepler's laws.

As a pure mathematician, Newton had no need of placing any restrictions on the constructs or imagined systems whose properties he wished to explore. But as a mathematical natural philosopher, he had as his goal to invent and elaborate the properties of only such constructs that seemed reasonable and that appeared to hold the possibility of being useful for natural philosophy, for explaining the world as revealed by experiment and observation. Newton was always first and foremost a mathematician, and so he could not entirely restrict himself to natural conditions. A mathematician's instinct always leads him to generalizations. We shall see below how Newton made just such generalizations of the conditions for Kepler's law and for Boyle's law.

But there can be no doubt that the main concern in the *Principia* is not with mathematical systems and constructs in general, but primarily with those that may either approximate or be equivalent to the experiential world of nature. By this I mean that they had to accord with the generally accepted principles of Newton's physics, that they would predict (or retrodict) the data of observation and experiment (or the laws based on those data), and that they would to some degree seem reasonably to be analogues of systems that do or might occur in nature. Of course, one cannot ever say that any given construct of this kind is equivalent to the reality of nature, for that would imply a knowledge of nature's reality and would obviate the need for the construct save in some such sense as the simplification of calculations. But Newton was greatly concerned in the *Principia* to know whether the conditions he explored were only mathematical or were possibly the conditions of nature. This feature is evident in the occasional "philosophical" scholium in bk. one and two of the *Principia*, in which Newton raises the question whether or not the situation under discussion might pertain to physics, if only to a limited degree, and not be restricted to a construct. We have seen examples of such scholia. But a particularly striking one occurs in the final sect. 14 of bk. one. Having recognized the 'analogy' that exists 'between the properties of rays of light' and the motion of certain 'minimally small bodies', New-

ton says that he '. . . decided to subjoin the following propositions
for optical uses, meanwhile not at all disputing about the nature
of the rays (that is, whether they are bodies or not), but only determining the trajectories of bodies, which are very similar to the
trajectories of rays [of light]' (scholium foll. prop. 96, sect. 14, bk.
one).

Here we see an example of both phase one and phase two. In
phase two Newton seeks to find out the degree to which the laws or
propositions of one or other of the constructs may be congruent
with, or closely approximate, phenomenologically determined
laws. In the *Principia* Newton does not explore phase two in full
save with respect to the system of the world: the motion and physical properties of sun, earth, planets, moons, and comets, and certain terrestrial phenomena such as tides, falling bodies, and the
shape of the earth. Once it had been seen that such congruences or
close approximations occur, or the conclusion had been reached
that the conditions of the construct may be modified so as to apply
to the explanation of nature, then the investigation could move on
to phase three: the use of the principles, laws, and rules found in
phases one and two in the elaboration of the system of the world.

Newton did not strictly limit such constructs to simplified conditions of nature, or to conditions that he believed might actually
occur in nature. But in almost every case the construct tended to
be somewhat similar to nature, however simplified, or it represented a natural condition with a change in the value or the power
of some term, or it embodied a natural possibility or potentiality
according to Newton's view of nature. Thus Newton might propose
a construct in which (perhaps only temporarily) one or more natural conditions, such as the gravitational interactions between the
planets, might be eliminated. But Newton never made use of a construct that went strictly against either his personal or the commonly accepted principles of nature: such as a system in which there
could be "physical" mass without the property of inertia. For example, Newton was aware that in nature the resistance of physical
media always is, *caeteris paribus*, some function of the speed of a
body. Newton accordingly considered cases in which the resistance
may depend in various different ways on the speed, but never did
he explore a resistance that would be independent of the speed or
that would decrease with increasing speed.

Knowing Kepler's harmonic law, that 'the periodic times are as the 3/2 power of the radii', he also explored the consequences of assuming that 'the periodic time is any power R^n of the radius R'.[1] Starting out with Boyle's law in the form of the density of a gas being as the compression, he found the centrifugal forces to be as $1/D$, where D is the distance between particles, and conversely. At once he generalized this result in a way that transcends the limitations of physical nature, considering that the cubes of the compressing forces are 'as the fourth power of the densities', and that 'the cubes of the compressing forces' will be as the fifth or even the sixth powers of the densities'. In the most general case, 'the compressing forces' are as the cube root of E^{n+2}, where E is the density, 'corresponding to a force of repulsion between the particles being inversely as any power D^n of the distance' (scholium to prop. 23, bk. two).

These examples exhibit nature simplified or extended, but they do not ever flatly contravene either the principles of nature according to Newton's belief or the phenomena of nature according to his knowledge. They thus differ from Descartes's apparent use of models in his *Dioptrique* (1637). There Descartes introduces three models to illustrate the transmission of light, each of which is a fundamental contradiction of his own principles of natural philosophy or of his view of nature. One model is a tennis ball moving at a finite speed, whose speed is altered as it goes from one medium to another;[2] whereas Descartes insisted that the transmission of light must be instantaneous. Another likens the propagation of light to grapes within 'a vat completely full of half-pressed grapes' immersed in wine; the vat has one or two holes at the bottom. This model is intended to exemplify the subtle material (the wine) that fills all of space and the 'heavier parts of the air as well as other transparent bodies'. Again, motion here is finite and not instantaneous (Descartes, 1965, p. 69). In yet another model, Descartes compares the motion of light to a blind man with a cane. There is no loss in transmission time: the blind man feels the sensation in his hand at the very instant when the cane strikes an object.[3] If the cane is rigid (and if not, there will be a transmission time), then this model does not preserve Descartes's distinction between motion and tendency or inclination (*conatus*) to motion, since the rigid cane cannot transmit a tendency or inclination to motion without at the same time transmitting the motion iself.

These models differ in their action from the world of nature, according to Descartes's own principles. They thus play a very different role in Descartes's thought from that played by the constructs or imagined systems in Newton's thought. In one respect Descartes's procedure resembles the use of models in classical physics, where the argument from analogy may provide useful information. For example, in classical physics, a model may be conceived of a gas made up of elastic particles in motion; certain conclusions may then be drawn in regard to energy, temperature, and so on. Similarly, Descartes uses his model of the tennis ball and racquet to derive the law of refraction, which was published for the first time in his *Dioptrique* (1637) (see Sabra, 1967, ch. 4).

In introducing these models in the *Dioptrique*, Descartes makes it clear that he is aware of 'the great difference between the stick of that blind man and the air or the other transparent bodies by the means of which we see' and that he has only been making a "comparison" (*comparaison*); the grapes in the wine is also such a comparison. In a letter to Morin (13 July 1638), Descartes referred again to the example of the blind man and the stick as a 'comparaison' which he said he had introduced primarily 'pour faire voir en quelle sorte le mouvement peut passer sans le mobile'.[4] Hence this model is presented for purely heuristic purposes: not to show what light or its transmission may be like, but rather to indicate that the kind of properties of motion he is invoking may occur in nature. As such, this use of models is like the evocation of electric and magnetic forces in an argument concerning gravitation: not to indicate that gravitation may be electrical or magnetic, or even have a similar cause or origin, but only to show that attractions do occur in nature. Descartes also uses the word *comparaison* in *Le monde*.[5]

Today's dictionary gives "simile" and "metaphor" as the fundamental synonyms of "comparison". A simile is 'an imaginative comparison between objects which are essentially unlike, except in certain aspects'.[6] This would equally apply to the use of models in scientific thought. The main difference between Descartes's *comparaison* and the models that have become characteristic of scientific thought is that Descartes wanted to illustrate a single property by means of a *comparaison*, whereas the use of models tends to reveal properties of nature that might not be discovered by direct experiment and observation or as consequences of a theory. An alternative use of models is in relation to a theory that is either not well established or fully acceptable or that presents certain concepts or principles that so go against conventional science that the author refers to his creation as a model rather than a theory. Thus Bohr presented his theory of atomic structure and spectral lines in relation to a "model" and Einstein did not refer to a "theory of photons".[7] But, as Mary Hesse points out, it would 'be odd today to speak of a wave model of sound'.[8] It was Descartes's goal to reduce complex phenomena to their 'simple natures', to entities of which we have certain knowledge, such as matter and motion.[9] Each *comparaison* made by Descartes illustrated a single particular property or aspect of light by means of a mechanical system. He apparently never envisaged that a single mechanical model might exhibit all the properties of light, probably because then it would replicate all the complexities of nature itself and thus not be helpful to our understanding.

In a real sense, then, Descartes's use of "models" may be akin to the way in which scientists and philosophers of science make use of "models" in their thought in our own day. Newton, as we shall see below, not only believed that matter is particulate, but also that the particles are either endowed with forces or have forces associated with them. So we may see why, when Newton considers a gas or elastic fluid that accords with Boyle's law, he may legitimately ask what the law of force is that produces this relation. But when he then went on to propose an explanatory system for Boyle's law, he was acting in a way that, as is the case for Descartes's *comparaisons*, is similar to today's usage. For the record shows that Newton rarely (if, indeed, ever) wrote with any true conviction about such particulate forces, and in this case (as we saw above) there are real problems, such as their terminating on nearby particles. Newton (scholium to prop. 23, bk. two) said specifically that 'whether elastic fluids [i.e., compressible gases] do really consist of particles so repelling one another is a physical question'. He had done no more than to demonstrate 'mathematically the property of [elastic] fluids consisting of particles of this kind' so that natural philosophers might 'take occasion to discuss that question'. The Cartesian *comparaisons* and the Newtonian "models" differ in a fundamental way, because in Descartes's *comparaisons* light is considered to be (or is compared to) a stream of moving particles or a kind of movement, whereas for Descartes light is only a *conatus* or tendency to motion. But for Newton there was the open possibility that his explanation of Boyle's law might correspond to the actual case in nature and insofar transcend the property of being a "model", as we would understand that term. We shall see in §3.11 that Newton attempted to construct systems to explain the properties of light that to some degree may partake of the character of "models".

Descartes, unlike Newton, gave a truly hypothetical character to his optics by introducing such *comparaisons* that were false according to his own principles. But he went further than that—in his *Discours de la méthode, Le monde,* and *Principia philosophiae*—when he informed his readers that he was introducing fables or novels (romances).[10] And at one point he even said that he was using false hypotheses.[11] At the very beginning of the *Principia*, Newton too may have seemed to have constructed an imaginary or fictitious universe. That is, he proceeds as if he had invented an imaginary system which absolutely transcends reality. This problem had to arise as soon as he began to write up his mature thoughts concerning force and motion and celestial mechanics, in the work that became the *Principia*. He was faced with the problem of imaginary system versus reality in the first of the three 'axiomata sive leges motus'. This law begins: 'Corpus omne perseverare in statu suo quiescendi vel movendi uniformiter in directum' ('Every body perseveres in its state of resting or of moving uniformly straight forward')—and then states the condition 'nisi quatenus a viribus impressis cogitur statum illum mutare' ('except insofar as it is compelled to change that state by forces impressed upon it'). In the real world, in which every body attracts and is attracted by every other body, there is no possibility that a given body will not have 'forces impressed upon it' and so be 'compelled to change that state.' In a sense, we can say that Newton was saying no more than that the first law may hold only for a purely imaginary or fictitious or hypothetical situation, either a universe with a single body in it and no force fields or a universe in which bodies do not gravitationally interact with one another.[12]

But to have the first axiom or law of motion be so characterized by pure hypothesis would give the Newtonian elaboration of dynamics a character wholly out of keeping with Newton's intentions and ordinary procedure. It would be more consistent with the spirit of the *Principia* to say that Newton is here proposing an extremely imaginary system, which in its pure state has only a limited analogy with the world of ordinary physics: In it there is a single body (or particle or mass point) that may move freely through a space of no resistance and is subject to the action of no external forces produced by other bodies and of no force fields. This is, in fact, the mathematical construct that Newton himself will shortly propose in prop. 1, where he starts out with just such a single body moving with pure inertial motion in a space free of resistance, in the absence of external forces or force fields. This is the wholly imaginary system that Newton is going to use, in other words, in order to illustrate the link between the law of inertia and Kepler's area law and to establish the significance and meaning of the area law. While this system cannot exist in nature, it can be approximated in our minds in the vast empty spaces far beyond the solar system, in which gravitational forces may be minimal.[13] But Newton himself does not suggest any such approximation to his imagined system.

It is a significant fact that on both occasions when Newton introduces this system, he at once introduces a further condition that changes it into the kind of construct which he uses normally in phase one of the style that characterizes bk. one of the *Principia*. Thus, in prop. 1, bk. one, Newton shows that under the initial conditions of this imagined system, the moving body or particle or mass point will trace out equal areas in equal times by a line drawn from it to any other point in space not on the line of motion. But then he introduces an external force, by whose action the moving body is given a single instantaneous blow or impulse, which alters both the direction and magnitude of the original motion; after a while there is another blow, and then another and another. Newton lets the time between successive blows decrease indefinitely, and in the limit there is a continuous force. The first alteration of the imagined system for prop. 1, by the introduction of a single blow of impulse which produces a change in motion (or momentum), corresponds to the alteration of the imaginary system proposed for axiom 1 (or law 1) by the introduction of axiom 2 (or law 2). The condition for the possibility of law 1 ('except insofar as it is compelled to change its state by forces impressed on it') becomes the condition of actuality in law 2, the statement of what happens when a force is indeed impressed on a body.

A single body in a universe with no forces and no resistances and with no other body with which it can collide is an extreme case of an imagined system in the *Principia*. It is so extreme, in fact, that Newton does not insist on it. In the physical examples intended to illustrate law 1 by showing that in nature there is a continuation of inertial motion, Newton does not invoke a particle in some remote space in the universe, far away from any other bodies and accordingly removed from gravitational forces of any significant or observable magnitude. Rather, the examples he gives involve circular (or curvilinear) motion. In them, there is a force acting, but it is centrally directed and so perpendicular to the tangential inertial motion. According to the rule for finding components of forces producing accelerations in any given direction ($F \times \cos \theta$), the component affecting the inertial motion ($F \times \cos 90°$) is null. Mathematical analysis has shown an instance of a long-term continuing inertial motion in the regularities of the solar system that have been observed for millennia.[14]

The reason that the foregoing example is extreme, going far beyond the conditions of the ordinary mathematical systems or constructs of phase one, is that the latter are usually mathematicizations of nature simplified and idealized. Neglecting perturbations, the earth–sun physical system behaves very much like the construct of props. 1 and 2 of bk. one of the *Principia*. The earth is so small and of such little mass compared to the sun that it may be considered as a particle moving about a fixed center of force. That is, the action of the earth to move the sun is all but negligible; or, the common center of gravity about which both the earth and the sun move in orbit is not only within the body of the sun but is very close to the sun's center. It is very different for the earth–moon system and even for the sun–Jupiter system; but the earth–sun system is a taking-off stage for the construct of a mass point moving about a center of force. But there is no such simple starting point for the system of pure inertial motion. Newton did not choose to give any example of a situation even approximating pure inertial motion, perhaps because of the awesome implication that pure inertial motion continues in a rectilinear path indefinitely–which implies at once properties of infinite or unbounded space which he may well have wished to avoid. In any event, he got over the difficulty by at once introducing (as we have seen: prop. 1, law 1–law 2) a further condition in which the indefinite or infinite extent of the motion would be curtailed.[15]

In one sense, Newton was flying in the face of the new science, which tended to start out from empirically established laws and properties. Galileo, for example, was less interested in constructing possible or imagined physical systems than he was in basing his

definitions and laws on nature herself (see §1.4). But Newton begins (phase one) with imagined systems or constructs such as a one-
body system with a central force field, and then proceeds to the
law of areas, which he held to be based on phenomena. Kepler had
put Newton's sequel to phase three first, giving a primacy to the
nature of the solar force and the principles of motion in his search
for planetary laws. As a matter of fact, Newton seems in his phase
one to have partially resembled the schoolmen of the fourteenth
century rather than the founders of the new science. They too had
constructed mathematical systems and had then explored the consequences of the conditions they had imposed. But there was a
major difference between them, since Newton always had a phase
two in mind whereas the medieval thinkers seem to have been not
at all concerned with the degree to which their mathematical systems, or the laws which they had derived from them, might or
might not be valid in explaining the physical world of external
nature.

3.8   *Newton's third phase and its sequel: the cause of gravitation*
        The great advantage of the three-phase Newtonian proce-
dure is that it separates the basic questions of science into separate
categories. In phase one Newton could explore the consequences
of any condition or conditions that he found mathematically interesting or stimulating–and he could do so just as his inspiration
drove him, without being blocked or deflected by questions whether certain forces or conditions of resistance do or do not ever occur
in nature (or might or might not do so). We have seen, in the case
of Huygens, how inhibiting it was not to have this freedom.
        One cannot overstress the importance of this lack of a premature
restriction for the creative effort of a scientific imagination such as
Newton's. When, in 1679, Hooke explicitly posed the problem of
planetary motions arising from a combination of an inertial component and a centripetal force, Newton–in response–did not stop
to consider whether there exists any known kind of force that could
possibly extend from the sun to the earth, or to Saturn and beyond;
or whether such a force was the result of pressure, or of a shower
of aethereal particles, or the effect of a vortex, or the result of an
aether of varying degrees of density. For Newton, these considerations became of paramount importance in relation to the force
acting on the planets only long *after* he had explored the mathe-

matical consequences of the conditions set by Hooke.[1] That is, Newton was able to consider the problem of planetary motion in its mathematical aspects, and only later, when he had found that his results conform with experience, did he have to face the physical (or, in his own terms, "philosophical") problem of what kind of entity this centripetal force could possibly be. When he thought that the simple system used in the beginning of bk. one of the *Principia* conformed to reality, there were many physical explanations of the planetary force that might have seemed possible (including some kind of vortex or set of vortices[2]); but once he had found that the planetary force is mutual, exerted by the sun on each planet and by each planet on the sun, and that this is the same force that keeps the moon in its orbit, and is the same as terrestrial gravity, then all of the known physical explanations failed.[3]

At that stage in the development of his thought, there were three major choices open to him. One was to assume that nature has endowed bodies with forces which may act on other bodies over great distances of empty space; this would contravene the accepted principles of the mechanical philosophy to which Newton had given his adherence and which would thus have to be modified. The second was to abandon the celestial mechanics he had developed, and to reject his own creation because it made use of "attraction", which was the kind of concept supposed to have been banished from science. The third was to accept the "fact" of universal gravity and to explore possible mechanisms or causes for its action, which make bodies tend to move toward one another as if by attraction.

At present, many scholars believe that Newton adopted the first of these positions. But the whole range of documentary evidence seems to me to declare otherwise. As I see it, Newton found that he could construct the system of bks. one and two of the *Principia* from a mathematical standpoint, in terms of a series of imagined systems or constructs whose physical reality or lack of reality was not primarily in question *at this phase* of the investigation. In phase two he found that certain forms of the basic construct (or system) led to an agreement with the phenomena to an extent that gave him confidence that the construct was not fictive; that is, it predicted or retrodicted the known phenomena and even new and as yet unknown effects that were later confirmed by observation. Phase three consisted of the elaboration of the system of the world, the application of the mathematical principles to natural philosophy. The results were stupendous. In his private world, and not in the public world of the *Principia,* he then devoted himself to an exploration of the cause of the gravitating force: the force that causes bodies to be heavy on earth and to be accelerated downward in free fall, the force by which the earth pulls on the moon to keep it in its orbit, the force exerted by the moon and sun to produce tides in the seas, and the force mutually exerted by the sun and planets on one another. Some of the properties of this force had been disclosed by the mathematical investigations in phases one and two and their applications in phase three: that gravity extends to great distances, that it diminishes as the square of the distance from bodies, that it is null inside of homogeneous spherical shells, that it is exerted by a homogeneous spherical body or a body composed of concentric spherical shells (on an external particle or body) as if all of its mass were concentrated at its geometric center, and that it acts on a body in proportion to its mass or quantity of matter and not its surface–and so differs from mechanical actions such as resistance to motion in fluids or the production of motion by pressure.

Newton's three-phase system has thus led him into a position which must have seemed perplexing. According to the accepted canons of natural philosophy, there was no conceivable way in which a force could act according to these properties.[4] And yet just such a force (as he later said) 'does really exist' and according to these properties explains the phenomena of the world. Phase three was followed by a sequel: the process of finding a cause of gravity and of understanding how gravity may operate. Phase three is equivalent to constructing a new natural philosophy in which the force of universal gravity is an essential ingredient. The sequel to phase three may even include the construction of explanatory mechanisms or quasi-physical models to account for or to explain gravitational attraction. The simplest way out would have been for Newton to assume that gravity is an essential property of matter, and to let it go at that, as Cotes did when writing his preface to the second edition of Newton's *Principia,* and as Bentley seemed to do (see §3.9). But Newton said again and again that he did not so consider gravity to be essential to matter, as impenetrability and inertia are.[5] We shall see him successively attempting to account for gravity by a kind of aethereal shower, by electricity, by a new kind of all-pervading aether of varying density. None of these attempted explanations worked fully and in detail. One reason for their failure is that they are all mechanical models of action, and we know today that gravity cannot be explained mechanically. They were never better than hypotheses–suppositions or speculations–and they just would not do. Newton never gave up this search, as we know from such documents as his tentative proposed revisions to the *Principia*, and the later queries planned for or published in the *Opticks*. In searching for a "cause" of gravity, Newton was in fact hoping to find some kind of causal mechanism to explain its action and to make its existence seem reasonable.

At the time of writing the *Principia*, he may have considered the most likely such explanation to be some kind of aethereal shower or a stream of aethereal particles. The evidence is found in the first edition of the *Principia*, in the only reference there to a possible cause of gravity or gravitational attraction. This occurs in the introduction to sect. 11, bk. one, when he formally introduces the two-body system. In this famous paragraph (discussed in §3.3), we have seen that Newton said that he would consider the mutual centripetal forces of bodies as 'attractions, although perhaps–if we speak in the language of physics–they might more truly be called impulses'. Impulses, or instantaneous percussive forces, arise from the action of some sort of particles that strike a body, as in a shower or stream of aether particles. From then on, Newton invented or tried out a series of explanatory "models" of physical action, none of which was fully satisfactory, and none of which ever transcended the stage of being more than just a "model", never attaining the status of what Newton could be convinced was truth or reality. These attempted explanatory "models" differ from the mathematical constructs or systems which characterize what I have called phase one. They are not mathematicized systems based on nature idealized and simplified, with given mathematical conditions of force and resistance, from which Newton draws the consequences or implications by the use of formal mathematics: geometry, algebra, proportions, application of the method of limits or fluxions, and infinite series. Rather, they fall into the category of imaginary or postulated mechanisms such as the motion of aethereal particles, the effect of electric effluvia or something like the effect of electric effluvia, the action of some kind of aether, the mediation of something that may be material or nonmaterial. As such they are similar to the models that characterize present thinking in the sciences and philosophy of science (these various attempts are discussed in §3.9).

In the first edition of the *Principia*, Newton made no statement concerning the possible cause of universal gravity, other than the above-mentioned reference to impulsion in the introduction to sect. 11, bk. one. In an unfinished *Conclusio* which he suppressed before the *Principia* was sent to the printer,[6] he discussed the attraction and repulsion of particles of matter, such as gross bodies are composed of, but he did not enter directly into the question of the cause of universal gravity.[7] Nor did he address this question in the drafts of the preface to the first edition.[8]

By the time of the second edition of the *Principia*, however, there was required a public statement, which appears in the concluding general scholium. It is here that Newton adopts a somewhat positivist point of view—but only "somewhat" in that he does insist that gravity 'really exists' ('revera existat') as well as that gravity is sufficient to explain the various phenomena of the universe. (Obviously, the phrase 'really exists' is anti-positivistic.) In one sense, we may see here direct overtones of the point of view that in phase one had enabled him to consider the mathematical consequences of a mathematical system or imagined construct based on the concept of a centripetal force without having to ask any physical questions about the force itself. Now, having shown in phase three that gravity exists and serves to account for the phenomena, he declares the validity of his system of the world, even though the cause of gravity remains hidden from us. Newton was never, however, a true positivist, because he himself never gave up the search for the cause of gravity and because he actually believed that such a cause exists and may be found; but he acts like a positivist to the extent that his system is said to be acceptable because it works, even though the cause of universal gravity may be unknown and even if gravity itself cannot be explained.[9] In the 'scholium generale', Newton's intention was not to set a boundary to scientific inquiry, although post-Newtonian scientists have tended to read that document as if it did set such a boundary.[10] But in that final scholium, Newton did set forth a standard of sufficiency for the acceptability of scientific systems, theories, or explanations that did not require an explanation of the forces or other causes of observed effects, and this standard was accepted as a matter of course by post-Newtonian scientists.

It was fortunate indeed that Newton was able to develop and to use his three-phase system; for he not only showed the way for the exact sciences to follow thereafter, but he saved himself from becoming hopelessly mired in a fruitless quest. By this I mean that after he had written the *Principia*, he did explore the question of what might possibly be the cause of gravity, and he continued to do so on and off for the rest of his life; but it never became an all-consuming passion to the inhibiting exclusion of everything else. He revised the *Principia*, readied the *Opticks* for publication, revised and expanded the queries and worked up propositions concerning the moon's motion, studied perturbation and tide theory, and so on. His published and unpublished writings do not show that a search for the cause or modus operandi of universal gravity was ever a major intellectual activity. His failure in this quest did not inhibit him from publishing either the *Principia* or the *Opticks*, with its queries as to the possible cause of universal gravity. As he said in the 'scholium generale', he had shown that universal gravity exists, and he had shown that universal gravity suffices to explain the phenomena of the heavens and the earth. This was the fruit of the first three phases. He was naturally curious about the nature of gravity, but in his eyes his system of the world was acceptable without such knowledge.

Not only was Newton himself unsuccessful in working out the cause or modus operandi of gravitation, but in terms of the aims he set himself no one else was ever able to do it either. Newton's own thoughts as to how gravity (and later universal gravity) might be produced went through a number of vicissitudes. In the early 1660s, he believed terrestrial gravity to be caused by a kind of "shower" of aethereal particles (see Westfall, 1971, pp. 330–331), and in 1679 he suggested in a letter to Boyle that gravity might be caused by a nonhomogeneous aether with a density varying according to a certain rule.[11] He had found experimental support for belief in an aether that could resist motion: a freely swinging pendulum was observed to slow down and come to rest in an evacuated vessel almost as quickly as it does in ordinary air. Newton interpreted this experiment as proof that there is an aether: something which remains in a receiver when the air has been pumped out, something that resists motion.[12] By the time of his solution of the problem of elliptical orbital motion according to an inverse-square force, presumably in 1679–1680 (that is, during or after his exchange of letters with Hooke), it was still possible for him to believe in gravity being caused by the pressure of a density gradient in the aether, or even by some kind of aethereal vortex. The reason, as has been mentioned, is that Newton had not yet come to the point of applying his third law or axiom; as yet, there did not have to be a mutual force between the earth and terrestrial objects, or between the sun and planets, or between the planets and their satellites. The change is documented in the revision of his tract *De motu*, sometime during or after December 1684 (see §5.6). From then on, the simple aethereal explanations would not do.

At some time before he wrote (or while he was writing) the *Principia*, Newton made another pendulum experiment, this time in air, which seemed to him to show that the resistance of the aether was either null or very small; presumably such an aether could not possibly produce any of the mechanical effects, such as gravitation, for which it had been intended.[13] In reporting this experiment in the *Principia*, Newton says that he is basing his presentation on memory, since the paper on which the results were recorded was lost (it has never been found among his papers). Newton gives no hint as to when these experiments had been done, but I would guess the most likely date to be after the composition of *De motu*, that is, after December 1684, probably when he was actually writing bk. two of the *Principia*. This would accord with the fact that Newton was still writing about vortices in relation to planetary motion in 1680 and 1681, as if there were no firm reason to reject the vortex-based explanations used by his correspondents Burnet and Flamsteed.[14] Thus Newton replied to Flamsteed in relation to the sun's heat (7 March 1681 N.S.) that 'the central parts' of 'the liquid matter swimming in the Sun' must 'become as hot as if the hot fluid matter surrounding it equalled the whole Vortex'. We are here more interested in Newton's unquestioning acceptance and echoing of Flamsteed's version of the vortex theory than his conclusion that 'the whole body of the Sun therefore must be red hot & consequently voyd of magnetism' (Newton, 1959–1977, vol. 2, p. 360). A few years later, Newton publicly attacked the concept of vortices in the *Principia*, showing in the conclusion to bk. 2 that they contravene Kepler's laws of areas. But he had believed in vortices in the late 1660s or early 1670s, when he had used the supposed vortical motion of an aether in the Cartesian manner to explain certain features of the moon's motion by what D. T.

Whiteside (1976, pp. 317–18) has called 'the pressure of the solar vortex upon the terrestrial one in which the moon rides its "planetary" path'. The fact that in 1680 and 1681 Newton was still writing as if the concept of an aether moving in a vortex was directly related to solar forces, indicates not only that he had not as yet made the pendulum experiments, but also that—even after the correspondence with Hooke—he had not as yet fully committed himself to ontologically independent solar or planetary forces as the only way to account for all of the observed motions of the planets and the moon.

The revisions to *De motu*[15] (made soon after November 1684) mention the aether as if it existed, even though its resistance seemed to be 'either null or . . . exceedingly small'.[16] Newton says that he has been considering 'the motion of bodies in non-resisting media' so that he 'may determine the motion of the celestial bodies in the aether'.[17]

Very soon after the revisions of *De motu*, Newton began to write the *Principia*. By this time his views concerning the aether have become a little less positive. He still does refer to the aether as such (by name) in a number of places,[18] but it is obvious that the aether concept did not play a significant role in the composition of the *Principia* itself. What is more notable is that although Newton sometimes refers to the aether as if he were an ordinary believer, at other times he discusses the aether as if he were doubtful as to whether the aether actually did (or could) exist. Thus he introduced the pendulum experiments 'because it is the opinion of some[!] that a certain aethereal and extraordinarily subtle medium exists that quite freely permeates all the pores and passages of bodies', and he said in the opening paragraph of the treatise (def. 1) that 'For the present, I am not taking into account any medium, if there should be any[!], freely pervading the interstices between the parts of the bodies'.[19]

In some parts of the *Principia* Newton mentions the aether, in others he refers to a subtle medium. There are some places where the aether is invoked only indirectly, by implication. I have referred above to the introduction to sect. 11, in which Newton comes to talk about mutual attraction rather than centripetal forces, and he says that 'in the language of physics' attractions 'might more truly be called impulses'. Impulses of what? The only answer that comes to mind would be impulses of some kind of aethereal particles, as in the earlier belief of the 1660s. That he must have had the aether in mind is reinforced by the conclusion to this same sect. 11, in which he lists among the possible causes of attraction 'the action of aether or of air or any medium whatsoever–whether corporeal or incorporeal–in any way impelling toward one another the bodies floating therein'.[20]

The belief that attraction must be caused by a medium between (and even permeating) gross bodies persisted even after the *Principia* had been published. In 1693, in a letter to Bentley (25 February), Newton said that 'Gravity must be caused by an agent acting constantly according to certain laws, but whether this agent be material or immaterial is a question I have left to the consideration of my readers' (Newton, 1959–1977, vol. 3, pp. 253sq; 1958, pp. 254sq). While Newton does not here commit himself to a material agent, he also does not exclude the possibility that such an agent would turn out to be material. But, whatever the agent, it would have to act 'constantly according to certain laws', as Newton and others supposed the aether would do. What is of greatest significance is that at the time of writing the *Principia* and very soon afterwards, Newton obviously did not believe the force of gravity to be an entity that could stand by itself or have an independent existence, for (as he said to Bentley) the idea 'that one body may act upon another at a distance through a vacuum without the mediation of any thing else by or through which their action or force may be conveyed from one to another is to me so great an absurdity that I believe no man who has in philosophical matters any competent faculty of thinking can ever fall into it' (Newton, 1958, pp. 302sq). In the context of this discussion, the word "material" may have been intended by Newton to mean having the properties of ordinary matter, primarily impenetrability and inertia or mass.

In the same year (1693) in which Newton was writing to Bentley, he also discussed gravitation, along with aether, in correspondence with Leibniz. 'Some exceedingly subtle matter', he wrote, 'seems to fill the heavens' ('At caelos materia aliqua subtilis nimis implere videtur') (Newton, 1959–1977, vol. 3, pp. 286, 287). It had to be 'exceedingly subtle' because of the pendulum experiments, which had set a limiting upper bound to the possible resistance an aether could offer to the motion of the body through it.[21] Newton even

went so far as to write to Leibniz: 'But if, meanwhile, someone explains gravity along with all its laws by the action of some subtle
matter, I shall be far from objecting (Newton, 1959–1977, vol. 3,
pp. 286, 287). And, in fact, at about this same time he embraced
with alacrity and enthusiasm an attempt by Fatio de Duillier to
account for gravity by a hypothesis based on the concept of the rectilinear motion of aether particles; Newton went so far as to say
that this was the only possible "mechanical" explanation of gravity
(see Hall & Hall, 1962, pp. 313, 315). Fatio's hypothesis of an 'aether
or such like subtile matter' had the additional virtue that from the
'rectilinear motion in all directions' of the aether 'there is deduced
the action of gravity in reciprocal proportion of the squares of the
distances'.[22]

Some time before 1702, in an essay on the theory of the moon,
however, Newton stated bluntly that there is no fluid medium in
space.[23] Then, for a time, Newton came to think that gravitation
might be caused electrically, apparently on the basis of some experiments performed by Hauksbee.[24] This supposition is expressed
in a final paragraph of the general scholium at the end of the *Principia*, written for the edition of 1713.[25] Newton's last thoughts on
the matter, at least in print, were a return to an aether, or an aethereal medium, as in the second English edition of the *Opticks* in
1717/1718.[26] This time the aether was "thin" rather than "thick",
and was homogeneous; Newton suggested that it might produce its
effects by means of a variation in density.

By 1685, when Newton had transformed the concept of a centripetal force acting on a body into a mutual attraction between
two bodies, the concept of the aether raised two major types of
fundamental questions. The first of these has to do with a vortex.
For it is a property of a vortex that it will tend to draw an orbiting
body toward the center whether there is a physical body at the center or not. In this way, the vortex theory denies the essentiality of
having two interacting bodies as a condition for gravitation. As
Newton said expressly in introducing sect. 11, bk. one, of the *Principia*, 'attractions . . . are directed toward bodies' and not toward
mathematical centers of force; and 'by the third law of motion,
the actions of attracting and attracted bodies are always mutual and
equal'. Huygens, however, argued–from the point of view of the
traditional vortex theory–that Newton was wrong. Huygens was

not at all 'convinced of the necessity of the mutual attraction of whole bodies, since' (as he wrote) 'I have shown that, even if there were no earth, bodies would not cease to tend toward a center by that which we call gravity'.[27]

But, vortex theory apart, there is another major kind of problem about explanations of gravitational attraction by means of an aether or an aethereal medium. The aether shower or motion of aethereal particles and the aether with a density gradient can explain how one body is urged toward another. Such a theory, for example, might very well show how a terrestrial object is pushed or pulled toward the earth, how the moon is pushed or pulled toward the earth, but not the converse. That is, aether theories generally fail to provide for a necessary *equal* and *opposite* force on each of two bodies: an apple and the earth. In addition to this qualitative problem, there is also the quantitative one of having aether considerations yield a resultant effect that is a force of attraction both directly proportional to the product of the two masses and inversely proportional to the square of the distance between them. And so it is no wonder that, in despair, Newton would (as recorded by Fatio de Duillier, 1949, p. 117) 'often seem to incline to think that Gravity had its Foundation only in the arbitrary Will of God' (Newton, 1959–1977, vol. 3, p. 70).

Newton tended to write about these various ways (dare we say models?) to explain gravity with some degree of tentativeness, at least in print, and we have no way of determining the absolute degree of his commitment to any of them.[28] From the point of view of the present chapter, however, it is important to observe that the law of universal gravity and its effects as delineated in the *Principia* are unaffected by the particular choice of explanation that Newton was adumbrating at any given time. Each mode of explanation is but a variant sequel to phase three, as I have called it, and insofar has no effect upon phase one (the construction of mathematical systems and constructs and the elaboration of their mathematical properties and consequences) or phase two (the investigation of the degree to which such systems and constructs agree with experiment and observation, or may need modification in order to fulfill such agreement). Being no more than a sequel to phase three, neither the devising of explanations for gravitation nor the search for a cause of gravitation or for its modus operandi was essential to the acceptance or rejection of the Newtonian *Principia*—at least for those who were willing to accept the Newtonian style in natural philosophy, or some variant of it.

3.9 *The Newtonian revolution as seen by some of Newton's successors: Bailly, Maupertuis, Clairaut*

In the age of Newton—roughly the first three-quarters of the eighteenth century—when it came to be recognized that science progresses through a series of revolutions, there were three primary candidates for the honor of having instituted revolutions in science: Copernicus, Descartes, Newton (see Cohen, 1976a). Galileo and Kepler, surprising as it may seem, were not credited by Jean-Sylvain Bailly with having inaugurated a revolution, although he greatly esteemed their contributions to astronomy. Bailly, who used the new concept of revolution in science more systematically than any other author of this period with whom I am familiar, did not explicitly attribute a revolution to Descartes, as d'Alembert had done,[1] but he praised Descartes's 'sublime idea of daring to reduce the general laws of motion for the universe to laws of motion for terrestrial bodies' (Bailly, 1781, p. xi). This was a wholly new idea, one belonging exclusively to recent times, he said, and it was due to Descartes. Of course, the vortices of Descartes proved to be a 'bad explanation of weight and of the system of the world', but at least these vortices had the positive virtue of being 'mechanical'. Descartes deserved the greatest possible credit, according to Bailly, because: 'He discovered that the same mechanical cause [*le même méchanisme*] must make bodies move in the celestial regions and on the earth's surface; even though he did not grasp the nature of this mechanical cause [*méchanisme*], one must not forget that this great new concept was the fruit of his genius.' And then he concluded: 'That which Descartes had proposed, Newton carried out. We strip Newton of none of his glory in rendering justice to Descartes' (ibid.). Others, such as the young Turgot, explicitly credited Descartes with having made (or inaugurated) a revolution in the sciences.[2]

For Bailly, as we saw in §2.2, Copernicus both destroyed an old system of the world and established a new one. He was 'the restorer of physical astronomy, and the author of the true world system'. A 'seditious mind', Copernicus 'gave the signal and the revolution occurred'. In yet another presentation, Bailly said that 'at this time [*époque*]' Copernicus 'made a great revolution and changed everything'.[3] Copernicus was responsible for a two-stage revolution, or two revolutions in one. The first was the elimination of the old Ptolemaic or geocentric system, the second the presentation of the new heliocentric system.

Bailly found the same features of a two-stage revolution exhibited in the advent of the Newtonian natural philosophy. In a characteristic presentation, Bailly first praised Newton for his modesty (a propos of the preface to the first edition of the *Principia*) and then went on to describe the revolution:

> Newton, more than anyone else, had to beg pardon for his lofty place. He had taken so extraordinary a flight and he came down again with such novel truths that he had to humor those minds that would have rejected these truths. Newton overturned or changed all ideas. Aristotle and Descartes still divided up the empire and were the preceptors of Europe. The English philosopher destroyed almost all of their teachings and proposed a new philosophy. This philosophy brought about a revolution. Newton achieved—but by more gentle and more appropriate means—that which conquerers who usurped the throne sometimes attempted in Asia: they wished to eradicate the memory of previous reigns so that their own reign would inaugurate an era [*époque*], so that everything would begin with them. But these arrogant and tyrannical enterprises were most often fruitless; they succeeded only insofar as reason and truth could obtain this advantage without false pretension! [Bailly, 1785, vol. 2, bk. 12, sect. 42, pp. 560sq.]

The use here of a full panoply of political metaphor is most striking: conquerors usurping the throne and wiping out all trace of their predecessors, and the contrast between violence or tyranny and reason or truth. But, again, it is to be noted that for Bailly a revolution in science is a two-stage action. Bailly (1785, vol. 2, bk. 13, sect. 1, p. 579) warned his readers, however, that although Newton's 'treatise on the Mathematical Principles of Natural Philosophy was destined to produce a revolution in astronomy', it was nevertheless true that 'this revolution did not occur all at once'.

Bailly was not content merely to state generalities concerning the Newtonian revolution in science. As he saw it, the key that in

Newton's hands unlocked the celestial mysteries was mathematics: geometry. First of all, Bailly said, the planets move in curved paths, and all curvilinear movement is the product of several forces; ergo, 'mathematics [*géométrie*] supposes two forces'. One is the 'force' in the heavenly bodies (uniform and constant)[4] and the other is 'placed in the sun, & capable of drawing toward it all the bodies which find themselves within its sphere of activity'. The force is not constant, and it must decrease as it spreads out; from simple geometrical considerations, it follows that this force can decrease only as the law of the inverse square of the distance. Geometry shows that under these conditions of force and motion, the planets describe areas with respect to the sun that are proportional to the times of description; they move in ellipses with the sun at a focus; and the periods of revolution are as the square roots of the cubes of the distances from the sun. Hence, 'the three consequences of this supposition are the three great phenomena observed by Kepler's genius' (Bailly, 1785, vol. 2, bk. 12, sect. 9, p. 486). Bailly does not follow the exact line of the *Principia*, but he has grasped the essential point: by mathematics–specifically geometry–Newton has found that the three phenomenologically or testably true laws of Kepler are consequences of the law of universal gravitation, which is thus demonstrated to be no mere imagined hypothesis, but rather a principle and system that actually conforms to the real world. Or, as Bailly expressed himself: 'What is supposed to make things move is what really makes things move; the demonstration was complete. Newton alone, with his mathematics [*géométrie*], divined the secret of nature'.

By mathematics (or "geometry") Bailly did not mean the classical geometry of Euclid, nor even the analytic geometry of Descartes and Fermat. He meant specifically the differential and integral calculus. He pointed out, quite correctly in my view, that:

> Newton formed an inseparable union between mathematics [*géométrie*] and astronomy; these two sciences [now] advance together and the separate progress in each of them is necessary for the progress of the other. The intimate knowledge of what goes on [*connoissance intime des choses*] depends on their accord inasmuch as one of them observes and the other one explains, inasmuch as mathematics predicts phenomena and astronomy

observes and confirms what has been predicted [Bailly, 1785, vol. 3, discours 6, pp. 326sq].

Then, after referring briefly to some of the post-Newtonian advances in astronomical science, he referred to the way in which, during this same time, 'mathematics [*géométrie*] has advanced by improving the two kinds of calculus invented by Newton'. One of this pair was the differential calculus, which 'left Newton's hands full-fledged'. The other, the integral calculus, was still incomplete. 'The complete method of the integral calculus would be a revolution in mathematics [*géométrie*] comparable to that of the application of algebra [to geometry] and to that of the invention of the differential calculus'. Despite this lack, 'three mathematicians, MM. Clairaut, d'Alembert, & Euler, Newton's successors, have been able—by pursuing a Newtonian path—to see better and further than he had' (Bailly, 1785, vol. 3, discours 6, pp. 327sq). Newton, as Bailly noted, had completely resolved only the two-body problem; with respect to the problem of three bodies mutually attracting one another gravitationally, Newton's genius could do no more than reveal to him 'the most palpable effects of this complexity'. Bailly compared Newton to a 'conqueror who has subjugated a vast empire', but who was not able by himself to make all parts submit to his rule: 'he imposed laws, and left it to the care and talent of his successors to make them known everywhere'. The 'great three-body problem' had been 'in effect' resolved by Clairaut, d'Alembert, and Euler, in a manner that is 'everywhere applicable, which is the base of all investigations of this kind, and which is the glory and distinctive characteristic of our century' (ibid.).

With rare insight, Bailly saw that 'the advantage of mathematical solutions is that they are general'. The argument that if the planets move according to Kepler's laws, they must be 'impelled by a force residing in the sun' depends only on mathematical or geometrical considerations and general principles of motion. No special physical properties of the sun appear in Newton's argument, which insofar differs from Kepler's, since the latter had invoked such special qualities of the sun as its magnetic force and the orientation of its poles. Accordingly, the identical mathematical argument shows that the satellites of Jupiter and Saturn, subject to the same laws of Kepler, must be equally 'impelled by forces residing in these two planets'. In other words, Jupiter and Saturn are to their satellite systems what the sun is to the planetary system, the only difference being one of extent and power. And the same is true of the earth and our moon (Bailly, 1785, vol. 2, bk. 12, sect. 9, pp. 486sq).

Bailly fully understood that to find the links between inertial motion plus centripetal force and Kepler's laws was 'a problem in mathematics [or geometry]': not of ordinary geometry (as has been just mentioned), but of 'a geometry [mathematics] which Newton had made suitable for these profound investigations' (ibid., sect. 5, p. 477). According to Bailly, furthermore, 'This mathematics was not in any case intimidated by the size of the orbits, of the variation in the speeds, of the enormity of the forces necessary to transport the ponderous masses of the heavenly globes'. The new science created by Newton, later called dynamics, 'considers force only insofar as it is manifested by its effects, the spaces traversed and the time expended. Science cares little whether the forces are weak or strong; it can consider a great many forces at a time, under a general abstract expression . . . (ibid., p. 478). Here then is seen the full force and virtue of Newton's mathematics [*géométrie*]; his methods 'are universal, . . . and are as great as nature which embraces all' (ibid.).

Bailly himself was perfectly willing to accept the concept and principle of a universal gravitating force, since so many phenomena were explained by its use: So many of the observed data and experiential laws could be derived by mathematics from the properties of universal gravity (Bailly, 1785, vol. 2, bk. 12, sect. 41, pp. 555sq). He was aware, however, that at first many scientists (notably in France) made a distinction between the Newtonian system as mathematical and as a true natural philosophy. Thus with respect to Maupertuis, who (according to Bailly) 'appears to us to have been . . . the first of our mathematicians to have used the principle of attraction', Bailly (1785, vol. 3, 'discours premier', p. 7) had to point out that 'at first he considered it only in relation to its calculable effects; he accepted gravitation as a mathematician, but not as a physicist'. That is, Maupertuis went along with the Newtonian mathematical system or construct (our phases one and two), but would not grant that in the system of the world (phase three) Newton was necessarily dealing with reality.

In fact, in a paper 'On the Laws of Attraction' (1732), Maupertuis had been very explicit on this point. 'I do not at all consider' (he wrote) 'whether Attraction accords with or is contrary to sound Philosophy'. Rather, 'Here I deal with attraction only as a mathematician [*géomètre*]'. That is, Maupertuis was concerned with attraction only as 'a quality, whatever it may be, from which the phenomena are calculated, considering it to be uniformly distributed through all the parts of matter, acting in proportion to the mass.[5]

Toward the end of his general introduction, and before introducing his purely mathematical commentary on, and expansions of, sects. 12 and 13 of bk. one of Newton's *Principia*, Maupertuis suggested two possible physical causes of gravitational attraction. One is that attraction arises from some kind of "emanation" from the attracting body in every direction in straight lines; another is that attraction is the effect of some foreign or external matter that pushes bodies toward one another. In the first case one can see easily that attraction must follow the law of the inverse square of the distance; in the second one can perhaps find why attraction occurs in this proportion (Maupertuis, 1736, p. 478).

Maupertuis, in other words, accepts the Newtonian style, and is willing, as "géomètre", to follow out the mathematical consequences of a law of gravitational attraction. Since the results accord with the phenomena observed in nature, Maupertuis then asks himself as natural philosopher whether there is such a force as a physical entity, or whether there may be some other reason why bodies act as if there were such a force. If such a force does exist, it must have a cause; and we may observe that his thought is still so embedded in the mechanical philosophy that he restricts himself to two material causes of this gravitational action: some emanation from within the attracting body or some kind of matter outside the body. As for himself, he is willing to abandon all physical causes, and he even asks (1736, p. 479): 'If God had wished to establish a law of Attraction in Nature, why would this law follow the proportion it seems to follow? Why should Attraction vary as the inverse ratio of the square of the distance?'

In the critical résumé that preceded Maupertuis's memoir (in the *Histoire de l'Académie Royale des Sciences* for the year 1732), this aspect of the Newtonian philosophy is developed at length. It was only because of the magnitude of Newton's 'great genius and great authority' that attraction has come back into physics, from which, it is said, 'Descartes and all his followers, or rather all Philosophers, had banished it by unanimous consent'. But it has returned a little disguised, not at all like the attraction of old; it is now 'only a name which is given to an unknown Cause'. The effects of this Cause 'are felt everywhere, effects that are calculated in order to know at least the manner in which their cause acts, while waiting for its nature to be unfolded.'[6] There can be no doubt that the author had read his Newton (or the Newtonian commentators) and had fully understood Newton's position, with which he was evidently in sympathy.

In a memoir by Clairaut (1747), in which we have seen him refer to Newton as having made a revolution (Clairaut, 1749, p. 329), there is an introductory discussion of the ways in which many readers of the *Principia* 'lost heart on first inspection, and flattered themselves that they had destroyed the Newtonian system without having followed the calculations and observations on which it is founded'. Such readers, furthermore, 'believed themselves able to avoid the troubles, by searching in Metaphysics for the means of proving the impossibility of attraction as a cause and as a property possessed by matter in itself' (Clairaut, 1749, pp. 329sq). But these critics had not fully grasped the essentials of the Newtonian style, or they did not agree with Newton that one could legitimately separate the results of scientific analysis from questions of the cause of gravitation, the possible mechanics of gravitational action, and purely metaphysical arguments as to whether universal gravity can exist. Following Newton, however, the scientist should not begin an enquiry with such questions as: What is force? What is the cause of motion? What is gravitation? But the critics of Newton, Clairaut found, did not yet understand how one can study gravitation—and force and motion—mathematically, without posing such questions. These critics, says Clairaut,

> did not realize that . . . they would have been refuted simply . . . by M. Newton, who avowed in his own words that he only used the term *attraction* while awaiting for its cause to be discovered; and in fact it is easy to judge by the treatise on the Mathematical Principles of Natural Philosophy that his only goal was to establish attraction as a fact [ibid., p. 330].

These comments of Clairaut are all the more remarkable in that in this very memoir—staunch Newtonian though he was—Clairaut found it necessary to consider that the moon's motion may require that the law of gravitation might have higher terms and thus not vary inversely as the square of the distance, as Newton had supposed.

3.10   *The Newtonian revolution in the perspective of history*
      This chapter has been largely devoted to a single theme: the Newtonian style as a key to the Newtonian revolution in science. The Newtonian revolution, of course, did not consist wholly of the introduction into science of the use of imagined systems and mathematical constructs as found in the *Principia*. Rather, that revolution was a radical restructuring of the principles and concepts of motion along the lines of mass, acceleration, and force; plus the elaboration of a system of the world operating in terms of the new dynamics, in which universal gravity is the governing force and inertia is a primary or essential property of matter. In terms of both the breadth of its scope and the profundity of the analysis, the *Principia* was unfolded in 1687 to a wholly unexpectant and unprepared audience who did not, in actual fact, know what to make of it or how to use it for some time.[1] Only gradually was it appreciated how deeply Newton had seen into the operations of nature, that he had recognized (for example) both the role of mass in inertial physics and the distinction between mass as a resistance to acceleration (what we call inertial mass) and as a determinant of force in a gravitational field (gravitational mass). Newton showed how one and the same universal force serves to account for the motion of planets around the sun, of both real and artificial satellites around planets, and of comets. This same force accounts for the tides, the equal rate of free fall of all bodies at any given place on earth (or elsewhere) and the change of that rate with change in latitude; and it produces the oblately spherical shape of the earth. It is difficult to think of any other scientific book that had ever been written which embodied so complete a change in the state of knowledge concerning the physics of the heavens and the earth.[2]

      Newton produced such an astonishing revolution in science by applying mathematics (geometry, algebra or proportions, fluxions, limit procedures, infinite series) to natural phenomena. It was Newton's example that such later scientific figures as Kant and Quetelet had in mind when they averred that a science's progress may be measured by the degree to which it becomes mathematical. The Newtonian style is of paramount importance because it made possible his mathematicization of nature's processes. And it is in this sense that the Newtonian style gives a key to the revolution in science associated with the *Principia*.

Newton was far from being the first scientist to build a mathematical system of nature. Ptolemy and Archimedes were distant (but only partial) predecessors; more immediate ones were Copernicus, Galileo, Descartes, and—above all—Kepler; while Huygens and Wallis were older contemporaries. But Archimedes had addressed himself to only a very limited range of nature, while Ptolemy's *Mathematical Composition*, as we have seen, embraced certain aspects of phenomena at the expense of ignoring others. As a result, Ptolemy's systems[3] for the sun, moon, and planets are essentially a set of geometric computing schemes or geometric models and appear not to have been intended as a representation of reality.

Pierre Duhem (1969) has called attention to the problem of devising computing schemes versus attempting to portray reality; he views the conflict between the two as a dominant theme in the history of the science of the heavens from Greek times to the seventeenth century. His thesis that there was a complete dichotomy down the ages between model-makers and realists is extreme.[4] Nevertheless, we may see the conflict between these two points of view coming to the fore with Copernicus's *De revolutionibus* (1543). This book had been printed in such a way as to give the impression that the author was presenting his new system as only a model, a hypothesis, or a scheme for computing solar, planetary, and lunar phenomena. Following the dedication of Pope Paul III, there is an introductory essay ('Ad lectorem de hypothesibus hujus operis'), in which Copernicus appeared to have said just that.[5] Kepler found evidence, however, that this essay was not a composition by Copernicus at all; it had been introduced into the book by Osiander (a Protestant clergyman who saw *De revolutionibus* through the press).[6] Kepler himself, as we have seen, was primarily a realist. He wanted to start right out with the causes of the motions of the planets and to find their true paths; he was not interested in any mere computing schemes.[7] And so he was especially delighted to find that Copernicus himself had been a realist, that

Copernicus had believed heart and soul in his own system and had not been the author of the prefatory essay on hypotheses.

Galileo was as convinced a Copernican as Kepler, but he did not either improve the schemes for calculation or seek for causes. Of course he believed in the reality of the Copernican system, and even concocted a reason (based on an explanation of the tides) why there had to be a rotation of the earth as well as a motion in revolution around the sun. Galileo did not, however, concern himself notably with the technical details of the Copernican system[8] so much as with philosophical and scientific arguments favoring a general heliocentrism and an anti-Aristotelian view of motion. It is, therefore, in his *Two New Sciences* rather than in his *Two Chief Systems of the World* that we may seek for a true precursorship of the Newtonian style as a step toward applying mathematics to nature. For example, Galileo had to deal with the realities of air friction or air resistance in relation to the motion of pendulums and the free fall of bodies. Since this reality situation was too complex and difficult for him to handle, he simplified nature as he found it by supposing a world of empty space in which there would be no effects of the air. He predicted, for example, that in such an imagined world, a coin and a feather would fall freely at the identical rate, or have equal accelerations.[9] On a smaller scale than Newton's, Galileo was thus considering a simplified case of physics as a stage toward reality. The slight difference in time of fall of a heavy and a light body dropped at the same time from a tower was attributed to air friction, the cause of the divergence of the idealized situation from reality. Pendulum experiments then provided evidence that air friction does indeed resist motion. Only in the case of an idealization or simplification of nature, and not in the real world of ordinary experience, are Galileo's laws of free fall and the parabolic trajectories of projectiles strictly valid.[10]

In the present context, these Galilean examples are of only academic interest—for the record, so to speak—since we have no reason to believe that Newton had ever read Galileo's *Two New Sciences*, and we do have many bits of evidence that indicate that he had not done so (see Cohen, 1967c). Important and fruitful research remains to be done on the whole question of the use of imagined systems and mathematical constructs in seventeenth-century physical science, both the systems and constructs that comprise a set of mathematically expressed conditions of force, resistance, and motion and those that embody physical systems or mechanisms to explain theories (and are like the "models" of today's scientists and philosophers of science). Such a study would no doubt point to possible sources of Newton's procedure, which he would have transformed, improved, and endowed with extraordinary new powers.[11]

Did Newton ever show any direct awareness that he was doing something new in his use of the Newtonian style? Not in so many words. But he was certainly aware that no one before him had found the many results he had discovered. He knew, of course, that some earlier scientists had guessed at the inverse-square law, and he even supposed that this law might have been known to the ancients.[12] But that did not mean that he took any the less credit for his invention: The inverse-square law may have been known to seers of a bygone day, but to have found it and to have proved that it caused elliptical orbits was something new in his own age, and to this degree he would not give Hooke or anyone else any credit for it. Furthermore, what was significant about the Newtonian achievement was not merely to guess or even to know the inverse-square law, but rather to use it to demonstrate elliptical orbits and to develop a system of the world based upon it. This could not have been done by experiment and observation, by induction, or by philosophical speculation, but only by mathematics. And the key to applying mathematics to the world of nature was the Newtonian style, in which by stages there could be added the conditions that would bring the original imagined system and mathematical construct into congruence with the realities of experience. The mathematics needed for this task was a new mathematics, the fluxional calculus, embodied in the continual use of limits and of infinite series,[13] and here Newton insisted that he was the sole inventor—the first inventor, not a rediscoverer of ancient methods (see Collins et al., 1856; Newton, 1715).

The degree to which the Newtonian style was revolutionary can be seen in the simple fact that so much of our exact science ever since has proceeded in a somewhat similar manner. I believe that there is a logical and simple penchant for a kind of positivism on the part of all who approach physical subjects as mathematicians, and for whom the exploration of the mathematical consequences of any system or of any set of conditions is equally fascinating; although certain ones are obviously of more importance than others, because they relate to nature as revealed by experiment and observation.

In the general scholium written for the second edition of the *Principia* in 1713, Newton expressed the quasi-positivist view which has inspirited much of the exact sciences from that day to ours, when he said that 'it is enough' ('satis est') that gravity exists and that from it we can deduce the motion of the heavenly bodies, terrestrial objects, and the tides. By this expression, Newton was arguing against the kind of criticism in which the whole structure of Newtonian celestial dynamics would be thrown out on metaphysical grounds, because of an abhorrence of "attraction" or doubts as to whether such a force can possibly exist. In 1717, in the second edition of the *Opticks*, Newton once again stated that he did not consider how 'these Attractions may be performed' (Newton, 1952, qu. 31, par. 1, 376). Repeating essentially what he had said in 1706 in the Latin edition, he pointed out, 'What I call Attraction may be performed by impulse, or by some other means unknown to me'. And he stressed the fact that he was using 'that Word [attraction] here to signify only in general any Force by which Bodies tend towards one another, whatsoever be the Cause' (ibid.). He had limited himself to the first stage of enquiry, in which 'we must learn from the Phaenomena of Nature what Bodies attract one another, and what are the Laws and Properties of the Attraction, before we enquire the Cause by which the Attraction is performed' (ibid.). But none of these statements, as has been said above, can be taken to imply that Newton himself was uninterested in the search for such causes, or that he himself had not instituted such a search.[14]

Newton's insistence that it is enough to be able to predict the celestial and terrestrial motions and the tides of the sea was, in fact, less a battle-cry of the new science than a confession of failure. For what Newton was saying in essence is that his system should be accepted in spite of his failure to discern the cause or even to understand universal gravity, because its results accord so well with the data of observation and experiment. The eventual acceptance of Newtonian celestial mechanics, in the absence of the knowledge of the fundamental cause, was in one sense a perversion of the philosophy Newton had expressed in the general scholium, because it inhibited any further search for a cause. But in another sense, modern science *has* been following the Newtonian principles, because Newton *did* believe that the first goal of mathematical physics (or of exact science) is to predict and retrodict the phenomena of nature.

Supplement to 3.10: Newtonian style or Galilean style

What I have called the Newtonian style often appears in the literature as a Galilean style. Many writers use the phrase "Galilean style" in physics in reference to Galileo's idealization of falling motion, his elimination of disturbing and complicating factors in order to formulate a simple mathematically stated law. As Ernan McMullin (personal communication) has put it, 'This law (in Galileo's view) holds exactly of the world to the extent that the complications are absent. The world obeys it precisely (this is why he is not a strict Platonist . . .). To find out how a particular system moves, one complicates it to the extent needful by allowing for the physical factors that have been left out.'

A major discussion of the Galilean style appears in Edmund Husserl's (1970) *The Crisis of European Sciences and Transcendental Phenomenology*, in a section on 'Galileo's mathematicization of nature'. But even earlier, in the immediate post-Newtonian era, it was recognized that the Newtonian style had roots going back to Galileo. In 1732, in a 'Discours sur les différentes figures des astres', Maupertuis discussed Newton's use of the term "attraction": Newton, he wrote, 'has often asserted that he only employed this term to designate a fact and not a cause, that he only employed it to avoid systems and explanations'. Maupertuis then went on to explain the procedure of a mathematician, who can study 'every regular effect, although its cause is unknown'. This style, according to Maupertuis, originated with Galileo who, 'without knowing the cause of the heaviness of bodies toward the earth', was nevertheless able to erect a beautiful and certain theory based upon this heaviness and to explain the phenomena that depend upon it (see Aiton, 1972, p. 202). Galileo, like Newton, was able to discuss problems of physics from a mathematical point of view without inquiring into the causes and nature of the forces. But the style that I have called Newtonian does not consist merely of determining the properties of physical systems without inquiring into causes. There is also required a systematic second and third phase in which causes are considered. The Newtonian style led, in the *Principia*, to relations between forces and the accelerations they produce and eventually to universal gravity. But Galileo was content to disclose kinematical laws of motion without moving on to dynamics. In short, in Galileo we may certainly find examples which are very much like the Newtonian style, but not carried through (as Newton was to do) to the third phase and the sequel.

Newton did not invent this style in the sense of creating something without any precursors. He was no doubt influenced by Barrow, who (in the manner elaborated by Newton in bks. one and two of the *Principia*) included "physical" or "philosophical" scholiums in his mathematical treatment of a physical subject such as optics. It would, in fact, be wholly inconsistent with the point of view I have advocated in this book to believe that the Newtonian style could have been formed in any other way than by a transformation of earlier versions or less highly advanced modes of scientific procedure. Among such predecessors I would certainly include Galileo, although I would have to insist that Galileo's style was at best a primitive antecedent of the Newtonian style; it was, furthermore, not used as systematically or extensively or as successfully as Newton's. Newton himself had probably not read enough of Galileo's writings (especially his *Two New Sciences*) to have been influenced by him in this regard.

I have used the expression "Newtonian style" because it seems to me to be an apt description of the style expounded in example in the *Principia*. It is a style which Newton developed and applied with telling effectiveness and it is a style, furthermore, that must be kept in mind as a key to Newton's procedure and hence to an understanding of the level of discourse in reading the *Principia*. It is the Newtonian style in the strict sense of being the style used by Newton in his *Principia*. This Newtonian style provided a means of combining the techniques of mathematics and the physics of experience that made the *Principia* possible. And it is the style of the *Principia*, elaborated and illustrated by Newton (and not by Galileo), that has—with certain later developments—largely been the key to the development of the exact sciences from that day to this.

3.11 *Optics and the Newtonian style*

A basic feature of the Newtonian style, as illustrated in the *Principia*, is the application of mathematics to a system or construct that is the mathematical analogue of a natural situation, though somewhat simplified and idealized. As further conditions are added to the construct, the latter becomes more nearly an analogue of nature. In this way a body of mathematical principles relevant to natural philosophy is built up and then applied to the world of physics, as in bk. three, on the system of the world. In this context of the Newtonian style, the word mathematics refers to the application of true mathematical techniques, such as algebra, the method of proportions, Euclidean geometry, the geometry of conic sections, elementary projective geometry, infinite series, fluxional calculus (at least fluxional arguments), the theory of limits. Mathematics in this context does not mean experiments yielding numerical results, numerical measurements, or a logical form of argument. Nor does mathematics mean that a work merely has the superficial form associated with mathematical treatises: definitions, axioms, and numbered propositions.

To test this dichotomy, consider Spinoza's *Ethica ordine geometrico demonstrata* (1677, posthumous), which attempts, as its title says, to present ethics 'proved in geometric order'. Hence it proceeds by definitions and axioms to propositions and their corollaries, following a long tradition of practice by Muslim, Hebrew, and Christian authors who used this form of argument in an attempt to identify 'the syllogistic form of demonstration with the Euclidean geometrical form'.[1] Such philosophical works would hardly be called "mathematical", and they are certainly not mathematical in the sense I have defined above for the *Principia*. Their contents are not subjected to analysis by the techniques and methods of mathematics, however much their authors sought to imitate the external form of Euclid's geometry.

Newton's *Opticks* is also cast in the external form of a mathematical work to the extent that it begins with definitions and axioms and proceeds to propositions. It is notable, however, that the propositions in the *Opticks* are not for the most part proved by logic in relation to the axioms, nor do the proofs invoke a sequence of references going back to the earlier propositions. More significantly, the propositions are not proved by the application of mathematical techniques. Rather Newton most often proceeds by giving 'PROOF by Experiment', and he tends to refer back to previous experiments rather than to the preliminary axioms. Hence, although Newton uses numbers (as in the results of experiment), his *Opticks* can in no legitimate sense be considered a mathematical treatise.[2]

Another way of stating this conclusion is that in the *Opticks* Newton does not proceed by using what I have been calling the Newtonian style. Yet we must not conclude that Newton considered optics as a subject which could not be developed by mathematics in the Newtonian style. In other writings, some published posthumously and others still in MS, Newton approaches the problems of optics in ways that are fundamentally different from the presentation that has come down to us in the *Opticks*. In particular, Newton uses mathematics extensively, not only in the geometry of catoptrics and dioptrics, but also in physical optics. In his inaugural set of lectures in Cambridge University (in mid-January 1670), Newton in fact gave a considerable discussion that—as the editor of the posthumous English edition (1728) put it—is not so much physical as 'purely geometrical'.[3] In the geometrical portions of these lectures,[4] Whiteside ('from a mathematical viewpoint') notably finds '[some] revealing illustrations, in an optical setting, of his technique for constructing the roots of algebraic equations by intersecting conics, of his deployment of limit-increments to effect a geometrical differentiation, and of his treatment of the extreme values of a given function . . .' (Newton, 1967–, vol. 3, p. 440).

Another remarkable feature of Newton's mathematicization of optical problems was 'his conclusive mathematical formulation, in culmination to two thousand years of slow empirical advance and Descartes' primitive, numerical theory, of the structure of the n-ary bow'. Other examples are his 'calculation of the chromatic aberration of rays (radiating from a unique point) which are refracted at a spherical interface' and 'his construction of a "new" catadioptrical telescope in which the mirror is a coated lens placed so that its chromatic distortion is minimized' (Newton, 1967–, vol. 3, p. 442).

In these lectures, especially sects. 3 and 4 of pt. 1, Newton develops his mathematical analysis by making certain arbitrary "physical" assumptions,[5] and his style or manner of proceeding suggests the approach to problems in the *Principia* in various ways. That is, one can discern the power of using imagined systems or constructs from which he draws consequences by actual mathematics, in what I have called the Newtonian style. Newton, in other words, did not conceive the subject matter of physical optics to be necessarily different in any basic way from physical mechanics, insofar as the use of mathematics in relation to physical problems was concerned.

And this may be confirmed for us by an examination of sect. 14 of bk. one of the *Principia*, where Newton considers a system of particles moving through force fields that have rather odd properties near the interface between any two mediums, and within certain mediums; as a result, a number of experimentally observed properties of rays of light can be mathematically derived from the conditions of this system.[6]

What is essentially the same system[7] appears in the *Opticks*, where it is conspicuous by being the only fully developed mathematical argument of this kind in the treatise.[8] It occurs at the end of prop. 6, bk. one, pt. 1, where Newton wishes to prove that 'The Sine of Incidence of every Ray considered apart, is to its Sine of Refraction in a given Ratio'. This proposition is proved by an experiment designed to show that when there is the same sine of incidence of various rays, there is a proportion between the sines of the angles of refraction of these rays. This is found to be true 'so far as by viewing the Spectrums, and using . . . mathematical Reasoning I could estimate'. Newton 'did not [however] make an accurate Computation', but he nevertheless concluded: 'So then the Proposition holds true in every Ray apart, as far as appears by Experiment'. Yet, 'that it is accurately true, may be demonstrated upon this Supposition. *That Bodies refract Light by acting upon its Rays in Lines perpendicular to their Surfaces*'.[9] Here then is a single isolated instance of the working out, in the *Opticks*, of the method which I have characterized as being the essence of the science of the *Principia* and hence the source of the Newtonian revolution in science.[10]

Are we then to conclude that the world of science saw Newton's revolutionary method exemplified equally in physical mechanics (and celestial dynamics) and in physical optics? Not at all. The reason is that Newton had no real success in this goal for physical optics, in the sense that is true of his work on the motion of mass points under the action of central forces. We may well understand, therefore, why it was when he wrote up and published what we know as the *Opticks*, the only full-scale application of mathematics in the Newtonian style which he put into print was the above-mentioned attempt to account for the law of refraction. Even this example could not have been wholly satisfying to Newton because of the difficulty in imagining a force in a medium that would act only perpendicularly or vertically.[11] The motion of particles under these conditions is elaborated in full, with the necessary mathematics, and applied to reflection and refraction in sect. 14 of bk. one of the *Principia*. It is to be noted, however, that even in this more fully developed presentation, Newton does not attempt to account for a broad range of optical phenomena.[12]

In order to discover Newton's attempts to use the Newtonian style in the domain of optics, it is accordingly necessary to venture beyond the printed *Opticks* and *Principia* and to explore the Lucasian optical lectures and the unpublished optical MSS. This task was undertaken by J. A. Lohne (1961, pp. 397, 398), who attempted to extract from Newton's MSS and his published writings a set of assumptions that seem to have been of determinative influence in his thinking about optics and in his performing and interpreting experiments, especially in the 1670s. Among them are concepts of globules of light and an all-pervading aether that is rarer in denser bodies (so that light globules move faster in denser bodies). An important condition[13] is that when a globule of light goes from a body or medium into another of higher density, the 'tangential velocity is unchanged, but the total velocity increases and is independent of the angle of incidence'; another is that "rubriform" (or red-producing) rays move faster in vacuo than blue-producing rays[14] (and hence also in every transparent medium or substance). Some of the mathematical features of Newton's optical writings have been elucidated by D. T. Whiteside, especially in vols. 4 and 6 of his edition of Newton's *Mathematical Papers* (Newton, 1967–); these do not, however, occur in the *Opticks*.

In an interesting and novel approach to Newton's optics (not *Opticks*), Zev Bechler has explored various "models"[15] proposed by Newton to explain color phenomena, among them what Bechler calls a 'mechanistic model of differential reflection of light corpuscles from bodies, whereby the colour of these bodies would be reduced to the laws of elastic collision'. Newton was here constructing a model for the behavior of light corpuscles or globules in the invisible or not-directly-perceptible realm in which they would follow exactly and rigorously the 'laws of the visible world'. This form of model construction is named by Bechler 'the "rule of normality", and the models constructed in accord with it [are named] "normal models" '. Newton supplemented this "rule of normality" in model construction, according to Bechler (1973), by a 'demand for rigorous mathematical reasoning', and he views this 'as the central contribution of Newton to the new methodology of science' and calls it 'the "mathematization of normality"'. Bechler finds that Newton proposed (or by implication can be seen to have used) a series of different models. For instance, there was what Bechler calls a "velocity model" (1669–1670), suppressed in 1675 and replaced by a "mass model", later refined and improved. He also finds various models in 1687, 1694, 1704, and 1706. These are all examples of Newton's wish for a mechanistic or "mechanical" explanation of dispersion. It is noteworthy that these "models" do not appear explicitly in the published *Opticks*, save for the summary (referred to above) in prop. 6 of bk. one (pt. 1). In particular Bechler analyzes and makes explicit the mathematical underpinning of Newton's first great paper on 'light and colours', which is to be found largely in the *Lectiones opticae*, although, as Bechler says, 'none of the worthies who read the paper knew of this massive mathematical basis, and Newton hardly even hinted at its existence'.[16] In view of the revelation of the corpuscular and mathematical presuppositions of Newton's thought (by such scholars as Bechler, Lohne, and Sabra), it would be difficult to maintain that this classic experiment and its interpretation can be understood on a simple level of experiment and observation.

Against this background we can see what Newton had in mind when he wrote that, 'A naturalist would scearce expect to see the science of those [i.e., colors] become mathematicall, & yet I dare affirm that there is as much certainty in it [i.e., the science of colors] as in any other part of Opticks' (1959–1977, vol. 1, pp. 96sq). This sentence occurs in that first great letter of 1672 on light and colors, which Newton sent to Oldenburg to be published in the *Philosophical Transactions*. Newton is saying that the science of colors (physical optics) is as certain as geometric optics, and the reason is—according to Newton—that what he has found is a 'most rigid consequence . . . evinced by the mediation of experiments concluding directly & without any suspicion of doubt' (ibid.). It is a fact of record, however, that this expression of mathematical hopes was omitted from the version published in the *Philosophical Transactions*.[17]

A somewhat similar statement was omitted from Newton's lengthy reply (11 June 1672; 1959–1977, vol. 1, 171sq) to Hooke's objections, published in the *Philosophical Transactions* with a note saying that Hooke's discourse (to which Newton was replying) 'was thought needless to be here printed at length'. This statement is worth considering in extenso, because in it Newton expresses his never-achieved hope of producing a mathematical theory of colors:

> In the last place, I should take notice of a casuall expression which intimates a greater certainty in these things then I ever promised, viz.: the certainty of *Mathematicall Demonstrations*. I said indeed that the *Science of Colours was Mathematicall & as certain as any other part of Optiques*; but who knows not that Optiques & many other Mathematicall Sciences depend as well on Physicall Principles as on Mathematicall Demonstrations: And the absolute certainty of a Science cannot exceed the certainty of its Principles. Now the evidence by which I asserted the Propositions of colours is in the next words expressed to be from *Experiments* & so but *Physicall*: Whence the Propositions themselves can be esteemed no more then *Physicall Principles* of a Science. And if those Principles be such that on them a Mathematician may determin all the Phaenomena of colours that can be caused by refractions, & that by computing or demonstrating after what manner & how much those refractions doe separate or mingle the rays in which severall colours are originally inherent; I suppose the *Science of Colours* will be granted *Mathematicall* & as certain as any part of *Optiques*. And that this may be done I have good reason to beleive, because ever since I became first acquainted with these Principles, I have with constant successe in the events made use of them for this purpose [Newton, 1959–1977, vol. 1, p. 187].

Such expressions of faith and hope were no doubt made in all sincerity, but they provide no warrant whatever for assuming that Newton had ever come near to realizing this ambition. Above all, we must be very wary of reading such statements out of context, and especially of assuming that the published *Opticks* in any way came up to such expectations.

In the 'hypothesis' sent by Newton to the Royal Society in 1675, in the second part, he produced a nomogram for determining ratios of thicknesses for which various colors appear in the production of Newton's rings (see Newton, 1958, p. 216). Since the nomogram was constructed by using certain regular sequences of numbers and proportions (including the 'Cube-Roots of the squares of the Numbers, 1/2, 9/16, 3/5, 2/3, 3/4, 5/6, 8/9, 1, whereby the Lengths of a Musical Chord to sound all the Notes in an eighth are represented'), Newton could conclude that his results had been 'mathematically demonstrable from my former principles'; he added at once that 'they, which please to take the pains, may by the testimony of their senses be assured, that these explications are not hypothetical, but infallibly true and genuine' (p. 223). In the *Opticks* he toned down this boast to read that 'as all these things follow from the properties of Light by a mathematical way of reasoning, so the truth of them may be manifested by Experiments'.[18] He had been referring here to the quantification which he applied to the color-effects seen in bubbles and other thin films. The basic notion was, as he said in 1675, that 'there is a constant relation between colours and refrangibility', so that 'the colorific dispositions of rays are also connate with them, and immutable'. A consequence is 'that all the productions and appearances of colours in the world are derived, not from any physical change caused in light by refraction and reflection, but only from the various mixtures or separations of rays, by virtue of their different refrangibility or reflexibility'. Hence, if the optical properties of all materials were known with respect to different colors, then it should become possible to compute color phenomena arising from direct reflection and refraction. And so Newton (1958, p. 225) boldly concluded that 'in this respect it is, that the science of colours becomes a speculation more proper for mathematicians than naturalists [i.e., experimental natural philosophers]'. In the *Opticks* he altered this conclusion slightly, by tempering it so as to have it read: 'a speculation as truly mathematical as any other part of Opticks'—which is similar to what he had said in his letter of 1672 to Oldenburg.[19] But in the *Opticks*, he now added a qualification: 'I mean, so far as they [i.e., 'the Production and Appearances of Colours in the World'] depend on the Nature of Light, and are not produced or altered by the Power of Imagination, or by striking or pressing the Eye' (Newton, 1952, p. 244). Yet even this extremely limited kind of mathematics was no more than an idle dream; Newton could not achieve even so much. One reason was the fact that he was unaware of the real potentialities of the wave theory of light and the possibility of producing colors by the interference of waves. Nor did he give due weight to the production of colors in objects by absorption rather than reflection. And he did not fully appreciate the niceties of additive and subtractive color mixtures as exemplified in mixing colored rays and pigments (see Biernson, 1972).

In any event, from the point of view of technical mathematics as commonly understood (the general application of mathematical techniques to physical phenomena), Newton's physical optics as actually expounded in the *Opticks* can hardly be denoted mathematical. It is only by confusing Newton's possible ambition and hope with real achievement that a serious scholar could have written that the '*Opticks* could have been called after Newton's earlier and greater book, *The Mathematical Principles of Light and Colour*, since that is its subject'. For the fact is that in this work, in an area which he most fully explored by his own experimental researches, he did not succeed in speaking in the same mathematical language of nature that he had employed in the *Principia*. From the special point of view of how Newton actually proceeded in the *Principia*, by successive stages of mathematics applied to increasingly more complex and more general systems or constructs, the *Opticks* represents a level of inquiry far removed from that of the *Principia* and displays the enquiring mind of an experimental natural philosopher rather than that of a mathematical physicist. No wonder that it has been considered a sign of Newton's supreme genius that he should have been both at once.

3.12 *The ongoing Newtonian revolution and the Newtonian style: mathematics and experience*

The Newtonian revolution was a turning point in the history of the exact sciences: mathematical astronomy and rational mechanics. But, as I have pointed out, this should not be taken to mean that these (and allied) sciences were in any possible sense "complete" or in a "final form" as they appeared in the *Principia*. Newton had taken a giant step forward, as may be seen by contrasting the level of discourse (the range, generality, complexity, and difficulty) and the actual subject matter of Galileo's *Discorsi* and *Dialogo* and Newton's *Principia*. This is also the case, although to a greatly lesser degree, with regard to Kepler's *Astronomia nova* or Huygens's *Horologium oscillatorium*. The greatness of Newton's achievement made dramatically clear what he had omitted from his considerations and also what he had done badly, imperfectly, or incompletely. For example, he had opened up the possibility of perturbation theory but had not made a truly significant contribution to it. Newton's stature is hardly diminished by an honest appraisal of his achievements and his failures, by the recognition that giants were still needed after the *Principia* in order to revise, improve, and complete the subjects he had treated and to exercise their collective genius to create all the other subjects that comprise classical exact science.

The scientists of the eighteenth century were well aware of the problems unsolved by Newton as well as those topics which he had pursued with such success. They were as cognizant of Newton's failures as of his impressive breakthroughs. We do not at present have an exact picture of the response to Newtonian science because the impression made by the *Principia* on the science of the eighteenth century has never been fully studied. What is needed is more than an examination of the influence of a general Newtonian philosophy or an inquiry into the stages and degrees of acceptability of the concept of gravitation; on these topics there are many monographic studies of excellent quality. To evaluate the impact of the *Principia* with precision, it is necessary to study in detail and in depth how individual scientists actually used Newtonian principles, methods, laws, concepts, and particular results. In such an inquiry, it does not suffice merely to see whether or not the three Newtonian laws of motion are used as such, although it may be of the greatest interest to enquire into whether (or in what form) Newton's second law may appear.[1] It would be particularly fruitful to see how the approach found in the beginning of bk. one of the *Principia* or in the propositions around prop. 41[2] may have been used by different writers on dynamics. In some cases the Newtonian principles, concepts, methods, or results will be found to have been adopted without change, but in others it will be found that there had to be significant transformations.

Such an investigation may show that a concept that might otherwise have seemed to us to be not very fruitful was apparently of real service. An instance may be seen in Newton's "vis inertiae". On page 3 of d'Alembert's *Traité de dynamique* (1743), he says: 'Following Newton, I call *force of inertia* [*force d'inertie*] the property which bodies have of remaining in the state in which they are; now a Body is necessarily in either a state of rest or a state of motion . . .'[3] This leads him to two laws:

1st law

A Body at rest will persist in rest unless an external cause draws it out of that state. For a Body cannot of itself [*de lui-même*] put itself in motion.

2d law

A Body once put into motion by any cause whatever must ever persist in motion that is uniform and in a straight line, if indeed a new cause, different from the former one which put the body in motion, does not act on it, that is to say, unless an outside force that is different from the cause of the motion [*cause motrice*] acts on this Body, and it will move continually in a straight line and will traverse equal spaces in equal times.

It is thus plain that d'Alembert is not following Newton in having a simple and single law of inertia, but rather–having accepted Newton's "force d'inertie"–he has gone back to the two separate laws of inertia in a form like that used by Descartes in his "leges naturae" in his *Principia philosophiae*.[4]

Another informative example is provided by Jacob Hermann's *Phoronomia, sive de viribus et motibus corporum solidorum et fluidorum libri duo* (Amsterdam 1716). Although this work is generally Leibnizian, it turns out to be replete with Newtonisms. Hermann used Leibniz's algorithm for the calculus, and he begins the treatise with an effusive tribute to Leibniz. So it is all the more remarkable that in 1716, just after the peak of the grim controversy over the discovery of the calculus, Hermann should refer to Newton in his preface as 'Summus Geometra Isaacus Newtonus' and call his book an 'aureum opus', a golden work.[5] In an introductory poem in honor of Hermann, reminiscent of the similar poem by Halley in honor of Newton at the beginning of the *Principia*, Nicolaus Westerman refers to Newton in one of the verses as follows:[6]

> Newtonus hospes divitis insulae,
> Sed nil habentis se magis aureum,
>   Hac primus ivit, Tuque forte
>   Nil populis dederis secundum.

This may be freely translated as follows:

> Newton, inhabitant of a wealthy isle,
> But an island having nothing more golden than himself,

> Was the first to go along this path; and you perhaps
> Give nothing of lesser value to the public.

Hermann states Newton's second law for continuous forces in differentials. Following Leibniz, Hermann calls Newton's centripetal forces "solicitations". If the mass of a moving body is M, and the speed acquired in a time T is V, so that the motion generated by a "solicitation" G acting uniformly during this time T will be MV, then, according to Hermann, $G = MV : T$. This is prop. 16 of bk. one; there follows a scholium in which it is considered that G does not act uniformly ("uniformiter") but difformly ("difformiter"), that is, G changes. In this case we are to consider an indefinitely small time dT ('. . . temporis tractum indefinite parvum dT') in which the movable acquires only an infinitesimal speed dV ('quo mobili tantum celeritas infinitesima dV acquiritur'); under these conditions $G = MdV : dT$, or $dT = MdV : G$ 'where G stands for the weight [pondus] or gravity [gravitas] however variable of a mass M' (Hermann, 1716, pp. 56–57).

Hermann uses the second approach to celestial mechanics found in the *Principia*, prop. 41 of bk. one rather than props. 1–11 et seq. Even more important, he follows Newton more closely in his definitions than he may have been aware. For example, having defined "vis motrix" as both "vis viva" and "vis mortua", and having introduced "solicitatio" and "vis activa corporum", he turns to "vis passiva":

> But there is also in bodies a certain passive force [*Vis passiva*], from which no motion nor tendency to motion results; but it consists in that reluctance [*Renixus*] by which it strives against any external force trying to bring about in bodies a change of state, that is, of motion or of rest. This force of resistance was called by a most significant name *Force of inertia (Vis inertiae)* by the very great Astronomer Joh. Kepler. This force of inertia reveals itself sufficiently in bodies at rest, for any body A striking against another but resting body B will lose some part of its force and motion, and the receiving body B will acquire some part of the force and motion of the striking body A. From this it is clear that the resting body B really has some passive force that must be destroyed and overcome by the force of the body A striking against it. Otherwise the impelling body A would have to have lost

nothing of its motion after the collision, since the resting body B, if it lacked the power of resisting, would not
be able to offer any impediment to the motion of the
other, so that both the impelling body A and the impelled
body B would have to move along after contact with that
very speed with which body A was going before the collision, which obviously is contrary to the phenomena
[Hermann, 1716, p. 3].

There are several significant observations that may be made about
this passage. First of all, it was Newton and not Kepler who referred
to a "Vis inertiae"; Kepler wrote of "inertia" or "inertia naturalis".
Second, Kepler never wrote of a "state" ("status") of motion, as
Newton (following Descartes) did. Third, it was Newton and not
Kepler who conceived of "inertia" (or "vis inertiae") as maintaining a body in its state of rest or of motion; Keplerian inertia merely
tended to bring a moving body to rest when and if an external moving force ceased to act. Fourth, the very phrase that this "vis re-
sistentiae" by a most significant word was called Vis inertiae has
been taken directly (almost word for word) from the *Principia* and
should have been followed by a reference to Newton and not Kepler; Newton, in def. 3, says (of the "vis insita") that 'nomine sig-
nificantissimo vis Inertiae dici possit'. Finally, the discussion about
resistance also follows directly from Newton, as does Hermann's
following statement of Newton's third law (Lex 3. 'Actioni contrariam semper & aequalem esse reactionem'), rendered by Hermann as 'Cuilibet actioni aequalis & contraria est reactio'. Most
important of all, Hermann follows Newton in the concept of mass
and its role in dynamics, of which we have seen an example above
in his reformulation of Newton's second law. Referring to "massa"
as "quantitas materiae",[7] he insists that 'the gravities or weights
of bodies are proportional to their masses' as 'the illustrious Newton' has proved by extremely accurate experiments on pendulums.[8]

These samples from Hermann's *Phoronomia* show how overtly
the Newtonian dynamics tended to infiltrate even the Continental
opposition.[9] What is more important, of course, is that all writers
on this subject eventually used a somewhat Newtonian style of dealing mathematically with the properties of imagined systems or constructs and separately exploring the additional properties that
would have to be added in order to have greater and greater con-

formity to nature. It was an essential part of this style that the investigation into the causes behind the forces be omitted altogether or relegated to a separate part of the discussion.[10] In the classical period, there is no doubt that this style reached its apex in Laplace's *Mécanique céleste*.

Just as Newton's great breakthrough, universal gravity, came as a result of his working on the solution of actual problems in dynamics and not just thinking in general terms about nature and the universe, so our knowledge of the real impact of Newton on the advance of science must come through the analysis of what working scientists were actually doing, rather than a study of even such gifted *hautes vulgarisations* as those of Pemberton, Voltaire, and Maclaurin. I showed, a few years ago, how a lengthy French summary of Newton's *Opticks*, prepared by the chemist Geoffroy, had been read at ten meetings of the Paris Académie des Sciences during the ten months from August 1706 to June 1707, thus acquainting the members with Newton's discoveries in relation to dispersion, color, and interference phenomena.[11] Recently, Henry Guerlac has been able to pinpoint the first successful reproduction in France of Newton's experiments with prismatic colors. He finds this date to be 1716–1717, when Dortous de Mairan repeated and verified these experiments, thus counteracting Mariotte's inability to confirm Newton's "experimentum crucis" (the experiment with refraction by two successive prisms to show that when a single color is separated from the spectrum produced by a prism, the second prism introduces no further dispersion).[12] This example of how a working scientist actually took the crucial research step of beginning with the repetition of a famous experiment is of more than ordinary interest. For it not only shows how the Newtonian technique actually entered the practice of French experimental optics; it also indicates that in the practice of experimental optical science any real influence of Newton in France became significant only many years after his election as an "associé étranger" of the Académie des Sciences in 1699.

I have referred earlier to the fact that for Newton himself, as for most of his successors, gravity acting "at-a-distance" over millions of miles was simply incomprehensible. Hence, as we have seen, there was a block against the wholehearted acceptance of his system of the world as a physical system. Indeed, it even must have seemed to many critical readers as if bk. three of the *Principia*, on the system of the world, was itself only a mathematical exposition, in which case there would not have been any real difference between bks. one and two and bk. three. In one sense there is a unity between bks. one and two, on the one hand, and bk. three, on the other, in that in all three books Newton does not consider the nature, cause, or mode of action (or transmission) of forces. But in bks. one and two the forces are abstract and arbitrary, whereas in bk. three the individual forces acting on the sun, planets, satellites, and all samples of matter are shown to be manifestations of gravity which Newton maintained (in the concluding general scholium) 'really exists'.

I believe that the outlook of post-Newtonian scientists, using a system based on the action of a universal force that they could not understand, was not wholly unlike that of present-day physicists with respect to quantum field theory. In a recent talk, Murray Gell-Mann (1977) said:

> All of modern physics is governed by that magnificent and thoroughly confusing discipline called quantum mechanics, invented more than fifty years ago. It has survived all tests and there is no reason to believe that there is any flaw in it. We suppose that it is exactly correct. Nobody understands it, but we all know how to use it and how to apply it to problems; and so we have learned to live with the fact that nobody can understand it.

It was much the same in the age of Newton for gravitational celestial mechanics. After a while, men of science like Euler, Lagrange, and Laplace used such concepts as universal gravity and force because this was the key to the successful mathematical solution of so many problems arising in physics and astronomy.[13] By the time of Laplace's *Mécanique céleste*, rational mechanics had developed to the stage where one could make an easy transition from mathematical systems to the physics of nature, without even a blink to philosophy. Thus in ch. 2 of bk. 1 of the *Mécanique céleste*, Laplace begins with 'the motion of a material point' (an obviously Newtonian mathematical construct), but shifts easily to aspects of terrestrial gravity, the motions of projectiles, and other aspects of observable physics (ch. 2, sects. 9–10, bk. one).

Of course, there were always physicists who were concerned with the philosophical problems of the nature of force–including such diverse figures as Boscovich, Mach, and Hertz. This too was in a

Newtonian tradition, however un-Newtonian their solutions to the Newtonian problem may have been, for it is directly related to what I have called the sequel to phase three. Newton's demonstration of the dynamical significance of Kepler's laws and the limits and conditions of their veracity, and his explorations of the motion of the moon, the perturbation of Saturn's motion by Jupiter, his explanations of tidal phenomena in the seas and the motion of comets, the computation of the masses of planets with satellites, and the study of the precession of the equinoxes as a consequence of the pull of the moon on the equatorial bulge of an earth that is an oblate spheroid—all of these and more must have collectively exerted a strong pressure on scientists to use the Newtonian system even without necessarily believing in it, without necessarily understanding how its central operative feature, universal gravity, could exist and could act as it must in that system.[14] Happily, a main feature of the Newtonian revolution in science was the Newtonian style, which provided a sharp division between the use of concepts like centripetal force and universal gravity that one could not understand and the search for the causes of such forces or the attempts to understand them.

Despite the impressive achievements of the *Principia*, Newton's failures could not but be evident to any first-rate mathematical physicist. Leibniz, Johann Bernoulli, Euler, Laplace were certainly not (in any simple and direct sense of the word) Newtonians. Even Clairaut, who held that Newton's *Principia* had created a revolution, found it necessary to warn against trusting Newton as our guide, and he made his reputation not so much for his commentary on Newton's treatise as for producing a genuine contribution to the theory of perturbations in which Newton had failed. Laplace, who had the highest praise for Newton's general achievement and who openly expressed his admiration for many parts of the *Principia*, could not help but observe that while Newton was 'fort ingénieux' there were many places where he was not 'heureux' in either his methods or his results. Newton, furthermore, had ignored certain topics of paramount importance, notably 'the mechanics of rigid and flexible and fluid and elastic bodies'. Clifford Truesdell (1970) finds that bk. two of the *Principia* offered a challenge to the geometers of the day: 'to correct the errors, to replace the guesswork by clear hypotheses, to embed the hypotheses at their just stations in a rational mechanics, to brush away the bluff by mathematical proof, to create new concepts so as to succeed where Newton had failed'. He concludes that 'rational mechanics, and hence mathematical physics as a whole and the general picture of nature accepted today, grew from this challenge as it was accepted by the Basel school of mathematicians: the three great Bernoullis and Euler', on the basis of whose work a succession of first-rate men during the eighteenth and nineteenth centuries 'constructed what is now called classical physics'. Furthermore, there are major subjects of mechanics—statics, energy considerations, rigid bodies— for which the *Principia* provided no direct illumination. How then could Newton's *Principia* have been of such primary importance in the eighteenth century?

I believe that the answer to this question is not to be found by merely looking for the ways in which a particular proposition or method was used or rejected or corrected. Nor do I believe that Newton's importance was merely in his use of such concepts as mass, centripetal force, universal gravity. Of course it was impressive to have demonstrated that many phenomena could be accounted for by universal gravity, and to have introduced a theoretical basis for the unification of the phenomena of the heavens and of our earth. But I believe that his stature and influence must be seen in more general terms than a measure of either his outstanding successes or his dismal failures, in relation to what Truesdell has called the 'program' that Newton 'laid down for us and illustrated by brilliant examples' in the *Principia*. Of course, the *Principia* contains many beautiful and ingenious solutions of outstanding problems, notably in particle mechanics and dynamical astronomy, and even his only-partial successes and his failures set forth new kinds of problems that were not even generally imagined a generation earlier. But above all Newton set forth a style of science that showed how mathematical principles might be applied to physics and astronomy (that is, to natural philosophy) in a particularly fruitful way. This may have been even more influential in the long run than his system of the world based on universal gravity. But, of course, what gave the Newtonian style, and the book in which it was set forth by example, a more than ordinary importance was not its degree of success so much as the primary nature of the subject to which it was addressed: the system of the world. As Lagrange wistfully remarked, and as Laplace agreed, there was but one law of the cosmos and Newton had discovered it.

In making so much of the Newtonian style, I do not wish wholly to brush aside the many fascinating and challenging statements made by Newton about how to proceed in the investigation of nature, and which have stimulated seemingly endless discussions by historians and philosophers. These include a general procedure for making experiments and drawing conclusions from them, views on cause and effect, statements concerning analysis and synthesis, a rule for obtaining principles that apply to *insensibilia* from those discovered for *sensibilia*, rules of simplicity, a precept concerning the illegitimate invalidation of induction by the invention of contrary hypotheses, the use of the method of induction itself, views concerning hypotheses and their possible uses (in certain limited varieties) in natural philosophy. These issues are of obvious importance to a complete understanding of Isaac Newton the scientist and they also illuminate his science. There have been historical investigations of Newton's method and practice of experiments, his concern with analysis and synthesis, his theories of matter, his experiments and his explanations of optical phenomena, his views on the aether and on action-at-a-distance, his chemical studies and his alchemical pursuits (both literary and experimental), his work on heat and heat transfer, and his opinions on God's active role in the physical world; these have revealed important aspects of Newton's mind and thought and their consequences in the development of physical science during the eighteenth and early nineteenth centuries. Such studies have led to a more exact understanding of the rise of certain experimental sciences such as electricity in the decades following Newton's death. It is a fact of history, however, that those scientists of the eighteenth century who saw a Newtonian *revolution* in the sciences found the revolution in the *Principia* and in the Newtonian style, and not in the *Opticks*, and certainly not in unpublished and private manuscripts that were then safely hidden from inquisitive eyes.

There is one final aspect of the Newtonian style that seems to be a feature of all mathematical science. I have in mind the problem of the primacy of mathematical theory over observational data that may seem to contradict or fail to confirm the theory. The natural reaction of any scientist is to try to save his theory, to "explain away" the discrepancy. When Newton recorded in the *Principia* (scholium concluding sect. 7, bk. two, third edition) his investigations of the resistances of fluids, he had to explain why certain experiments made in St. Paul's had not given the desired result. He found an explanation for the discrepancy in the "fact" that hollow balls made of hogs' bladders would not always 'fall straight down, but sometimes oscillated to and fro while falling'. One of the balls was 'wrinkled and was somewhat retarded by its wrinkles'. In some cases, the poor result cannot so easily be explained away. Even then, I believe it is rare for a scientist to give up at once a cherished theory or hypothesis in that way that Huxley once described in his romanticized image of a 'beautiful hypothesis' slain by an 'ugly fact'.[15]

Of course, Newton would have been very pleased and satisfied if experiments or observations would have ever been in close agreement with theory. But the point is that his own experience had all too frequently been to find a discrepancy rather than such agreement. In his lunar theory, he found himself off by a factor of 2 and in another part of it the answer was off the target by about 30 percent. One can well understand the temptation to fiddle with the numbers, to "fudge" the results. And this was a way out all the easier in those days when the canons of experiment had not yet been fully established and when it was not yet a universal practice to publish complete tables of all the data. In 1966, at a meeting in Austin, Texas, to celebrate the two-hundredth anniversary of Newton's "annus mirabilis", Clifford Truesdell called attention to the fact that in the second edition of the *Principia*, Newton 'introduced the fiction of the "crassitude of the solid particles of air" so as to insert what would nowadays be called a "fudge factor" yielding the desired numerical result from a recalcitrant theory': The problem at issue was the calculation of the velocity of sound. R. S. Westfall (1973) published an article on 'Newton and the fudge factor', in which he declared that 'Not the least part of the *Principia*'s persuasiveness was its deliberate pretense to a degree of precision quite beyond its legitimate claim'. The three examples he has explored are the overly exact correlation of free fall on earth (g) with the motion of the moon, the determination of the velocity of sound, and the mode of computing the precession constant. We have the suggestion that possibly the advance of the lunar apogee should be added to the list.[16] In some cases Newton chose numerical values out of a possible range so as to make the theory (in Cotes's words) 'appear to the best advantage as to the numbers', but in other cases he simply introduced an arbitrary factor to make the theory and the data agree. It would seem to me that in each such case Newton was acting as if he so fully believed in his mathematical theory that he was willing to alter or juggle numbers as if in disdain, obviously not being willing to have his theory be killed by an 'ugly fact'. The "new science" of the seventeenth century demanded confirmation of theory by observation and experiment, and Newton went along with this necessity—but when the two were in conflict, it was not necessarily the basic theory that was rejected (although it might be revised or fiddled with), but rather the data or the numerical calculations.

An expression of the mathematical delight in theory that does not depend on agreement with practical applications or with experiments and observations may be found in Lagrange's *Méchanique analytique* (1788). In the preface, he writes (translated in Hobson, 1912):

> We have already various treatises on Mechanics, but the plan of this one is entirely new. I intend to reduce the theory of this Science, and the art of solving problems relating to it, to general formulae, the simple development of which provides all the equations necessary for the solution of each problem.
>
> I hope that the manner in which I have tried to attain this object will leave nothing to be desired. No diagrams will be found in this work.
>
> The methods that I explain require neither geometrical, nor mechanical, constructions or reasoning, but only algebraical operations in accordance with regular and uniform procedure. Those who love Analysis will see with pleasure that Mechanics has become a branch of it, and will be grateful to me for having thus extended its domain.

In this purely mathematical treatise there is not only an absence of figures, but also an absence of all numbers. There are no numerical data of experience in this work, nor even practical numbers for the determination of constants of integration. It is a general work on dynamics conceived as a purely mathematical subject and as such reminiscent of bk. one of Newton's *Principia*.

I shall conclude this section with a final example: Albert Einstein. In a lecture given in 1933, he boldly declared that reason and not empirical data is the basis of any 'scientific system'. He stressed the 'free inventions of the human intellect' and said:

> I am convinced that we can discover, by means of purely mathematical constructions, those concepts and those lawful connections between them which furnish the key to the understanding of natural phenomena. Experience may suggest the appropriate mathematical concepts, but they most certainly cannot be deduced from it. Experience remains, of course, the sole criterion of physical utility of a mathematical construction. But the creative principle resides in mathematics. In a certain sense, therefore, I hold it true that pure thought can grasp reality, as the ancients dreamed.[17]

Gerald Holton has called attention to an even more remarkable expression of Einstein's belief in the primacy of mathematical theory. The events are related by a student, Ilse Rosenthal-Schneider, who was with Einstein in 1919 when he heard the first report from Arthur S. Eddington that the British expedition to observe a total eclipse had found the first verification of general relativity: There *is* a bending of light rays from distant stars passing through the gravitational field of the sun. What was Einstein's reaction? According to his student,

> Once when I was with Einstein in order to read with him a work that contained many objections against his theory . . . he suddenly interrupted the discussion of the book, reached for a telegram that was lying on the windowsill, and handed it to me with the words, 'Here, this will perhaps interest you.' It was Eddington's cable with the results of measurement of the eclipse expedition. When I was giving expression to my joy that the results coincided with his calculations, he said quite unmoved, 'But I knew that the theory was correct'; and when I asked, what if there had been no confirmation of his prediction, he countered: 'Then I would have been sorry for the dear Lord—the theory *is* correct.'[18]

In this aspect, the Newtonian style is seen to be more a universal characteristic of the science and its practitioners than of the man Newton. The Newtonian style was not a completely original creation of Newton's, but a transformation in which he brought to a high level a tradition going back to Greek antiquity, one that had been undergoing a series of radical or significant transformations during the seventeenth century. I believe it to be a sign of Newton's genius that he was able to grasp the potentialities of this style[19] and to transform it so effectively in the working out of his mathematical philosophy of nature applied to problems of dynamics and the mechanics of the system of the world.

PART TWO

Transformations of scientific ideas

4

The transformation of scientific ideas

4.1   *A Newtonian synthesis?*

The words "synthesis" and "revolution" abound in the literature concerning the formation of modern science. The previous chapters contain testimony to the ways in which Newton's great work has been conceived to have been "revolutionary" ever since the age of Newton. The Newtonian achievement, however, is often categorized as a "synthesis" rather than as a "revolution": as in such phrases as "The Newtonian Synthesis" or "The Great Synthesis" (for example, see Ginzburg, 1933, p. 369*a*; Whitehead, 1922; Butterfield, 1957, p. 106; Rosen, 1973; Gillispie, 1960, pp. 88, 144, 335, 510; Koyré, 1950*b*). The writers who use this word do not define exactly what they intend by it. For instance, in a brilliant essay on Newtonian science entitled 'The Significance of the Newtonian Synthesis', Alexandre Koyré (1950*b*) not only did not define what he meant by "synthesis"; he hardly used this word in the body of this text, and never in so general a context as the title would suggest.

It would appear that the Newtonian "synthesis" occurs in the literature of the history and philosophy of science as a convenient name for the Newtonian achievement taken as a whole, or the Newtonian system of the world, or the Newtonian natural philosophy. There is usually an implied sense of Newton's having put together the contributions (or, possibly the incomplete contributions) of such predecessors and contemporaries as Copernicus, Kepler, Descartes, Galileo, Huygens, and Hooke, and also John Wallis and Wren. As a statement of gross truth, no one could quarrel with a declaration of Newton's indebtedness to certain predecessors, but all such statements lack elements of fine structure and insofar are not very illuminating. Clearly, Newton profited by coming *after* the illustrious men listed above; not even a Newton could have produced his great work had he lived before them. Ergo, there seems to be no possibility that a Newton (in the sense of a creator of the Newtonian natural philosophy) might have arisen in the times of Galileo or Kepler, or that either Galileo or Kepler might have been such a Newton. To follow such men, rather than to have been their contemporary or predecessor, may have been a necessary condition for our Newton, but hardly a sufficient one: witness the failure of Huygens to have created a Newtonian system (see §3.4).

The term "synthesis", in my opinion, is a misleading word to use in the historical analysis of science, because it tends to mask the creative way in which any scientist uses the work of his contemporaries and predecessors. Derived from the Greek word σύνθεσις (or composition), "synthesis" in this context means the 'putting together of parts or elements so as to make up a complex whole'–as opposed to "analysis".[1] Thus the historian William Lecky (1879, vol. 1, p. 168) wrote of a 'system which would unite in one sublime synthesis all the past forms of human belief'. Here, then, is one of the most current senses of the word: another is related to this one, save that it means combining together by ingenuity and artifice either something that would never have come into a single whole by itself or that had hitherto only been produced by nature alone. One aspect of this second sense is used commonly today in reference to artificially produced materials (often called "manmade"), the products of the laboratory or the chemical plant and not of natural processes on or in the earth or in our atmosphere.

As we shall see below, in a number of examples, neither of these two meanings truly gives an accurate description of the process of Newton's creative thought. He did not create an artificial system to be superimposed upon nature and he certainly did not merely combine in a synthetic "stew" the principles of Copernicus, Kepler and Galileo, Descartes, Hooke, and Huygens. Rather, he carefully selected certain ideas (concepts, principles, definitions, rules, laws, and hypotheses) and *transformed* them, giving each of them a new form which only then was useful to him. Let us insist, then, that if there is such a historical event as a "synthesis" in scientific thought, then the synthesizer must be highly selective, in the first instance choosing what he finds good or useful and rejecting the rest. But in many cases, the "synthesis" becomes possible only after the selected ideas will have become transformed in such a manner as to become essentially new and different ideas. In some cases they become obviously new and different, since there is a resultant contradiction of the prototypes. In other cases, they are applied in such novel ways (or to such novel situations that go far beyond their authors' intentions and restrictions) as to constitute new ideas altogether. In yet other cases, they become the starting points for developing new ideas that may only faintly resemble the originals. In short, I see two separate creative processes: transformation and synthesis. Synthesis directs our attention to the act of putting various parts together into a whole. Transformation is the prior process in which ideas are put into such a condition that a synthesis is possible.

One should perhaps make a distinction between "the Newtonian synthesis" and "Newton's synthesis". Gerald Holton (1978) suggests that the first of these be used for 'the successful development of physical science from the late seventeenth century' and the second for 'Newton's own achievement'. The gulf between the two is wide and deep. For example, Newton himself was never able to encompass physical optics within the force-mass-motion framework of the *Principia*, nor did he ever succeed in producing a system of physical optics in what I have called the Newtonian style which had worked so well in the *Principia*. And it is the same with respect to chemistry (or alchemy) and theories of matter. Hence "Newton's synthesis" would seem to be limited to a part of rational mechanics, rather than be inclusive of all of statics and dynamics. We have seen that "Newton's synthesis" excluded much of dynamics itself, the most obvious omission being the consideration of rigid bodies and almost all of the physics of deformable bodies. Nor did Newton concern himself with the point of view being developed by Huygens, Hooke, and Leibniz that embraced the concept of energy. I would find the chief contribution of Newton to "the Newtonian synthesis" to have been the demonstration of the power of the Newtonian style in mathematizing physics and then relating the mathematical principles to natural philosophy. Of course he also contributed to the "Newtonian synthesis" such crucial concepts as mass and the concept and law of universal gravitation plus its exemplification in the solar system.

"Newton's synthesis" has at least two senses. One is the unification within a single scientific structure of subjects previously held to be separate or previously not seen to be closely related. Thus Newton showed that the falling of bodies on earth, the phenomena of tides, the motion of the moon, and the motions of planets and their satellites are all part of a single system of physics and are all effects of the same force of universal gravity. It is well known that Newton sought to make a synthesis, however unsuccessfully, of the physics of gross bodies and the physics of the particles of which they are composed. The other sense of "Newton's synthesis" is the production of a system of physics by synthesizing concepts, laws, and principles of Galileo and Kepler and possibly of Descartes and others. It is in this process that "transformation" is of such importance that it would be more strictly accurate to refer to Newton's synthesis of concepts, laws, and principles "originating" with Galileo and Kepler rather than concepts, laws, and principles "of" Galileo and Kepler.

In what follows, instances are given of a number of different types of "transformation" of scientific ideas, not only those made by Newton but also by those very men whose ideas he transformed in turn. Examples are Newton's transformation of Keplerian "inertia" (by which bodies come to rest when the motive force ceases to act) to a kind of "inertia" (or "force of inertia") by which a body maintains its "state of motion" in the absence of an external force; Kepler's transformation of Aristotle's physics of local motion by applying Aristotelian terrestrial physics to the heavenly bodies, for which there was supposed to be a wholly different set of Aristotelian principles of motion.[2]

One way of seeing that "Newton's synthesis" could hardly have been a mere putting-together of the science of his predecessors is to observe that the *Principia* exhibited the express falsity of certain basic or fundamental principles of their science. Among them are the following:

Copernicus: the solar system does not have its center in the true sun, but in a fictitious or "mean" sun (actually the center of the earth's orbit), with respect to which all planetary orbits are reckoned; the planetary orbits are compounded of circles on circles (epicycles on deferents or other epicycles).

Kepler: the three planetary laws are "true" descriptions of the motion of the planets; a solar force, exerted on those bodies, diminishes directly as the distance and acts only in or near the plane of the ecliptic; the sun must be a huge magnet; because of its 'natural inertia', a moving body will come to rest whenever the motive force ceases to act.[3]

Descartes: the planets are carried around by a sea of aether moving in huge vortices; atoms do not (cannot) exist, and there is no vacuum or void space.

Galileo: the acceleration of bodies falling toward the earth is constant at all distances, even as far out as the moon; the moon cannot possibly have any influence on (or be the cause of) the tides in the sea (see §1.4).

Hooke: the centripetal inverse-square force acting on a body (with a component of inertial motion) produces orbital motion with a speed inversely proportional to the distance from the center of force: this speed law is consistent with Kepler's area law (see §3.1).

Since "Newton's synthesis" must thus deny the general validity of ingredients it has allegedly fused together, it follows that Newton must have been extremely careful in selecting those parts of the science of others that might be useful. Furthermore, even such useful concepts, rules, methods, and systems would require serious transformation before being melded or blended together. A "true" natural philosophy or system of the world can hardly be an amalgam of falsehoods. And yet, from a strictly logical point of view, even such a basic set of ingredients for "Newton's synthesis" as Kepler's laws proved to be strictly false in the system of the world expounded in bk. three of the *Principia*.[4] It is, of course, undeniable that Newton's system does embody some altered or transformed versions of these earlier systems, doctrines, or principles. For instance, in the *Principia* Copernicus's system becomes heliocentric as well as heliostatic and the planetary and lunar orbits become elliptical—both innovations made by Kepler.[5] Also, in the *Principia* it is shown that the planetary orbits are not exactly ellipses.[6] Galileo's doctrine that the acceleration of bodies falling freely downward toward the earth is the same for all bodies and is everywhere constant is transformed in the *Principia* to an acceleration on earth that is the same for all bodies at any given place but that more or less varies with terrestrial latitude and that out in space near the earth varies inversely as the square of the distance (see §5.1).[7] And so on. In my view, each such case reveals the special creative force of Newton's mind in the individual act of transformation he made; the supposition of mere synthesis without transformation seems to me to imply a creative act on a much lower level. Furthermore, transformation suggests at once the importance of recognizing potentialities of useful transformation in some idea or set of ideas. And, as shall be seen below, there is a special quality to some ideas that enables them to become susceptible of useful transformations and thus to serve in the further advancement of science. Accordingly, an examination of such transformations is a necessary condition for understanding the Newtonian revolution in science.[8]

4.2   *Transformations of scientific ideas*

The creative intellectual process whereby Newton transformed certain ideas[1] of his predecessors and contemporaries is not an idiosyncratic feature of the Newtonian revolution in science. It is my belief that all revolutionary advances in science may consist less of sudden and dramatic revelations than a series of transformations, of which the revolutionary significance may not be seen (except afterwards, by historians) until the last great step. In many cases the full potentiality and force of a most radical step in such a sequence of transformations may not even be manifest to its author.[2] It is my thesis that cognizance of such a series of transformations, while explaining the last stage and making it seem less miraculous than would otherwise be the case, should in no way lessen our admiration and esteem for the extraordinary qualities of a great discoverer or radical innovator, however much he may prove to be (in the special sense adopted here) merely the author of the last of a sequence of transformations.

A great intellectual leap forward may, by this mode of analysis, be reduced not merely to but one stage in a succession of discoveries, but may even prove to have been part of a sequence so long that the concept of a single originator who began it all becomes meaningless. To be sure, a long chain of this kind may not seem quite as satisfying as a single miraculous stroke, but such psychological problems relate more to the spectator than to the succession of events he beholds. I would observe, furthermore, that there is much convincing empirical evidence that what I have called the last or final stage generally requires a mind of truly heroic creative proportions: a Kepler, a Galileo, a Newton, a Clerk Maxwell, an Einstein, a Rutherford, a Niels Bohr, or a Fermi; a Harvey, a Linnaeus, a Haller, a Darwin, a Pasteur, or a Claude Bernard.

Historians of science are often criticized for their constant belittling or debunking of the great figures, for showing that Galileo did not invent the new science of motion out of whole cloth and was not the original author and creator of the experimental method and, hence, of modern science as we know it. What often seems to be a particular grievance for some critics, many of whom are practicing scientists, is the conscious tendency of historians to seek out prior instances of the occurrence of scientific concepts, methods, laws or rules, experiments and observations, and even theories. On the surface, it would appear that today's historians of science *are* engaged in finding feet of clay wherever they can. Do not the writings of historians show that Galileo was *not* the first discoverer of the principles of uniform and of uniformly accelerated motion? And do they not declare that Galileo did *not* make all of his discoveries by experiments? It has been shown, furthermore, that even the definition Galileo used for uniformly accelerated motion and his statement for the primary theorem of the equivalence of uniformly accelerated motion and a uniform motion at the mean speed were well known by the end of the fourteenth century (Clagett, 1959; Maier, 1949). Even the diagram used by Galileo in proof of the mean-speed theorem seems to have been found with a minor variation in a text by the fourteenth-century French thinker, Nicole Oresme.[3] And, as has been mentioned earlier, it can no longer be maintained that Galileo had no predecessor in declaring free fall to be an example of such uniform acceleration. Do such revelations by historians shrink Galileo's stature or belittle his monumental achievement? Not at all! In a sense, these new revelations make Galileo's achievement at once more plausible and far greater in magnitude than anyone had ever before even imagined.

Let us put aside for the moment the question of plausibility and reasonable historical belief. As to dimensions of greatness, I believe—paradoxical as it may sound—that the traditional point of view held by scientists actually denies its major premise. Lagrange once said wistfully that Newton was the most fortunate of men since there is only one system of the world to establish and Newton did this. Laplace is said to have repeated this opinion with a new emphasis (Delambre, 1812, p. xlv, summarized in Brewster, 1855, vol. 1, p. 319n). Lagrange was apparently implying that he—alas!— could not discover a cosmic law although he had the genius to have done so, since there is only one such law and Newton happened to have lived already and to have discovered it. And Laplace was saying in effect that if Newton had not lived when he did (and if, therefore, Newton had not discovered the law), then Lagrange would have done so; so that there was no possibility whatever that he (Laplace) could ever have discovered it, having been born doubly too late! Laplacian probability was against him. Most scientists do not object to such statements by Lagrange and Laplace, and yet surely it is a real diminution (and certainly not an aggrandizement) of Newton's stature even to imply that Newton's primary quality as a scientific innovator was to have antedated Lagrange and Laplace. And, in general, the theory of sudden, unanticipated, precursorless (almost as if "miraculous"?) creation in science—to have invented something wholly new rather than to have been a final transformer; that is, to have been a complete innovator or first man rather than the last man in the succession—seems to be a "belittling" theory insofar as it always must suggest that the definition of genius consists primarily in the good luck of having been born before certain discoveries had been made.

Contrast this point of view with the one being advanced here. I hold that a quantitative measure of Galileo's genius may be found in the actual number of ideas we find him adapting and then using, which were "in the air", which were to some degree known to his contemporaries and even to his predecessors. These many men did not have the transcendent genius to put all the pieces together, to reshape them in the process, and to add some further new ideas and to create the new science of motion. But Galileo recognized the potentialities in certain ideas then current that awaited transformation and that could become inspirited with new life, when linked to Galileo's own ideas, and made to serve a new purpose.[4] Benedetti, Tartaglia, Cardano, Soto, Beeckman, even Descartes and Kepler, did not find the Galilean laws of falling bodies and of projectile motion. The very existence of such first-rank contemporaries, to whom the information available to Galileo and used by him was also available, may be taken as evidence of the special quality of genius that was required to transform those ingredients and to go on to create not one but in fact 'Two New Sciences' ('mechanics' and 'local motion'). Similarly, the science of the *Principia* was not created by Wallis or Halley or Wren or Hooke, or even Leibniz or Huygens. What better testimony than the existence of these men could we have as to Newton's greatness?

The concept of transformation is to some extent merely an explicit formulation of the practice of all historians of ideas, and especially of all historians of scientific ideas. Nevertheless, making transformations of scientific ideas explicit will tend to highlight the truly creative aspects of the ways in which the concepts, theories, methods, experiments, and even laws of one scientist are used by another. The very term implies that the scientific creative process is not merely a scissors-and-paste selection and assembly from available documents and other sources, embodying the possible recognition of gems in a bed of dross. The concept of transformation goes even beyond the view of a radical confrontation or juxtaposition of several sets of ideas, and a subsequent choice or selection among them.[5] For "transformation" at once suggests not merely the recognition of what may be useful, but an insight into the potentialities of ideas that are not obvious and that can be realized only by the transformations produced by the force and activity of a scientific mind with a high order of creative genius.[6]

The word "transformation" is more than an expression for change in general; it is used in mathematics to describe the alterations that may occur in geometrical figures (curves, lines, or systems of curves and lines) or spaces and spatial elements under certain processes whereby one space is changed into another. A simple example of this kind of transformation occurs whenever a plane is projected upon another (nonparallel) plane. In such a projection, a circle may be transformed into an ellipse: as is seen whenever bright sunlight falls on a shuttered window with a small circular hole (or on a hole in the surface of a tent) and an image of the hole is projected on the floor, transformed into an ellipse. In such mathematical transformations, particular attention is always directed to constancies, the "invariants" of the transformation. Thus two intersecting lines may be transformed into a pair of curves in such a way that the point of intersection of the two lines is still an intersection, but of curves. Or, there may be a point-to-point transformation of a plane ring into itself; a famous theorem, stated by Henri Poincaré and proved by G. D. Birkhoff, explores the condition of there being two points which remain the same, or are invariant under the transformation.[7]

By analogy, it is suggested to the historian of ideas to keep on the alert for possible invariants in the transformation of scientific ideas. These may, in some instances, be concepts or laws; but the invariants may prove to be only the names of such concepts and laws, serving as archeological remains to guide the inquiring historian to the sources of a given transformed scientific idea.[8]

Synthesis assumes that revolutionary scientific innovation is like the completion of a huge jigsaw puzzle, possibly one in which some of the key pieces may be missing and require invention. Transformation, by presenting each anterior concept, theory, law, or principle as the *occasion* of an innovation, focuses attention on the *cause*, the possible reason why only one of the many scientists to whom the scientific idea was known produced the transformation in question. In this way, the doctrine of transformation may help to clarify the actual stages of the scientific creative process.

4.3   *Some examples of the transformation of scientific ideas: Darwin and intraspecific competition, Franklin and the electrical fluid*

In §4.5 the Newtonian concept and principle of inertia will be seen to have been part of a long sequence of transformations going back at least to Galileo, Gassendi, Kepler, and Descartes. And in Ch. 5 one of the main themes is Newton's transformation of Kepler's laws of planetary motion. The two examples that follow, one from the nineteenth century and one from the eighteenth, illustrate two very common types of transformation. The first shows how a scientific idea, competition for survival among species, becomes transformed following the stimulation a scientist receives during the course of his reading; in this case Darwin was reading Malthus. The second example shows how a scientific idea (a subtle elastic electrical fluid) arises from the transformation of a generally accepted or widely diffused notion (the Newtonian aether or 'Aetherial Medium') during the course of attempting to explain a series of experiments. In both examples, the scientist is attempting to analyze and account for a set of phenomena, and not simply transforming at will certain received scientific ideas in order to find out what their consequences might possibly be.[1]

These two examples are highly typical of the creative process in science and in many aspects of the arts. Again and again we may see Newton developing new ideas as he makes annotations or comments on his reading. These initial transformations then go through a whole further set of transformations until the final intellectual achievement takes form. This may be seen most clearly in the transformations leading up to the first stage of the differential calculus, begun in the transformations of ideas and methods that Newton had encountered in reading Schooten's edition of Descartes's *Géométrie* (see Whiteside, 1964a; Newton, 1964–1966, pp. ixsq; 1967–, vol. 1, passim). In other cases, the immediate stimulus or occasion is less clear, as in Newton's transformation of the second law of motion for impulsive forces into a law for continuous forces: that is, the recognition that if an impulsive force produces a single instantaneous change in momentum, then a continuously acting force must produce a continuous change in momentum. This last transformation (see §4.4) required Newton's concepts of time, of limits, and of mass.

A striking example of the transformation of a scientific idea is to be found in Charles Darwin's concept of natural selection in relation to Charles Lyell and Thomas Malthus. The creative factors in relation to Malthus and Darwin have been much debated, and a crucial bit of information has been the date on which Darwin actually read Malthus, now firmly established by his 'Notebooks on Transmutation of Species'.[2] In an analysis of this problem published in 1971, Sandra Herbert has shown that prior to Darwin's reading of Malthus (the dated entry on Malthus is for 28 September 1838), he was fully aware of Lyell's concept of a 'universal struggle for existence' in which the 'right of the strongest eventually prevails'.[3] But, as Sandra Herbert points out, 'Lyell is not really speaking of competition between individuals of the same group to represent that group in nature'; the result is that 'the kind of selection always uppermost in his mind was . . . the competition between various species and races, to maintain their place on an earth with limited amount of life space'. What prevails in the 'struggle for existence', then, is not the 'strongest' individuals but the 'strongest' species. Lyell's concentration was thus 'interspecific rather than intraspecific'. It is apparently for this reason that Darwin ('who accepted Lyell's presentation of competition without protest') did not 'come to natural selection sooner' and was certainly 'not thinking in that direction at the time he read Malthus'.[4]

Malthus called Darwin's attention forcibly to the 'terrible pruning . . . exercised on the individuals of one species', according to Herbert, and thereby 'impelled Darwin to apply what he knew about the struggle at the species level to the individual level'. He saw 'that survival at the species level was the record of evolution, and survival at the individual level its propulsion'. In short, Lyell's 'concentration on competition at the species level' apparently numbed Darwin 'to the evolutionary potential of the "struggle for existence" at the individual level'. Thus Herbert concludes that Malthus should be considered 'as contributor rather than catalyst' to the 'new understanding' achieved by Darwin, after 28 September 1838, 'of the explanatory possibilities of the idea of struggle in nature'.

From the point of view presented in this book, it would be said that Darwin transformed Lyell's concept of an interspecific competition or 'struggle for existence' to an intraspecific one, a transformation from the species to the individual level that led on to an explanatory principle of some kind of selection factors that determine survival and reproduction. If Darwinian natural selection is based on three elements—'individual variability, the tendency toward overpopulation, and the selective factors at work in nature'—then we can see how crucial this transformation was as a stage in Darwin's creative thought. And, furthermore, we can now make precise the exact role of Malthus, not in adding yet another factor to a supposed Darwinian synthesis, nor in supplying Darwin with a mathematical law of population increases, but rather in directing Darwin to transform Lyell's concept into a struggle among individuals by bringing him 'to concentrate on the competitive edges to nature—predation, famine, natural disaster—as they played upon the individual differences of members of the same group'.[5]

In this example, finally, we may see certain conditions favorable for a transformation to occur: the existence of a concept with potentialities for useful transformation, and contact with a mind capable of making the transformation. But above all, these necessary conditions must be augmented by the condition of a working creative scientific mind wrestling with real scientific problems, not merely speculating on a philosophical level.[6] Darwin was evidently deep into problems of overpopulation; and he was developing his thoughts about the causes of variation and the rules of inheritance, both of which would prove to be relatively unfruitful and to lose their apparently crucial importance 'once he had natural selection to rely on'. Concerned with the struggle for existence, and the occurrence of variation within a single species (and not just the more abstract problem of the variations in species), Darwin was in the most favorable condition imaginable to apply Malthus's considerations to the struggle for existence among all the varying individuals comprising a species and thus (by transforming Lyell's concept under the pressure of Malthus's findings concerning the constraints on a single species) set off on the intellectual road to Darwinian natural selection.

A somewhat different kind of transformation may be seen in the development of Benjamin Franklin's concept of the electric fluid. Franklin and his co-experimenters in Philadelphia were doing a series of experiments in which there were spark discharges or transfers of the electrical "influence" from one person or inanimate object to another. Naturally enough, in the context of scientific thought of the 1740s, it seemed as if there were some material or quasi-material substance or "subtle" fluid that might be in motion.[7]

Franklin himself was steeped in the tradition of Newtonian experimental philosophy, which he had encountered in reading the queries of the *Opticks* and various books on Newtonian experimental science by such authors as Boerhaave, Hales, and Pemberton.[8] Newton had conceived of an aether that enters into bodies, filling the interstices between their solid parts. This aether, furthermore, was said to be possibly "elastic", expanding to fill up all kinds of spaces because of its composition: possibly a collection of minuscule particles mutually repelling one another. That is, Newton introduced the supposition 'that *Aether* (like our Air) may contain Particles which endeavour to recede from one another' (Newton, 1952, p. 352). Another property assigned to this aether was having different density in various circumstances and in relation to different kinds of bodies. In qu. 22 of the *Opticks*, furthermore, Newton introduced the subject of electrical effects to explain how the aether could be so rare and yet potent. If anyone should ask how so potent a medium as the aether 'can be so rare', said Newton,

> Let him also tell me, how an electrick Body can by
> Friction emit an Exhalation so rare and subtile, and yet
> so potent, as by its Emission to cause no sensible Diminu-
> tion of the weight of the electrick Body, and to be ex-
> panded through a Sphere, whose Diameter is above
> two Feet, and yet to be able to agitate and carry up Leaf
> Copper, or Leaf Gold, at the distance of above a Foot
> from the electrick Body [Newton, 1952, p. 353]?

When Franklin began to think of a matter that might produce

electric effects, this Newtonian aether (or 'Aetherial Medium') would have come to his mind at once, along with the notion of electrical exhalations.

We can easily trace how Franklin's concept of an electrical fluid or electrical matter resulted from a transformation of the Newtonian concept of aether. The first full statement concerning his new theory was embodied in a paper entitled 'Opinions and Conjectures concerning the Properties and Effects of the Electrical Matter, arising from Experiments and Observations made at Philadelphia, 1749'.[9] This paper began with the proposition (1) that the electric matter consists of 'extremely subtile' particles since it can easily permeate all common matter, even metals, without 'any perceptible resistance'. Here Franklin used the term 'electrical matter' for the first time. Although he indicated a cause for belief in its 'subtility', he took its atomicity or particulate composition for granted. Franklin did not refer to any particular experiments to prove that electricity passes through conductors and not just along their surfaces; he was content to say that (2) if any one should have doubts on this point, a good shock taken through his body from a Leyden jar would convince him. (3) The difference between electric matter and 'common matter' lies in the mutual attraction of the particles of the latter and the mutual repulsion of the particles of the former (which causes 'the appearing divergency in a stream of electrified effluvia'). In eighteenth-century terms, electrical matter—like the Newtonian aether, whose particles 'endeavour to recede from one another'—is a particulate, subtle, elastic fluid. (4) The particles of electric matter, though mutually repellent, are attracted strongly by 'all other matter'. (5) Therefore, if a quantity of electric matter be applied to a mass of common matter, it will be 'immediately and equally diffused through the whole'. In other words, (6) common matter is 'a kind of spunge' to the electric fluid. A sponge absorbs water more slowly than common matter absorbs electric matter since the 'parts of the spunge' are impeded in their attraction of the 'parts of the water' owing to the mutual attraction of the 'parts of the water' which the sponge must overcome. Generally, (7) in common matter there is as much electric matter as it can contain; therefore, if more be added, it can not enter the body but collects on its surface to form an 'electrical atmosphere', in which case the body 'is said to be electrified'. All bodies, however, (8) do not 'attract and retain' electrical matter 'with equal strength and force', a phenomenon presently to be explained, and those called electrics *per se* 'attract and retain it strongest, and contain the greatest quantity'. (9) That common matter always contains electrical fluid is demonstrated by the fact of experience that a rubbed globe or tube enables us to pump some out.[10]

The new properties of the transformed concept include the attraction between particles of this fluid and the particles of ordinary matter; the natural quantity of this fluid in bodies; the particular laws of the distribution of this fluid in bodies of different shape. The theory led to the still-continuing nomenclature of electrical science: "plus" or "positive" for bodies that have (presumably) gained an excess of fluid over their normal quantity, and "minus" or "negative" for those which have lost some of their normal quantity and are in a deficient state. Two positively charged bodies mutually repel one another because the excess electrical fluid of one repels the excess of the other; and there is, similarly, a reason for attraction between two bodies of opposite sign. Since electrification results from the transfer or redistribution of electrical fluid, it follows that the occurrence of any charge must be accompanied by the simultaneous occurrence of a charge of exactly the same magnitude but of opposite sign; and that when charges are destroyed or are annulled, equal quantities of charges of both signs must mutually cancel out one another. Franklin's "law of conservation of charge", for which there is no counterpart in the original Newtonian concept of aether, is a sign of the degree of creative innovation in Franklin's transformation.[11]

4.4   *Some transformations of ideas by Newton, primarily the transformation of impulsive forces into continually acting forces and the formulation of Newton's third law*

In Newton's day, a primary meaning of force was the impulsive or momentaneous action that occurs when one object strikes or is struck by another. Such percussive forces were the subject of intensive study in the seventeenth century, by such men as Galileo, Baliani, Marcus Marci, Descartes, Wallis, Wren, and Huygens.[1] Descartes had announced a set of rules for the outcome of impact which were notorious for their having been demonstrably false (see Koyré, 1965, pp. 76–79). Clarification of this problem required a proper distinction between what we call elastic and inelastic collisions, and an understanding of the vectorial nature of momentum. In solving the problem of impact, Wallis, Wren, and Huygens independently discovered the law of conservation of momentum.

In the scholium to the laws of motion in the *Principia*, Newton gives an instance of the application of 'the first two laws and the first two corollaries', together with the third law, in the way in which 'Sir Christopher Wren, John Wallis, S.T.D., and Christiaan Huygens, easily the foremost geometers of the previous generation, separately found the rules of the collisions and reflections of hard bodies, and communicated them to the Royal Society at nearly the same time'.[2] Newton then devotes considerable space to the laws of impact, including the distinction between the collisions of elastic and inelastic (or partly elastic) bodies and the confirmation of the rules by experiments made with colliding pendulum bobs.

This percussive type of force was fundamental in Newton's philosophy, as may be seen in the fact that it is the force that appears in the second law of motion, in its formal expression in the *Principia*:

> *Law 2*: Change of motion is proportional to the motive force impressed and takes place along the straight line in which that force is impressed.[3]

Newton says that the motive force may be impressed either 'all at once' in a single blow, or 'by degrees and successively' in several blows whose net effect is thus the same as that of the single blow which is their sum.[4] This is the second law of motion[5] in the "impulsive" form: We would state it in today's language of physics as a proportionality between an *impulsive* force and a *change* of momentum. It may thus be contrasted with the more familiar second law, which states a proportionality between a *continuous* force and a *rate of change* of momentum.

In the *Principia*, Newton never states the continuous form of the second law as an explicit axiom, although he uses it again and again. For example, in the proof of prop. 24, bk. two, Newton states that 'the velocity that a given force can generate in a given mass in a given time is as the force and the time directly and the mass inversely', which reads almost as if it were a textbook statement of the second law for a uniformly acting force. Although this accelerative (or continuous) version of the second law does not appear *expressis verbis* as an axiom or law of motion, it is in fact to be found among the definitions. In defining three "measures" of centripetal force, Newton has to deal with the accelerative version of the second law because a centripetal force (such as gravity and magnetic and electric attractions) acts continuously, or at least appears to act continuously.[6] And so Newton had to introduce the time of action whenever he was considering continuous force. For instance, one measure of centripetal force is 'the accelerative quantity' (def. 7), 'proportional to the velocity that it [i.e., the centripetal force] generates in a given time'; another is the 'motive quantity' (def. 8), proportional to the momentum 'that it generates in a given time.' The 'accelerative' measure is thus the 'motive measure' per unit of mass; and Newton points out that 'accelerative force is to motive force as speed is to [quantity of] motion [or momentum]'. He explains this in terms of an easily recognizable form of the second law of motion:

> For quantity of motion [or momentum] arises from [and is thus proportional to] the speed and quantity of matter conjointly; and motive force [arises from and is thus proportional to] the accelerative force and the quantity of matter conjointly.

It is of great interest that here Newton says in words what we would write as an equation $(F = mA)$. In the preceding two definitions, he introduces the factor 'in a given time' for a continuous force.[7]

Thus, we have apparently two different second laws of motion: one for continuous forces, which we today would write $d(mV) = k_1 F dt$, and another for impulsive forces, which we today would write $d(mV) = k_2 \Phi$. Newton called both types of "force" by the same name, but I have given one the symbol F and the other Φ, not only to differentiate one from the other but to indicate that they have different physical dimensions—an aspect that appears also in the different constants of proportionality, k_1 and k_2. Whereas we today tend to reconcile these two equations (or laws) by defining Φ (the "impulse") as $F dt$, whereupon $k_1 = k_2$ as long as we stick to a single set of units, Newton assumed both that the continuous form of the law follows from (or is implicit in) the definitions and that it can be derived in a special way by a transformation of the impulsive form of the law.[8]

Another way of saying this is that in the *Principia* Newton does proceed from a succession of impulses or blows to a continuously acting force, but he also considered that it was a consequence of dt being "given" that the second law could actually embrace both

$d(mV) \propto F$ and $d(mV) \propto F \cdot dt$ and also $d(mV) \propto \frac{1}{2} F \cdot dt^2$. It would then be only our modern post-Newtonian prejudices that would require a distinction to be made between these forms of the law and even necessarily between F and $F \cdot dt$ (cf. Cohen, 1970*d*, apps. 1, 2).

Before turning to this transformation, let us observe that as a result of Newton's conception of time in relation to his theory of fluxions, these two laws are effectively equivalent. I have mentioned earlier (§§3.1, 3.2) that Newton's conceptions of time in dynamics and in pure mathematics are analogues, and that the power of his physics of motion may have been due in large measure to his having developed a useful concept of time in mathematics. We saw that for Newton, time is 'absolute, true, and mathematical' or 'relative, apparent, and common' – as he put the matter in the scholium following the definitions in the *Principia*. This distinction was a natural one to any astronomer, who must be aware of the difference between mean time and local apparent solar time: an artificial and absolute time, regular and uniform, and a local or common time. And, since this absolutely flowing time, which occurs in Newton's mathematics and also analogously in his physics of motion, is measured (as Newton says) by a velocity; it follows that there must exist an absolute velocity; and such an absolute velocity necessarily defines and requires an absolute space. Much has been written on Newton's concepts of absolute space and time, including the suggestion that his belief in an absolute space may have been caused by his psychological needs, without taking cognizance of this basic aspect of Newton's mathematical dynamics and dynamics-based mathematics.

In the *Principia* the very first proposition on motion displays a transition from discrete impulsive forces to continually acting forces. This is the proposition to which reference has been made several times, in which Newton proves that the action of a centrally directed force on a body with inertial motion will produce motion that accords with the area law. Newton does so in three stages. First, he shows that in pure inertial motion, the moving body will trace out equal areas with respect to any point not on the line of motion. Second, he supposes that at regular intervals the body is given a blow from an impulsive force, always directed toward the same point P.[9] This converts the rectilinear inertial path into a polygon, the sides of which still determine equal areas (in these

equal times) with respect to *P*. Then, Newton says, 'let the number
of triangles be increased indefinitely . . . ; their ultimate perimeter
. . . will be a curved line; and so the centripetal force . . . will act . . .
continually, while any areas described . . . , which are always proportional to the times of description, will be proportional to those
times in this case also.' This transition from a succession of impulses to a continually acting force is a kind of hallmark of Newton's physics.[10] It appears in his manuscripts of the 1660s,[11] when
he first found the law of centrifugal force, and it was his key to that
discovery; it occurs again, much as in the *Principia*, in the preliminary tract *De motu*, which he wrote in 1684.[12] In the opening
section on physics in the *Principia*,[13] this proposition serves as a
kind of declaration that Newton will use the method of limits as a
primary tool throughout the book.

There are some rather significant aspects of this mode of approaching continually acting forces. First, if a centripetal force is
conceived as the limit of a sequence of impulses, then the essence
of the gravitational force is not so much a continuum of pressure

Figure 4.1. Let a body move with uniform inertial motion along the
straight line *ABCDE*... Since the motion is uniform, the distances
AB, BC, CD, DE, ... are traversed in equal times. Let the points
A, B, C, D, E, ... be connected to some point *P* (not on the line
of motion) by lines *AP, BP, CP, DP, EP*, ... forming triangles *ABP,
BCP, CDP, DEP*, ..., all of which have equal bases *AB, BC, CD,
DE,*... Erect a perpendicular from the line *ABCDE* ... to *P*,
which is the common altitude of all the triangles. Since the triangles
have equal bases and the same height, they have equal areas. Since
the bases were determined by equal times, the areas swept
out by a line from the moving point to *P* are proportional to the
time, or are equal in any equal times. This is Newton's discovery
of the relation between the law of inertia and the area law.

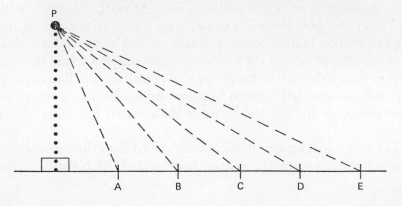

as the sum of individual blows, whose total effects are computed mathematically in the limit. This mathematical view of individual gravitational impulses was transformed into a physical hunch when Newton said in the introduction to sect. 11 of bk. one of the *Principia* that if he were to 'speak in the language of physics', the mutual gravitational forces (or centripetal forces) 'might more truly be called impulses' (see §4.3). In qu. 31 of the *Opticks*, first published in Latin in the *Optice* of 1706 and then in English in the second edition of 1717/8, Newton again hinted that 'What I call Attraction may be performed by impulse, or by some other means unknown to me' (Newton, 1952, p. 376). The aether, which he seems to have hoped would yield an explanation of the cause of gravity, was (in its last manifestation in the *Opticks*) particulate; as were, presumably, the electrical effluvia which at one time he also thought might give him the key to gravity.

A second aspect of the Newtonian limit-mode of transition from individual impulsive forces to continually acting ones is that it represents a transformation of the physics of impacts or collisions, which had become a staple of seventeenth-century physical thought, into the physics of attractions, or of forces that act as if they attract. To base forces acting at a distance on forces of contact, such as blows or percussive impulses, enabled Newton to give his physics a sounder ontological base in phenomena than if he had started out with attractions. For in the action of impulsive forces one may see both the physical event or cause (the blow given by a tennis racquet, or by a billiard cue, or by the impact of one billiard ball on another) and the effect in a change of momentum. In the case of major centripetal forces (as in the orbital motion of planets, planetary satellites, and comets), one sees only the effect (a change in momentum in any given period of time, Δt or dt, however small), and has to suppose there is a cause. This transformation from the physics of the known (contact forces of collisions and physical blows) to the physics of the unknown (gravitational forces that appear to have to act at a distance) was a tremendous intellectual leap forward that enabled Newton to transcend the limits of the terrestrial physics of his age and to forge the celestial dynamics of the future.

In Ch. 5 it will be seen that Newton's third law of motion played a considerable role in the transformation that led Newton to the concept of a universal force of gravitation. Of the three laws of motion, this may have been the most original with Newton. There is a rather long list of transformations leading up to Newton's first law, or the principle of inertia.[14] His own contribution to the first law was largely in relation to his conception of mass and his introduction of the technical term "inertia" in what was to become a traditional sense.[15] It was the very same concept of mass that gave Newton's second law its distinctive Newtonian character and made possible Newton's quantification of the relation between an impulsive force and the change in momentum it produces, and between a continuous force and the rate of change in momentum (or the net change in momentum in any given time) that it produces. But no one before Newton had clearly conceived of mass (quantity of matter)—the equivalent of inertia as measured by the resistance to a change of state—as the regulating constant in the second law, whether for impulsive or continuous forces.[16]

How original was the third law, that action is always equal and opposite to reaction? By now, we should expect the answer to lie in the degree of radicality of Newton's final transformation. The third law seems to have arisen in the first instance from the principles of collision: when a body A strikes a second body B, the "action" of A on B is accompanied by an equal and opposite "reaction" of B on A. As a generalization from this restricted law to a law for continually acting forces of pressure, and then to a law for attractions of any sort, Newton's third law represents a gigantic intellectual leap forward,[17] a transformation that probably makes the third law the most truly original of the three. Allen Debus (1975) Alan Gabbey (1971), and R. W. Home (1968) have traced some of the antecedents of Newton's third law[18], and, to the degree that this law arose from considerations of impact or collisions, it goes back in some essence to at least Descartes. But as a general principle of nature, applicable to attractions as well as impulses, it was Newton's own, though obviously a transformation. In particular, the leap to the general law required (as was the case for the second law) a transformation from a principle for impulsive forces to one for continually acting forces. We shall see, in Ch. 5, that the full and clear recognition of the general power and consequences of the third law did not become evident to Newton until the revision of the tract *De motu* (no earlier than December 1684), at the time that he took the first major step toward universal gravitation.

In stating the second law in the *Principia*, Newton gave no physical example to show how reasonable it is, or that it is founded in experience. In the scholium following the laws, however, he did note that it had been used by Galileo in deriving the law of falling bodies and the parabolic path of projectiles, which—although an incorrect historical statement about Galileo[19]—shows that he had physical examples in mind. These Galilean examples, be it noted, illustrate the second law for a continuous force, i.e., weight. For the first law, Newton gave evidence from experiment and astronomical observations to show that the law is true.[20] For the third law, he gave physical examples: a stone pressing on a finger that presses down on a stone, a horse being drawn back by a stone that it pulls. But in the scholium following the laws of motion, he did give experimental proof in relation to collisions and he discussed examples (or thought-experiments) to illustrate, as he said, that this 'law holds good also in attractions, as will be proved in the next scholium'.

Newton's interest in collisions (or "reflections") may be traced back to the 1660s (in the *Waste Book*[21]). In 1666 or 1667 he wrote an essay on this subject which he called 'The Lawes of Motion'.[22] A few years later, in 1669, John Wallis, Christopher Wren, and Huygens wrote and published their papers on this subject, in which the law of conservation of momentum was a central feature.[23] In the scholium following the laws of motion, Newton wrote of how these three geometers 'found the rules of the collisions and reflections of hard bodies' and how 'Wren proved the truth of these rules before the Royal Society by means of an experiment with pendulums'. Newton made additional experiments, which he reported in the scholium, to 'take account of both the resistance of the air and the elastic force of the colliding bodies'. He also made experiments with semielastic bodies, such as 'balls of tightly wound and strongly compressed wool'. Thus he could answer the objection that 'the rule which this experiment was designed to prove presupposes that bodies are either absolutely hard or at least perfectly elastic, of which sort none are found in bodies of natural composition' (*Principia*, scholium to the laws of motion).

These experiments were a kind of confirmation of the rules of impact. But it is to be noted that, as Whiteside has pointed out, Newton had attained in January 1665, or soon afterwards, '. . . an exact insight into the problems of both elastic and inelastic collisions between bodies moving not only in the same straight line but in different ones' (Newton, 1967–, vol. 5, pp. 148sq, n. 152). At this time he wrote out what may be his earliest step toward the third law,[24] at least for problems of impact or collision. In his *Lectures on Algebra*, under the date of 1675, he gave as problem 12, 'Given the sizes and motions of two spherical bodies moving in the same straight line and colliding with each other, to determine their motions after recoil.'[25] The first of the conditions or stipulations he made for solving this problem was 'that each body shall suffer as much in reaction as it impresses in its action upon the other' ('ut corpus utrumque tantum reactione patiatur quantum agit in alterum'). The combination of equal and opposite 'reaction' ('reactione') with the acting ('agit') presages the eventual law 3 ('Actioni contrariam semper & aequalem esse reactionem').

But the real transformation was to extend the general rule for collisions, expressed clearly and unambiguously in terms of action and reaction in the *Lectures on Algebra*, to attractions. In the *Principia*, in the scholium to the laws, this is justified by a clever analysis or thought-experiment, invoking corol. 4 to the laws, that the 'common center of gravity of two or more bodies does not change its state either of motion or of rest as a result of the actions of the bodies among themselves', from which it follows that 'the common center of gravity of all bodies acting upon one another (excluding external actions and impediments) either is at rest or moves uniformly straight forward'. The way Newton "demonstrates" the third law for attractions, in the scholium to the laws, has two parts, each of which is a thought-experiment.

In the first of these thought-experiments, Newton supposes that two bodies, *A* and *B*, attract each other and that an obstacle is put between them so as to prevent their coming into contact. If one body (say *A*) were to be more attracted toward the other (*B*) than that body (*B*) toward the first body (*A*), then the total force of *A* pressing on the obstacle would be greater than the similar force of *B*, with the result that the obstacle would be subject to the action of an unbalanced force and would move (according to law 2) with an accelerated motion in the direction from *A* to *B*. But, as Newton points out, the isolated 'system' of *A, B,* and the obstacle must necessarily (according to the first law of motion) 'persevere in its state of resting or of moving uniformly forward' and hence it is necessary that 'the bodies will urge the obstacle equally and on that account be equally attracted to each other'. This conclusion, Newton says, he tested with an experiment in which a lodestone and a piece of iron were floating in separate vessels on the surface of water.[26] It should be observed that Newton's theoretical reasoning here is incisive and original and that the experiment does confirm the theoretical result. But this is not an example of beginning with phenomena, with the data of experiment, and generalizing the results by induction.

In the second thought-experiment, Newton imagines the earth to be cut into two unequal parts by a plane. He then supposes that the larger of the two segments is cut by a second plane parallel to the first one, in such a way that the two outer segments are equal. This situation is shown in fig. 4.2. Newton argues that the middle part, *B,* 'will not preponderate by its own weight toward either of the extreme parts' but will 'be—so to speak—suspended between both, in equilibrium, and will be at rest'. But the outer part, *A,* 'will press upon the middle part with all its weight and will urge it toward the other extreme part', *C.* Therefore, the 'force' by which *A* and *B* together will tend toward the third part, *C,* will be 'equal to the weight of' part *A* only, which (since *A* is an exact duplicate of *C*) is the same as the weight of *C.* Whence Newton concludes that 'the weights of the two parts of the sphere', *A + B* and *C,* 'toward each other are equal, as I set out to demonstrate'. The argument is somewhat vitiated by assumptions concerning 'gravity between the earth and its parts', for it has not yet even been suggested that gravity is a universal force (a concept that will not be introduced until well into bk. two), and it is accordingly not clear what meaning is to be given to the "weight" of *A,* of *C,* and of *B +*

Figure 4.2

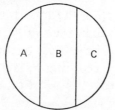

A when weight thus far signifies the propensity of a body to be "heavy" toward the earth. I would rather put my faith in the concluding sentence of the paragraph, that if 'those weights [i.e., of the parts into which the first—or any other—plane divides the earth] were not equal, the whole earth, floating in the free [i.e., nonresisting] aether, would yield to the greater weight and in receding from it would go off indefinitely'.

In any event, this thought-experiment can hardly qualify as an example of how a principle is either 'derived from Phaenomena' or rendered 'general by Induction' (see §4.5, especially n. 16). Nor is the extension of the third law from impulsive forces of collision to attraction an example of what is usually considered to be induction, since it is not a generalization of a number of specific instances[27] so much as an extension or transformation of a restricted principle to wholly new areas, in which case it would be an example of what has been called "transdiction" (see Mandelbaum, 1964, ch. 2). Possibly, however, Newton may have included such "transdiction" in his concept of induction, since "transdiction" is the aim of rule 3, along with ordinary induction. That is, Newton wishes (by what we would call "transdiction") to determine from the properties of gross bodies what must be the properties of their component particles. But he also asserts from the gravitation of all bodies toward the earth and of the moon toward the earth (all in proportion to their masses)—and from the gravitation of the sea toward the moon, of the planets toward one another, and of the comets toward the sun[28]—that (by induction) 'all bodies gravitate mutually toward one another'.

Following the above two thought-experiments, with their empirical sequelae, Newton turns to the application of the third law to the analysis of a balance (with a weight suspended from the ends of its two arms) and the classic simple machines. 'By these examples', he says, in conclusion, 'I wished to show . . . how widely the third law of motion extends and how certain it is'.

The third law is thus revealed to have been the result of a transformation (or series of transformations) that began with Newton's early considerations of the rule of impact for the special impulsive forces of percussion[29] and that ended up as a highly original and abstract principle that applies to every sort of force whatever—continuous as well as impulsive, forces of contact as well as forces of attraction—and hence is of the greatest generality. The first two laws have a quality of originality in the sense that "inertia" is "equated" to "mass" in the first law and that this same "mass" becomes the factor determining either the change of velocity or the rate of change of velocity for a force that is respectively impulsive or continuous. The real originality in the third law may perhaps be seen in the fact that it has often been misunderstood. For example, it has often been supposed that the third law is related to a condition of equilibrium, on the erroneous supposition that the "action" and "reaction" act on the same rather than on different bodies.

4.5 *Newtonian inertia as an example of successive transformations*

The fundaments of Newton's *Principia* are the introductory axioms, on which the following propositions are said to depend. These are three in number. The first states the law or principle of inertia, the second is a form of what is generally known today as "Newton's second law" (or just "the second law"), while the third embodies the principle of action and reaction. The second and third laws have been discussed in the previous section; it remains to examine the transformations by means of which Newton arrived at the first axiom or law of motion.

The concept of inertia and the system of physics based upon it was one of the real novelties in seventeenth-century physical science. In particular, the new dynamics (the science of force and motion, christened by Leibniz a few years after the *Principia* had been published[1]) is different from all considerations of motion from antiquity through the sixteenth century because of this new principle. Basically, seventeenth-century scientists conceived that a body could maintain itself in uniform rectilinear motion without requiring a "mover" or external force.[2] Thus there was a kind of equivalence between uniform rectilinear motion and rest, each of them being a "state", to use the expression that, following Descartes, has become standard in the language of physics.

Whereas Aristotelian physics had conceived motion to be a "process"[3] and so unlike rest, which is a "state", the new physics would allow that a special kind of motion is also a "state". I shall not here go into the prehistory of the concept of inertia, nor even the possible origins of Descartes's concept of a "state" (or *status*) of motion (this is the subject of a separate study, in progress), but I shall begin with Descartes's *Principia philosophiae*, since there can no longer be any doubt that it was from reading this book that the youthful Newton found and took over both the principle of inertia and some of the language used by Descartes to express it (see nn. 7–9; also Koyré, 1965, ch. 3, and Herivel, 1965a, ch. 2, pt. 2). Thus both Newton and Descartes use the same phrase, 'quantum in se est', in describing this new principle (see Cohen, 1964b). There are other similarities that are at once apparent on comparing the language used by Descartes and that used by Newton in his *Principia*, notably in Newton's statement of the first law and the antecedent definition 3 (which, in fact, anticipates the enunciation of the law itself) (see Koyré, 1965). We may observe, however, that the very designation 'Axiomata sive Leges Motus' used by Newton in the *Principia* seems to be a conscious or unconscious transformation of Descartes's 'Regulae quaedam sive Leges Naturae' in his *Principia*. Furthermore, as has been mentioned, the very title of Newton's *Principles of Philosophy* (as he himself was wont to call his book) must have been a transformation of Descartes's *Principles of Philosophy (Principia philosophiae)*, so as to make it become *Philosophiae naturalis principia mathematica*.

In the *Waste Book*, in the 1660s, Newton entered two axioms, as follows:

Axiomes and Propositions
1. If a quantity once move it will never rest unlesse hindered by some externall caus.
2. A quantity will always move on in the same streight line (not changing the determination nor celerity of its motion) unlesse some externall cause divert it [U.L.C. MS Add. 4004, fol. 10; printed in Herivel, 1965a, p. 141].

Then he began a new series of axioms, starting from this one:

Ax: 100. Every thing doth naturally persevere in that state in which it is unlesse it bee interrupted by some externall cause, hence axiome 1st and 2d . . . A body once moved will always keepe the same celerity, quantity and determination of its motion [ibid., fol. 12; printed in Herivel, 1965a, p. 153].

One has only to compare these axioms with the statements in Descartes's *Principia* (pt. 2, secs. 37, 39, which I give below together with an English translation; Descartes, 1974, vol. 8, pp. 62sq) to see where Newton's axioms come from:

[*Postil*]

Prima lex naturae: quod unaquaeque res, quantum in se est, semper in eodem statu perseveret; sicque quod semel movetur, semper moveri pergat.

The first law of nature: that everything whatever,—so far as depends on it,—always perseveres in the same state; and thus whatever once moves always continues to move.

[*Postil*]

Altera lex naturae: quod omnis motus ex se ipso sit rectus ...

The second law of nature: that all motion by itself alone is rectilinear ...

[*Text*]

Harum prima [lex naturae] est, unamquamque rem, quatenus est simplex & indivisa, manere, quantum in se est, in eodem semper statu, nec unquam mutari nisi a causis externis.

Of these the first [law of nature] is that everything whatever, insofar as it is simple and indivisible, remains, so far as depends on it, always in the same state, nor ever changes [in its state] save by external causes.

... concludendum est, id quod movetur, quantum in se est, semper moveri.

... it is to be concluded that whatever moves, will—so far as depends on it [*i.e.*, so far as it can, in and of itself] —always move.

It should be noted that in axiom 100, Newton not only uses the word "state" (which is not present in axioms 1 and 2) but also refers to "external cause" and "persevere". In the *Principia*, where Newton has combined his original designation of "axiom" with the Cartesian "law", the first law is stated as follows:

Lex 1

Corpus omne perseverare in statu suo quiescendi vel movendi uniformiter in directum, nisi quatenus a viribus impressis cogitur statum illum mutare.

Law 1

Every body perseveres in its state of resting or of moving uniformly straight forward, except in so far as it is forced to change that state by impressed forces.

In the first two axioms in the *Waste Book*, Newton has two separate laws, one for the continuity and the other for the rectilinearity, corresponding to the postils in which Descartes set forth two separate laws within this same distinction. By the time of the *Principia* these have coalesced into a single one.[4]

In a chapter on transformations, I cannot help but observe that a real transformation has occurred in Newtonian scholarship with respect to the role of Descartes in the development of Newton's scientific ideas. It was not too long ago that Descartes was conceived in relation to Newton's science primarily as the enemy who had to be attacked and destroyed before the new natural philosophy could be established. His name is nowhere mentioned in the *Principia*, but bk. two concludes with a thumping argument against vortices, which are shown to lead to results which are completely contradic- tory to astronomical phenomena and so are a hypothesis which 'conduces not so much to explaining as to perturbing celestial mo- tions' (scholium at the end of *Principia*, bk. two). Newton ended bk. two by saying that 'the way these motions are performed in free spaces without vortices can be understood from book 1', some- thing that 'will now be shown more fully in the system of the world', that is, in bk. three. Furthermore, in the preface written by Cotes to the second edition, considerable space is still given to the attack on vortices.[5] In Mme de Chastellet's translation of the *Principia*, it is said that the chief purpose of bk. two was the attack on Descartes, a sentiment repeated by Lagrange.[6]

The last decade and a half have witnessed a considerable change with respect to Descartes's influence on Newton. In 1962 A. Rupert Hall and Marie Boas Hall published a long essay by Newton, writ- ten in the 1660s, showing how deeply he had studied Descartes's *Principia*;[7] although this essay ends in a critique of several Car- tesian positions, one can see in reading it how important Descartes's book was in the formation of Newton's thoughts about motion. In 1965 a brilliant study by Alexandre Koyré on Descartes and New- ton explored the importance of Descartes's principle of inertia for Newton, especially calling attention to the concept and word "state" (*status*) in relation to motion (Koyré, 1965, ch. 3). D. T. Whiteside, editing Newton's mathematical juvenilia, found that many of Newton's early thoughts, which would eventually lead him to his magnificent achievements with regard to the calculus, were forged while reading Descartes's geometry, in the edition of Schooten.[8] In 1965 John Herivel explored the influence of Des- cartes on Newton in relation to his study of some early Newtonian texts on dynamics (Herivel, 1965a, ch. 2, pt. 2). And I myself showed a number of links between Newton and Descartes, par- ticularly in relation to certain expressions and concepts which Newton took over from Descartes, including "quantum in se est".[9] This phrase goes back to Lucretius, who used it in relation to what may be considered an anticipation of a kind of inertial idea in the motion of atoms, and was synonymous for "naturaliter", which—in

English translation—occurs in axiom 100 in the *Waste Book*, in what is one of Newton's earliest expressions of a principle of inertia.

One further transformation may be noted. We have seen that for Descartes the principle of inertia was expressed as a "regula" as well as a "prima lex naturae". Descartes's usage is carried over into the *Principia*, since Newton designates the principle as both an "axioma" and a "prima lex".[10] But Newton's law is a "prima lex motus" rather than a "prima lex naturae". There is a twofold transformation: the change of the name from "law of nature" to "law of motion" and the change in status of the laws of motion to make them "axioms". As to the first of these, Newton may have been influenced by the series of papers by Wallis, Wren, and Huygens on the laws of impact and the conservation of momentum, published in the *Philosophical Transactions* in 1669 under the title 'Laws of Motion'.[11] This would not be surprising, since we know how important these papers were in Newton's development, both the second law of motion and the third law of motion taking their rise from these principles.

But if there was any such influence, it would have been in strengthening his resolve rather than as a primary source of this expression. For even before these papers had been read and published, Newton wrote a paper entitled 'The Lawes of Motion', which deals with the motion of 'solitary bodyes' and then with the way 'Bodys are Reflected', or the problems of collision.[12] Newton's essay, which exists in two states, was apparently composed in the mid-1660s, most probably in 1666, but possibly in 1667,[13] well before 1669, when the articles on collisions were published in the *Philosophical Transactions*. In both cases (i.e., Newton's essay and the papers published in 1669) the application of the phrase "laws of motion" to the problems of impact or collision has Cartesian overtones, and was in this sense a transformation of the Cartesian "leges naturae," strictly speaking, since the final group of Descartes's laws of "nature" in his *Principia* had been devoted to the rules of collision. This example may serve as an illustration of the fact that to the extent that innovations are transformations, it is common to find the same transformations being made independently. In this case, there is a transformation being made of Descartes's "leges naturae" into a set of "laws of motion" in the same context: the laws of collision or of impact ('choc').

In 1684, in the tract *De motu*, written just before the *Principia* (published in Rigaud, 1838; Ball, 1893; Hall & Hall, 1962; Herivel, 1965*a*; Newton, 1967–, vol. 6), as in a paper in English on the motion of bodies according to Kepler's laws (of which the beginning is somewhat similar: published in Ball, 1893; Hall & Hall, 1962; Herivel, 1965*a*), Newton stated the laws or principles as Hypotheses. In a revision of *De motu*, however, each such Hypothesis (with some textual changes in successive versions) is changed to a Lex. The same designation (Lex) occurs in another tract, of which we have only a fragment, written between *De motu* and the *Principia* and entitled *De motu corporum in mediis regulariter cedentibus* (published in Herivel, 1965*a*). In the latter, six laws are grouped together under the general heading of 'Leges Motus' as Newton would do in the *Principia* for three such laws.

One significance of the transformation of Descartes's "leges naturae" to the Newtonian "leges motus" is that the latter is more exact in the sense that Descartes's laws, like Newton's, deal more with motion than with nature at large.[14] Newton's *Principia*, furthermore, is more restrictive than Descartes's *Principia*, since it deals only with "mathematical" principles of "natural" philosophy rather than with principles of philosophy in general. Even more important, Descartes had believed (as he said both in his *Principia* and in *Le monde*) that the principle of inertia was a universal rule or law of nature that derived from a higher divine principle of conservation; essentially he was saying that inertia results from the fact that the motion that God had created at the beginning of the world could never of itself be destroyed or wither away and so would continue forever.[15] Newton gave this principle a wholly different ontological basis, although he was probably sympathetic with the idea that all motion originated with God. His designation of "axiom", however, is a way of saying that these were the unarguable bases from which his system of dynamics would be derived, with the obvious overtones of the axioms of Euclidean geometry. Just before the publication of the second edition of the *Principia* in 1713, Newton referred to 'the first Principles or Axiomes which I call the laws of motion' in a letter to Cotes. Concerning them he said, 'These principles are deduced[16] from Phaenomena & made general by Induction: which is the highest evidence that a Proposition can have in this philosophy'. Although this statement has often been repeated to show that Newton founded the *Principia* on induction, it hardly squares with the facts. The first law of motion had not been 'deduced from Phaenomena', nor learned by induction; Newton found it in Descartes's *Principia*. Furthermore, in a strict sense, it applies only to an imaginary world, as has been pointed out in the previous chapter, and so is not really even based strictly on phenomena.

But Newton did offer some phenomenological evidence in support of the first law by considering the inertial component in long-lasting rotations and orbital motions (see § 4.4, n. 20). We have seen, additionally, how the third law took its rise from Newton's studies of collisions; insofar as the law is a statement only concerning impact, it was backed up by some evidence of experiment.[17] Newton's experiments with elastic and semi-elastic colliding bodies did confirm the third law of motion for collisions other than those of "hard" bodies, in the mode he described to Cotes; but the important (and the truly original) part of the third law was the generalization which Newton made so that he could apply it to 'the mutual & mutually equal attraction of bodies', which Newton said 'is a branch of the third Law of motion' (1959–1977, vol. 5, pp. 396sq). In the above-mentioned letter to Cotes, Newton told him that he might 'see in the end of the Corollaries of the Laws of Motion' (that is, in the scholium following those corollaries) how 'this branch is deduced from Phaenomena'. He did not say specifically, however, that 'this branch' had been 'made general by Induction'. In fact, as has been explained in §4.4, the chief argument for the mutuality of attraction (in the scholium) is a pair of hypothetical or thought-experiments, with a subsequent physical illustration for one of them. This is hardly an example of a principle being 'deduced from Phaenomena' or being rendered 'general by Induction'.

Nevertheless, it is significant that the Newtonian axioms were closely related to and confirmed by experience. The traditional or then-current sense of "axiom" in given in Harris's *Lexicon technicum* (1704), where an axiom is said to be 'such a common, plain, self-evident and received Notion, that it cannot be made more plain and evident by Demonstration'. This could hardly have been the case for the Newtonian laws of motion, since all three were so non-self-evident that they had been unknown throughout all the previous centuries. As for the law of inertia, most people before the seventeenth century believed—as many have afterwards, including many people today—that without a "mover" or a "moving force" a body will stop moving and come to rest. The law of inertia, in particular, contradicts common sense. Even Kepler and Galileo had believed in a different kind of "inertia".[18] Possibly this is the reason that only in the case of the law of inertia did Newton give experiential examples to prove that it is demonstrably true and not merely true by supposition. Unlike the physical examples given for the third law (which illustrate rather than prove it), these are offered to show that the first law is valid. It was a real novelty to construct a system of dynamics based upon "axioms" that could be demonstrated either by experiments and observations or by deriving consequences that could be so tested.[19]

The most important transformation that Newton made of the Cartesian principle of inertia was to associate it with "quantity of matter", which Newton made precise as "mass". He thereby introduced permanently into the language of the physical sciences the term "inertia" (from the Latin for laziness or inactivity), which had originated with Kepler. To achieve this, however, Newton had to transform the Keplerian concept of "inertia" (for Kepler, a property of matter which would bring bodies to rest whenever the force producing their motion ceased to act) into something that would keep bodies in whatever state they were, whether a state of rest or a state of moving uniformly in a straight line.[20] Newton did not, however, encounter the expression "inertia" by reading Kepler directly. For it is a fact of record that Newton read very little of Kepler's writings. But he did encounter the terms "inertia" and "natural inertia" in the Latin edition of the correspondence of Descartes, adjacent to another letter on a similar subject, one to which Newton actually refers in the essay on Descartes's *Principia* which he wrote as a young man.[21] Apparently, when Newton wrote the *Principia*, he did not even know that the name and concept of "inertia" which he had transformed had originally been introduced into physical science by Kepler. Neither in the *Principia* nor in any published or MS writing of Newton's prior to 1713 with which I am acquainted, is the term "inertia" ever associated with Kepler's name. Nor, as a matter of fact, is Kepler's name mentioned in this connection in the above-mentioned volume of Descartes's correspondence.

After the second edition of the *Principia* was published, however, we know that Newton had encountered the Keplerian notion, since he wrote into his own copy of his *Principia* a note for a future edition, reading (in translation): 'I do not mean the Keplerian force of inertia by which bodies incline to rest, but a force of remaining in the same state of rest or of motion'.[22] Newton could well have encountered references to Kepler and inertia in Leibniz's *Theodicy* (1710) of which Newton's personal copy still survives, with pages turned down in indication of his having read it (see Cohen, 1972). He also could have encountered references to Keplerian inertia in David Gregory's textbook of astronomy (1702), in which Newton's essay on the moon's motion was printed, along with a considerable discussion by Newton on the wisdom of the ancients (which was included as if it had been written by Gregory himself for his preface).[23] In the Leibniz-Clarke correspondence (published in 1717), furthermore, there is a note very much like the one Newton entered in his personal copy of the *Principia*. It reads:

> The *vis inertiae* of matter, is that passive force, by which it always continues of itself in the state 'tis in; and never changes that state, but in proportion to a contrary power acting upon it. 'Tis that passive force, not by which (as Mr. Leibnitz from Kepler understands it), matter resists motion; but by which it equally resists any change from the state 'tis in, either of rest or motion.[24]

Since Newton was closely associated with Clarke at the time of this debate-by-correspondence (see Koyré & Cohen, 1962), it is not surprising that this sentiment should so closely resemble Newton's private note.[25]

It is to be noted that in the paragraphs quoted above, both Clarke and Newton use the expression "force of inertia", which occurs in the *Principia* in relation to the way bodies continue in motion. This "force of inertia" is said by Newton (*Principia*, def. 3) to be only another and 'most meaningful name' for an 'inherent force of matter' ('materiae vis insita'), which is related to the property of 'inertia of matter'. This "force" is always proportional to a body's mass, says Newton, and 'is not at all different from the inertia of its mass, except in the manner of conceiving it'.

There has been much puzzlement concerning this "force", some of which may be resolved by recognizing that the concept and expression of "vis insita" was a fairly common (though by no means universal) one in Newton's day. It occurs again and again in Kepler's writings,[26] and may be found in certain dictionaries such as that of Goclenius,[27] where it means a "natural power". It is used by Henry More, an important author in the early development of Newton's thought. His volume on *The Immortality of the Soul* was read carefully by Newton, who made notes from it.[28] More important, however, is its occurrence in the handbook of Aristotelian philosophy by Magirus, which was summarized in some notes made by Newton while an undergraduate in Cambridge.[29] Newton was merely using (and so transforming) a common expression, "vis insita", to express what he conceived as a source of a body's continuing in a state of motion, and hence having some "natural" or "inherent" internal "power" by which it would, in a more general way, resist any change in state. Thus the "vis insita" causes a body at rest to resist any effort by a force outside the body (or an external force) to put it into motion, and equally causes a body in motion to resist any attempt by an outside force to alter that state of motion: by slowing the body down or speeding it up, by changing the direction, or by bringing it to rest. This is, in fact, the Newtonian version of inertia, a transformation of the Keplerian concept by associating it with the Cartesian law of persevering in a state. This is made clear by Newton's statement in def. 3 that the "vis insita" (taken from the current usage) may be known by a 'most significant name, vis inertiae'. This power, despite Newton's names of "vis insita" and "vis inertiae", is simply proportional to the mass of a body ('Haec semper proportionalis est suo corpori'), and it in no way differs from the property of inertia (or inactivity) of mass ('neque differt quicquam ab inertia massae') save that it is conceived in a different way ('nisi in modo concipiendi').[30]

The critical reader may find it odd that Newton, who systematized the new inertial physics into dynamics (to use the name given to this subject in 1691 by Leibniz), should have found it necessary to write of the inertial property of bodies in terms of a "force". But although Newton wrote of this as a "force", exerted by a body 'during a change of its state, caused by another force impressed upon it', and even conceived of this as a "force" of resistance, he made it perfectly clear that this was very different from the kind of "force" that acts in statics and different also from the kind of "force" that produces (or may produce) a change in a body's state.[31] Rather, from the point of view of dynamics, this internal "force" (whether it is to be called a "force of inertia" or "inherent force") is not really a force at all. A body changes its state, whether a state of rest or of uniform rectilinear motion, according to def. 4, by means of a force externally "impressed" upon the body, and such an "impressed force" is defined by Newton to be 'an action exerted on a body [and hence from outside the body] to change its state either of resting or of moving uniformly straight forward'. Here, again, in the case of "impressed force", Newton is using the traditional language of an older physics, but he is doing so in a new way. Whereas in the older physics an "impressed force" is something put into a body in order to keep it moving, Newton says that "impressed" force rather consists 'in the action [i.e., of changing a body's state] alone and does not remain in a body after the action has ceased'. He is thus differentiating his views and those of the writers of an earlier age. And he makes this clear by saying that 'a body perseveres in every new state by the force of inertia alone' (Def. 4).

In the *Principia* there is no confusion in practice from this dual use of "force". Newton does not, for example, ever use the parallelogram law to vectorially compound the "force" of inertia and the usual external forces that change a body's state—whether percussions, pressures, or centripetal forces—although he will compound or vectorially combine a motion *sustained* by the "force of inertia" and the motion *produced* by such external forces as percussions, pressures, or centripetal forces.[32] As Clarke made absolutely plain in the statement quoted above, the force of inertia of matter is that 'passive force, by which it always continues of itself in the state 'tis in; and never changes that state . . .' Newton himself said the same thing in qu. 31 of the *Opticks*: 'The *Vis inertiae* is a passive Principle by which Bodies persist in their Motion or Rest, receive Motion in proportion to the Force impressing it, and resist as much as they are resisted. By this Principle alone there never could have been any Motion in the World' (Newton, 1952, p. 397).

The scholar cannot help but be interested in the fact that Newton still continued to use the traditional word "vis" or "force" in relation to a body's ability to resist a change of state or to maintain whatever state it may be in (or may have gained), almost as if he could not fully separate himself from the long tradition in which all motion was thought to require a mover or a moving force. This example, in fact, shows us the invariances in intellectual transformations and the continuities that remain despite the gap between an old and a revolutionary new science. The presence of "vis insita", transformed into a "vis inertiae", is like an archeological sign directing the critical reader to the prehistory of a new subject and the continuity within the change. But it would be at peril to any reader's understanding of Newtonian science to confuse this concept with a true dynamical "force", of the kind which in Newtonian terms may be "impressed" on a body in an action which changes its state. And in the present analysis, what is most significant is not that Newton uses "force" in relation to inertia, but that for Newton this "force" of inertia or inherent "force" (which is not a force at all in the sense of the dynamics of the *Principia*) is said by Newton to be 'always proportional to' the mass of a body.[33] It must be stressed that the only difference, according to Newton, between the "inertia of mass" and the "force of inertia" is 'in the manner of conceiving it'. But whether we conceive of it as some kind of "force" which does not produce a change in momentum or merely (to use Newton's alternative) as 'the inertia of its [i.e., the body's] mass', the significant point is that it is associated with and 'always proportional to' the mass. It is mass that keeps a body going in inertial motion, and that is the source of a body's resistance to a change of state, whether a 'state of resting or of moving uniformly straight forward'. It is this same mass (or inertia) that determines (or is the measure of) a body's resistance to acceleration by an external force (a true dynamical force), and as such enters in the second law of motion. In the second law, mass appears either as the constant of the proportionality of an impulsive force to the change in momentum it produces or as the constant of the proportionality of a continuous force to the acceleration it produces.

Hence Newton not only transformed the concept of inertia (Keplerian) and the concept of inertial motion (Cartesian), but he did so by associating both with his new concept of mass, which was itself a transformation of the traditional concept of "quantity of matter", which had often been associated with a body's weight.[34] From Newton's time onward, "mass" has been a central and fundamental concept in all physics. And it is a measure of Newton's genius and his extraordinary insight that he saw the need of exploring the two aspects of mass as they appeared in the *Principia* (as we shall see in Ch. 5), one the inertial aspect of mass appearing in the first and second laws of motion,[35] and the other the gravitational aspect of mass appearing in the law of universal gravity, or the determinant of a body's response to the action of a gravitational field in which it may be placed.[36]

4.6 *Some general aspects of transformations*

In the presentation thus far, I have indicated how, in the course of thinking about phenomena, or attempting to account for phenomena, or working out an explanation of experiments, a scientist may transform a generally received idea or a particular scientific idea he has encountered. There is little that can be said at present about the relative frequency of transformations arising from a triggering stimulus (as in Darwin's reading of Malthus) and the general interaction of the problems posed by ongoing scientific research with an idea or set of ideas that may be on the subconscious as well as the conscious level. History provides all too few documented specific acts of scientific discovery, of which Darwin's transformation of the struggle for survival from an interspecies to an intraspecies concept may rival Archimedes' discovery of the principle that bears his name while taking a bath. Newton recorded the event of discovery when he saw an apple fall in the garden, but it is not entirely clear as to what it was that he discovered on that famous occasion.[1]

Darwin correctly read and understood Malthus and then went on to transform a scientific idea of Lyell's which Darwin had also read correctly. But sometimes the transformation occurs in an incorrect reading.[2] I have referred earlier to Dalton's erroneous interpretation of prop. 23, bk. two, of Newton's *Principia*, in which Dalton happily but quite wrongly assumed that Newton had proved that the molecules (or 'particles') of gases repel one another with a force varying inversely as the distance (see §3.3). Another such example occurs in the *Principia* itself, where Newton states his belief that: 'Galileo found, by means of the first two laws [of motion] and the first two corollaries, that the descent of heavy bodies is as the square of the time and that the motion of projectiles takes place in a parabola, in accord with experience, except insofar as those motions are a little retarded by the resistance of the air' (scholium, foll. corol. 6, to the laws of motion).

With regard to Galileo's having known the first law, or principle of inertia, there is some room for debate,[3] but it is certain that he did not conceive this law in both the strict and universal sense in which it occurs in the *Principia*. As for the second law, Galileo could not possibly have known it since he did not have the concept of mass which provides the proportionality between 'force' and 'change in [quantity of] motion' (or change in momentum).[4] And

Newton implied much more than the simple and bare fact that acceleration occurs in the direction of action of an external force; for he explains exactly how he conceives Galileo to have discovered that 'the descent of heavy bodies is as the square of the time':

> When a body falls, uniform gravity, by acting equally in individual equal particles of time, impresses equal forces upon that body and generates equal velocities; and in the total time impresses a total force and generates a total velocity proportional to the time. And the spaces described in proportional times are as the velocities and times conjointly, that is, as the squares of the times.[5]

Whether we accept Galileo's own presentation in his *Two New Sciences* as a true record of his steps toward discovery of the law of falling bodies, or assume some other road to discovery on the basis of manuscript fragments, there can be no warrant whatever for Newton's reconstruction of Galileo's thought processes. The best that can be said for Newton's analysis is that he was but imperfectly acquainted with Galileo's writings and paid him the compliment of transforming him into a Newtonian (see, further, Cohen, 1967c).

While the foregoing example is of interest in showing how a transformation may arise from an imperfect acquaintance with (or reading of) the scientific ideas of a predecessor, it is not of the same quality (with regard to significance) as the Daltonian transformation of Newton's prop. 23. For the latter led Dalton forward in his development of the atomic theory, whereas the former was a transformation in retrospect, written long after Newton had himself formulated the first two laws of motion. It may be observed at this point that all too many accounts of the historical development of science are based on conscious or unconscious transformations of the ideas of the past; stress is apt to be placed on aspects of the science of the past that are in harmony with present science, or a deliberate attempt may be made to read today's scientific ideas into the concepts of the past. An outstanding example of this practice is the attribution of the concept of the electron to Franklin, because he theorized that the 'electrical fluid' is composed of fundamental particles that repel one another;[6] another is the attribution of the concept of entropy to Sadi Carnot, because he apparently distinguished between 'chaleur' and 'calorique' (see, e.g., La Mer, 1954, 1955; Kuhn, 1955). Yet another may be the eighteenth-century transformation of Copernicus into the inaugurator of a revolution in astronomical science (see Cohen, 1967c).

Whether such transformations, when made by historians, are any less justified than when made by a creative scientist forging ahead in the advancement of science is a question that would go far beyond my limited intentions in this study.[7] But it is not possible to escape observing that certain scientific ideas have potentialities for such transformation whereas others do not. There are many instances in the history of science in which such transformations go so far beyond any possible intention of the original author that he would—in all probability—have denied any alleged parenthood of the progeny. A dramatic example is provided by the "Platonism" of Galilean science. One response to the thesis of Alexandre Koyré (1943) concerning Galileo and Plato has been to insist that the Plato that seems to inspirit certain aspects of Galileo's thought is so far removed from the "true" Plato as to be an unrecognizable travesty. This may, accordingly, represent either Galileo's ignorance of the true Platonic doctrines or his willful distortion of a figure he may have lauded primarily because this was a way to assert his own anti-Aristotelian position. After all, Galileo, in my opinion, was acting very deliberately when he chose to refer to himself in his *Two New Sciences* as 'our Academician', with its direct overtones of Plato's Academy.[8] From the point of view adopted here, it is easy to reconcile Plato's point of view and Galileo's. Plato had a primary concern for pure numbers and number relationships, and for pure geometrical forms. He accordingly rejected an astronomy based on the observation of the motions of the celestial bodies because it shows irregularities and variations, whereas a true astronomy should show the perfections of uniform circular motions that characterize such "perfect" bodies. In my opinion, Galileo transformed this point of view, to the degree to which he was actually acquainted with it, into an ideal science that combines the principles of numbers and geometrical forms (conceived "Platonically") with results of actual experiments and observations: a remarkable transformation that actually embodies the creation of modern science to a degree that is shared with Descartes and Kepler but not at all with Bacon. The result of Galileo's transformation is to some real degree a departure from the canons of Platonism and even the currents of later neo-Platonism, since it uses the considerations of ideal forms in an un-Platonic way to arrive at the laws and principles of nature as revealed by experiment and observation.[9]

Accordingly, in evaluating scientific ideas in relation to their potentialities for transformation—both possible and actual (or realized)—it must be taken into account that some ideas may have far greater potentialities than others. I do not know whether such distinctions or hierarchies necessarily follow upon any inherent logical or purely internal qualities of scientific ideas—a topic worthy of exploration. But it is a historical fact that some ideas are closely linked to the ways in which science develops and even become powerful forces in that development, whereas others simply do not. And some ideas may undergo successive transformations and continue to live on for a long time in science, such as atom, energy, and impetus (transformed into inertia transformed into inertial frame), whereas others (aether transformed into the imponderable fluids of heat, electricity, and magnetism) have a measured but fruitful existence and survive only as archeological remains in the scientific language: "perfect" gas, "good" conductor, heat "flow" or heat "capacity", and so on.

What may be especially of interest to the historian of ideas is that the end-product of a transformation, or a series of transformations, may even be opposed to the original doctrine. Who would ever have thought, on reading Plato, that the *Republic* or the *Timaeus* might ever possibly serve the formation of a mathematical physical science based on direct experiment and observation! Such negative potentiality, the ability to serve a purpose quite opposite to any beliefs of an author, was discussed for the first time in relation to literature and society (so I am informed by George Steiner) in a famous pair of letters of Friedrich Engels concerning Balzac.

In April 1888, writing to Mary Harkness (who had sent him a copy of her novel *City Girl*), Engels stated his preference for Balzac ('whom I consider a far greater master of realism than all the Zolas, past, present, or future') over authors of 'purely socialist' novels (*Tendenzromane*), and he said that Balzac shows how 'realism may creep out even in spite of the author's views'. In his view,

> Balzac was politically a legitimist; his great work is a
> constant elegy on the irreparable decay of good society;
> his sympathies are with the class that is doomed to extinc-
> tion. But for all that, his satire is never keener, his irony
> never more bitter, than when he sets in motion the very

men and women with whom he sympathizes most deeply—
the nobles. And the only men of whom he speaks with
undisguised admiration are his bitterest political
antagonists, the republican heroes of the Cloître Saint
Méry, the men who at that time (1830–36) were indeed
representatives of the popular masses.

That Balzac was thus compelled to go against his own
class sympathies and political prejudices, that he *saw*
the necessity of the downfall of his favorite nobles and
described them as people deserving no better fate; that he
saw the real men of the future where, for the time being,
they alone were to be found—that I consider one of the
greatest triumphs of realism, and one of the greatest
features in old Balzac.

The same point is made in a letter to Minna Kautsky (26 November 1885), in which he said 'that the bias should flow by itself from
the situation and action, without particular indications, and that
the writer is not obliged to obtrude on the reader the future historical solutions of the social conflicts pictured'.[10] George Steiner
has referred to the Marxian 'concept of dissociation—the image of
the poet as Balaam speaking truth against his knowledge or avowed
philosophy'.[11] He suggests that Engels may even have gotten this
position from Marx, especially with reference to Marx's 'counter-
ideological enjoyment of Balzac and Scott'. In any event, whatever
the source, we may indeed see a "Balaam effect" being introduced
by Marx and Engels to describe what Steiner calls 'the case of a
writer whose implicit insights and revelations conflict with his
overt statements and intent'. In Balzac, Marx and Engels (the latter
especially) would see 'an "outward conservative" whose genius
compelled him, against his conscious purpose, to reveal the decay
and revolutionary potential of French society' (Steiner, personal
communication). Here we may see an implicit doctrine of transformation that causes a literary work to have potentialities of an
ideological kind that far transcend the author's own intentions.

The doctrine of transformation which I have been invoking for
scientific ideas is well developed in the area of literary history. All
imaginative writing, whether poetry, fiction, or drama, is generally
recognized as representing a transformation of the author's experience, whether the conscious or unconscious experience of daily
life or the vicarious experience of reading, hearing tales or con-

versations, going to the theater, and so on. In some cases, historians have traced the sources of images, concepts, and themes.[12] A classic example of such literary detective work, closely akin to the job of the historian of scientific ideas, is John Livingston Lowes's study of Coleridge's sources in his readings, a work aptly entitled *The Road to Xanadu* (1927). To any reader of Lowes's work, what is most impressive about 'Kubla Khan' and 'The Ancient Mariner' is not the variety and extent of Coleridge's reading that ended up in the poems, but rather the ways in which Coleridge transformed rather prosaic bits of information, culled from newspapers, travelers' accounts, and even the *Philosophical Transactions* of the Royal Society of London, and produced the haunting images of his famous poem.

Lowes has not only highlighted the transformations made by Coleridge of factual bits of information into poetic images, he has also shown that many apparent examples of poetic license that had at one time seemed to transcend the boundaries of scientific credulity may be shown by research to have been originally factually based. A famous example is Coleridge's image of 'The hornèd Moon, with one bright star / Within the nether tip.' At first he had used the image of a star 'almost atween the tips' (Lowes, 1927, pp. 180–184). Was this possibly a free exercise of the poetic imagination, so wild a transforming of possible experience as to raise the eyebrows of any astronomer, to whom the 'hornèd Moon' climbing 'above the eastern bar' is the waning moon which rims with light a limb of the rest of the opaque solid moon? To see a star within the crescent would thus seem to contradict the simplest and most basic facts of astronomy and optics.

And yet, as Lowes found out, the source of this image was a pair of articles in the *Philosophical Transactions* of the Royal Society of London, one of science's most respected periodicals. Coleridge was quite familiar with a communication published by Cotton Mather in the *Philosophical Transactions* in 1714–1716 and written in 1712, in which he reported 'a Tradition . . . that in November 1668, a Star appeared below the Body of the Moon within the Horns of it'. Here, then, is the source of that first image of a star 'almost atween the tips' (Lowes, 1927, p. 180). Just three years before he wrote 'The Ancient Mariner', furthermore, there appeared in the *Philosophical Transactions* (for 1794), two independent sets of observations of the appearance of such a "star" under the title,

'An Account of an Appearance of Light, like a Star, seen in the dark Part of the Moon, on Friday the 7th of March, 1794'. One of the observers wrote of his certainty concerning 'this spot appearing WITHIN the circumference of the moon's circle'; and the other reported 'a light like a star, and as large as a middle sized star, but not so bright, in the dark part of the moon' (Lowes, 1927, p. 509, n. 46; the word "within" is printed in capitals in the page of the *Philosophical Transactions*). Coleridge had almost certainly encountered an account of these two communications, presented in the *Philosophical Transactions* by the Astronomer Royal (Nevil Maskelyne), since they were reviewed ('with their arresting titles given in full, and with substantial extracts') in the *British Critic* (June 1795), of which Coleridge was then 'an attentive reader' (Lowes, 1927, p. 181). Coleridge may also have known a communication on this same subject by Sir William Herschel (*Philosophical Transactions*, 1787) concerning a bright spot seen in 1783 in the dark area of the moon, which 'seen in the telescope resembled a star of the fourth magnitude as it appears to the natural eye' (ibid., p. 510, n. 49). So much for those critics of Coleridge who suggested 'that on that spectral ocean the sun and moon did not obey the ordinary laws of the universe, and that the defiance of them was intentional' (ibid., n. 50).

Lowes himself has used this example some years before writing *The Road to Xanadu*, and before he knew the facts in the case, to illustrate how the poet's images may be so apt and so well presented as to appear to have 'a semblance of truth' and so to procure 'that willing suspension of disbelief for the moment, which constitutes poetic faith' (Lowes, 1927, pp. 181 and 510, n. 51). Now his research had uncovered records of an observation that gave a warrant for the possible actuality of a 'startling phenomenon, [which] however explained, was not unique'.[13] A few years ago I mentioned this episode to my colleague and former teacher, Fred L. Whipple, a specialist in the theory of comets and meteors. I used the story to show that what had been dismissed as poetic license in defiance of obvious scientific fact had proved to be based on reports in a sober scientific journal. On hearing the story, Whipple was intrigued with it, but not for historical so much as for very practical reasons. He was interested in the unsolved problem as to whether craters on the moon might possibly be caused by large meteors striking the moon's surface; accordingly, his scientific imagination immediately saw the possibility of a historical test of a current scientific speculation or theory. If the appearance of this "star" or flash of light had been located accurately on the surface of the moon, then possibly a comparison of lunar maps made before and after the event might show the existence of a new (or not previously noticed) crater at the same place where the "star" had been seen. Unfortunately, this opportunity for the history of science to serve the advancement of present-day science came to naught, since it requires a higher degree of accurate detail than appears in those older records and lunar maps.

This example may illustrate the importance of historical research in order to ascertain whether a poetic image is a free creation of the literary imagination or (as in this case) the transformation of a scientific observation, so presented as if to deny its empirical progenitor. Indeed, what is most surprising is that Lowes himself had originally considered this image to be even more of a transformation than in fact it actually was. We may see in Coleridge's images how scientific facts and observations may serve as sources of transformation for nonscientists as well as scientists. While on the subject of transformations other than those that occur in the development of scientific ideas, it is not amiss to recall that a foundation of psychoanalysis is the transformation of wishes and anxieties of the unconscious into symbolic experiences expressed in dreams. Further, there are many examples in which the advance of science has notably depended on the use by scientists of ideas or concepts which are taken from outside the domain of science and have been transformed.

The expression of Coleridge's genius, in his transformation of what he encountered in his reading into poetic images and symbols, may serve to remind us that the transformation of scientific ideas also reflects the particular creative talent of the scientist in question.[14] It may indeed be argued that the maximum effect on the growth of science exerted by the transformation of an idea occurs in an encounter between that idea and a mind of transcendent genius: at the right time. That individual creative force displays itself in the first instance in an act of recognition, which must be a prior condition to every transformation. Most likely, the recognition and the transformation are not independently separate sequential acts. The recognition requires a preparation of mind, in the sense used by Louis Pasteur in relation to the role of chance in scientific discovery, a condition in which the mind is alert (whether consciously or unconsciously) to the possibilities or potentialities of any idea in the solution of a problem at hand.[15] Many men and women may be privy to the same ideas and be interested in solving the same problem or set of problems in science, but only a Galileo, a Newton, a Darwin, a Morgan, or a Fermi recognizes a potential source of transformations leading to a solution.

This situation is similar to the archeologists' experience of encountering primitive stone tools at Choukoutien, near Peking. These tools are so primitive that they would hardly have been recognized as tools at all, but would have been considered only crude nondressed or nonworked stones, had they not been found in association with hominid remains, of *Homo pekinensis*.[16] Only the trained eye of the practiced archeologist, alerted by the presence of human-like bones and teeth, would have made the imaginative mental transformation of these stones into the tools they were.[17] Without such transformation, these stones might not have shown their signs of human or hominid action and use; an imaginative transformation was apparently required to reveal that these stones had been partially worked, however crudely or primitively, so as to become truly tools. Just as the trained eye and insight of the archeologist transforms (in his mind's eye) some ordinary-looking stones into primitive human tools, so the 'prepared mind' and insight of the creative scientist enables him to grasp a chance acquaintance with a fact of experiment or observation, a method, a theory, or a concept, and to transform it into a new and productive source of knowledge or understanding. The doctrine of transformation of scientific ideas thus not only identifies the creative act in the development of science, but evokes the special quality of preparation and insight that enables a particular scientist to recognize the idea or ideas having the potentiality of transformation.[18]

It follows at once that a creative act in science must be seen much less as a unique performance than is usually assumed. That is, we should not be astonished to find that more than one scientist may make the same transformation (or a similar one), since the ideas to be so transformed are available to one and all. As science was becoming more and more professionalized and institutionalized in the nineteenth and twentieth centuries, there was an ever more efficient communication of scientific ideas to an international scientific community that was constantly growing. It is a natural consequence of the doctrine of transformation that the probability of multiple so-called independent discoveries would accordingly have constantly increased, and that scientists would come to accept such multiple discovery as the rule rather than the exception—a fact that has been fully demonstrated by Robert K. Merton (1957, 1961). Here again, however, the concept of transformation leads us from the gross features to the fine structure of scientific discovery. Different simultaneous discoverers may achieve the same end (really, a similar end) by different routes, by a different set of transformations. This aspect of the phenomenon is plainly shown by a comparison and contrast of the formation of the idea of natural selection by Darwin and by Alfred Russel Wallace.[19] We are led, by an analysis of the particular transformations made by these two men, to a contrast between their ideas, whereas too often stress is laid upon a comparison. The same phenomenon occurs even more strikingly in the transformations leading Helmholtz, Mayer, and Joule to the principle of conservation of energy (see Kuhn, 1955; Elkana, 1970). And the examination of the actual transformations once again leads us to the fine structure of history in the contrasts, as opposed to the gross features revealed by comparison.

Finally, I shall mention but not explore the existence of nonfruitful transformations of scientific ideas. I certainly do not advocate a Whiggish history of science, although it is true that in this book I have largely concentrated on incidents and episodes in the forward movement of science. Many a path that did not lead to advances in science has been paved with unfortunate transformations. For there can be no doubt that poor or useless or unfruitful ideas are as much the result of transformations as good or useful or fruitful ideas and on that account are worthy of our study and attention in the attempt to understand the creative scientific mind at work. But of course there is a natural hierarchy, and I cannot believe that Newton's alchemical notions or his theological beliefs merit as close attention, page by page, as the *Principia*. For example, would scholars be as interested as they are in Newton's alchemical "active spirit" if he had not written the *Principia*?

4.7   The transformation of experience

Closely allied to the transformation of ideas is the role of ideas in the transformation of experience. I do not refer here to the simple ordering of experience by the imposition of intellectual constructs (or ideas, or concepts), but rather the projection of ideas onto experience in a way that determines what one sees or concludes.[1] Leonardo Olschki has shown by dramatic example that 'what Columbus saw on landing in the West Indies' was conditioned to a high degree by a long tradition of travel literature. While Columbus was 'very meticulous and exhaustive in giving a very detailed report of the appearance of the natives, their customs and peculiarities', even 'depicting their life and habits with a keen and expressive realism' (Olschki, 1941, 1937), he was vague and bookish about the geographical features and 'the natural aspects of the newly discovered islands' and—like generations of explorers before him, with whose travel accounts he was familiar—he tended to exaggerate the 'dimensions and proportions'.[2] To us it seems curious that there should be such a difference between Columbus's close attention to the human beings he encountered and his apparent indifference to (and hence exaggerations about) physical nature, but Olschki has shown that in this Columbus was no different from the many travelers and explorers over many centuries, going back to medieval times. In the language of the present chapter, we would conclude that the visual experience of Columbus and other explorers of his day had been transformed by his reading so as to accommodate itself to tradition. He saw the human life around him with the eyes of an observer, but every other aspect of the new experience was transformed into a pattern set by the travel literature, or by literary tradition.[3]

A somewhat different pattern of transformation of experience occurs with respect to Newton's identification of seven colors in the visible spectrum of sunlight. A number of hypotheses have been advanced to explain how Newton could possibly have claimed to have seen seven colors in the spectrum (red; orange; yellow; green; blue; indigo, or what we would call violet; and violet, or what we would call purple), since 'anyone who honestly looks at the spectrum cannot see more than six distinctly different colors'; 'there is no indigo' (Biernson, 1972). Of course, in a continuous spectrum there are no absolutely distinct separations of the colors, each one shading by gradual degrees into its neighbor, so that any observer can arbitrarily divide the spectrum into seven or more colors at will. But Newton's choice of seven colors has caused interpreters to assume that he was consciously or unconsciously superimposing upon his observations some intellectual constraint. Or, to put it another way, he had transformed the results of experiment so as to see seven colors where most observers would have seen only six (red, orange, yellow, green, blue, purple–or violet).

This transformation probably arose from Newton's interest in musical theory, and the Pythagorean division of the "octave" into seven parts or intervals.[4] Recently, still another source has been proposed: the color circle of a painter (Biernson, 1972, sec. 3). Newton's own color circle has been shown to make sense if one supposes that it applies to the so-called subtractive color mixtures that result when mixing paints or powders;[5] it does not, in fact, apply to the so-called additive mixtures that are obtained by mixing differently hued light beams (produced by prisms and diaphragms or slits). It has been remarked that Newton probably 'obtained his color circle from a painter' and had in mind 'seven distinct hues: red, orange, yellow, green, blue, violet, and purple. He assumed that all these colors must be in the spectrum and so deceived himself into believing that he could also see seven distinct spectral colors (Biernson, 1972).

The example of the colors in the spectrum shows how directly observed experience may be transformed in the mind of the beholder by consciously or unconsciously held preconceptions. An even more striking example of this phenomenon occurs in Leonardo da Vinci's anatomical studies of the heart. Leonardo wrote many precepts concerning the importance of the direct experience of nature. For instance: 'To me it seems that all sciences are vain and full of errors that are not born of Experience, mother of all certainty, and that are not tested by Experience'. Again, 'All true sciences are the result of Experience which has passed through our senses'. And he even contrasted those who cite 'the authority of certain men held in great reverence by their inexperienced judgment' with himself, his works being 'the issue of simple and plain experience which is the true mistress' (Richter, 1952, pp. 1–5). And yet he described and even made a schematic drawing of pores in the septum that separates the left and right ventricles (or the interventricular wall), and another drawing showing how the blood can pass from the right to the left ventricle. This is what he says: 'The extraction of blood out of the right ventricle, which penetrates, through wide porosities, the wall interposed between the right and left ventricle, which porosities narrow themselves with pyramidal concavities till they pass into imperceptible pores through which the viscous blood penetrates and goes on subtilising itself to great subtlety' (*Quaderni d'anatomia*, vol. 4, fol. 11ᵛ; in Keele, 1952, p. 75).

Now, there are no such pores, and Leonardo is here giving almost a direct paraphrase of Mondino's account of the same subject, with one major difference (as K. D. Keele, 1952, pp. 75, 57–59, has observed): Leonardo 'emphasizes that the pores become "imperceptible" except at their openings from the right side of the septum'. But why did he imagine that such pores exist and why did he sketch them into his drawings of the heart? The answer is that those pores were necessary to the then-ruling Galenic physiology. In this case, to quote J. Playfair McMurrich (1930, p. 156), 'Leonardo allowed tradition to master observation'. Or, as George Sarton put it (preface to McMurrich, 1930, p. xix): 'Galenic prejudices were part of the very atmosphere which he was breathing; they were beyond the need of scrutiny or dispute. And so it was that this keen observer saw things not only with his own eyes, but sometimes with those of Galen!'

This transformation of experience, so as to make it conform to Galenic preconceptions, may be contrasted with Vesalius's studies of the interventricular wall of the heart. Vesalius's *De fabrica* (1543) had been written because Vesalius had found that Galen's writings on the bones (which Vesalius had been asked to edit for a new printing) were not in agreement with the results of dissection. And so it might be thought that Vesalius would have seized upon this example in order to show the inadequacy of Galen's knowledge and understanding of the human body. Not at all! In the first edition of his book he merely pointed out that there are pits[6] on the interventricle wall and observed that 'none, so far as the senses can perceive, penetrate from the right to the left ventricle'. This led him to conclude: 'We are thus forced to wonder at the art [*industria*] of the Creator by which the blood passes from the right to the left ventricle through pores that elude the sight.' By the time of the second edition, however, he had become bolder. Now he said that he had 'not come across even the most hidden channels by which the septum of the heart is pierced', even though 'such channels are described by teachers of anatomy who have absolutely decided that blood is carried from the right to the left ventricle'. It was not long ago, he wrote, that he 'would not have dared to turn aside even a nail's breadth from the opinion of Galen, the prince of physicians'. Nevertheless, 'the septum of the heart is as thick, as dense, and as compact as the rest of the heart', and he could not discern 'in what way even the smallest particle can be transferred from the right to the left ventricle through the substance of that septum'. Hence, when 'these and other facts are considered, many points concerning the arteries come forward about which doubts may reasonably arise'.[7] Vesalius was thus conscious of the Galenic preconception, but did not allow it to transform his observations; despite more than a millennium of assertions that such pores do (and must) exist, Vesalius could record that he had found none.

It is an interesting question as to why Leonardo should have sketched in the pores required by the Galenic physiology while Vesalius did not. Possibly the reason is that Vesalius was conscious of Galenic science, which he had had to study and master as a medical student, and that Vesalius had come to be doubtful of the absolute truth of Galen's writings when he compared, chapter by chapter and line by line, what Galen wrote and what he himself saw on making dissections of human cadavers. Leonardo, on the other hand, had never been formally schooled in Galenic science and had picked up what he knew primarily because it was all around him; thus he did not specifically contrast Galen's writings with his own empirical evidence, as Vesalius had to do in the preparation of an edition of Galen on the bones. Even apart from differences in individual personality, there is a further separation of these two men insofar as the eye and hand of the artist render nature on his two-dimensional canvas or sketchpad in its general outlines and holistic impressions, while the eye and hand of the anatomist tend to concentrate on each small detail.

The case of Leonardo and the septal pores shows how direct observation may be so transformed, and even contradicted, as to yield information which is not actually present. Or, to put it another way, Leonardo's drawings of the septal pores show us how the direct experience of observation may be so transformed as not to deny a theoretical position that is well established. But there is another negative aspect to the transformation of experience, which results in the nonobservation of phenomena that do not accord with preconceptions. I believe that this is the only reasonable explanation of the apparent historical fact that in all of Western Europe there is no record of the appearance of the supernova of A.D. 1054, which we know today only through its debris or what remained after it exploded, as the Crab Nebula. Apparently the men and women of eleventh-century Europe "did not see" this supernova in the heavens; perhaps the reason is that they had been brought up on a belief that the heavens are incorruptible and unchangeable. But in China and Japan, where no such prejudices existed, the appearance of this incredibly bright new star was duly recorded. It is, in fact, only one of a number of occurrences of novae and supernovae to have been observed and recorded in China during 'that period between Hipparchus and Tycho Brahe during which the rest of the world remained in almost complete ignorance of the fact that "new stars" sometimes appear in the heavens' (Needham, 1959, p. 172). Sunspots, also observed in China, are another group of 'phenomena [which] were regularly observed by the Chinese for centuries, which Europeans not only ignored, but would have found inadmissible upon their cosmic preconceptions' (ibid.). One can only add that these same cosmic preconceptions ('The heavens were thought to be perfect') apparently also inhibited Europeans from "seeing" the phenomena.[8]

We may actually document the way in which a new star or nova is "not seen" by reference to the supernova of 1572. Tycho Brahe was an exception, but even he had difficulty in convincing himself that there was a new star in the heavens. His experience shows dramatically the inhibiting effect of accepted dogmas and prejudices. At this time the doctrine that change—coming into being and passing away—cannot occur in the celestial regions was still widely accepted by thinking men and women. As Tycho Brahe explained, in his tract *De nova stella* (1573),

> It is agreed by all philosophers, and the facts themselves
> clearly declare it to be the case, that in the aethereal re-
> gion of the heavenly world no change occurs with respect
> to generation and corruption [i.e., coming into being
> and passing away]; but the heavens and the aethereal
> bodies that are in the heavens are not increased and are
> not diminished, and are not changed in number or size
> or brightness or in any other respect, but always remain
> the same [Brahe, 1913–1929, vol. 1, p. 16].

Accordingly, a changing object or a new object that appeared in the heavens would be considered to be located below the sphere of the moon, in the region of the four Aristotelian elements (earth, air, fire, water), where change was supposed to occur. Such an object could not be a planet or a star, but would—according to Aristotelian principles—be a comet or a "fiery meteor". Tycho Brahe's experience in relation to the new star of 1572 is particularly revealing as to the influence of the prevailing philosophy, indicating the difficulty in reconciling the obvious appearance of a new bright star with the dogma that no change can occur in the heavenly regions and that, therefore, there cannot be a new star in the heavens.[9]

Tycho first saw the new star on the evening of 11 November 1572. He was returning home, after having been busily engaged in chemical (or alchemical) experiments and, as was his habit, he looked up at the stars shining in the clear night sky. Directly overhead, in the constellation of Cassiopeia, he noticed what appeared to be a new and strange or unusual star that far exceeded the other stars in brightness. Since he had, from his boyhood days, known all the stars that are visible in the heavens (a knowledge which, he says, is not difficult to attain), it was evident 'that in that place in the sky there had never been a star before, not even a minimally small one, much less one of such conspicuous brightness'. Tycho was astonished, and he later admitted that he was not ashamed to have doubted the evidence of his own eyes. Because he could hardly believe what he saw, Tycho turned to some servants who were attending him and asked them whether they could see the star in the place he showed them. Even though they said they could, Tycho still needed additional reassurance before being willing to admit that there was a new star in the heavens. Since some peasants (or farmers) happened to be passing by, Tycho asked them if they too could see the star. They agreed that they were able to see a very bright star at the place pointed out by Tycho. Then, and only then, did Tycho cease to doubt the evidence of his senses (Brahe, 1913–1929, vol. 2, p. 308).

In attempting to give the reader some sense of the extraordinary experience of seeing a new star in the heavens, Tycho could only compare it to 'a miracle'—'either the greatest of all miracles that have occurred in the whole realm of nature since the beginning of the world, or one that must necessarily be put into the same category as those attested to in holy writ: the stopping of the Sun in its course in answer to Joshua's prayers and the darkening of the Sun's face at the time of the crucifixion' (ibid., vol. 1, p. 16). Then Tycho reminds his readers why this star was so miraculous. Its coming into being in the heavens was an event transcending 'the ordinary Laws of Nature' and if or when it ceased to be seen this would again be an event transcending 'the same Laws of Nature'. No wonder Tycho could not at first believe the evidence of his senses. Tycho's experience affords an admirable illustration of the supreme and almost insuperable difficulty of transforming a sense impression into the concept of a physically caused event when that cause directly contradicts the general philosophical or scientific precepts of the age. Other astronomers or observers would simply have doubted the evidence of their senses and would have either unconsciously dismissed or denied the appearance of the new star.

An alternative for Tycho would have been to conclude that the data of sense perception did correspond to a physical object, but not a star. In this case, the bright object would have been interpreted as a comet, since it was then commonly believed (as Aristotle had taught) that comets and meteors occur in or near the earth's atmosphere, or at least below the sphere of the moon, a region where change can occur, but not out in the celestial spaces of planets and stars where change cannot occur. Tycho was fully aware of this possibility, which he rejected on a number of grounds. The seen object did not have a tail, it twinkled like a star, and it did not move in a direction contrary to the daily rotation of the heavens. Even more important, careful measurements with a new sextant that he had just constructed showed that the star had no discernible parallax and so had to be located out beyond the sphere of the moon, in the region of the stars. He concluded that 'this star is not some kind of comet or fiery meteor, whether these be generated beneath the Moon or beyond the Moon, but is a star shining in the firmament itself—one that has never previously been seen before our time, in any age since the beginning of the world' (Shapley & Howarth, 1929, p. 19).

Tycho has also recorded for us how men of learning of his acquaintance were ignorant of the new star's existence. Early in 1573, he says, he paid a visit to his friend Johannes Pratensis in Copenhagen. Although Pratensis was a leading figure in the intellectual circles of the university, he had not even heard that there was a new star shining in the heavens. Although this star had a brightness equal to that of Venus at its maximum, it had not been seen or discussed among the faculty of the university. Pratensis could scarcely give credence to Tycho's announcement, which is all the more remarkable in that—as he later informed Tycho, and as Tycho duly recorded in his *Progymnasmata*—he was lecturing in the university on book two of Pliny's Natural History, in which Pliny reported that Hipparchus had seen a new star, an occurrence which had led him to construct his famous star catalogue. Tycho told his readers all about these events and the alleged incident of Hipparchus having 'noticed a star different from all others previously seen, one that was born in his own age'. But Tycho evidently did not fully trust Pliny's story, as he showed by using a conditional clause: 'if Pliny is to be trusted' ('*si Plinio adhibenda est fides*') (Brahe, 1913–1929, vol. 1, p. 16).

Soon after Tycho's arrival in Copenhagen, he and Pratensis were invited to dinner by the French envoy, Charles Dancey, who was very much interested in astronomy. (In 1576 Dancey would, at his own request, perform the ceremony of laying the cornerstone for Tycho's observatory 'Uraniborg' on the island of Hveen.) When Tycho declared that there was a new star in the heavens, Dancey thought that Tycho was joking and could not be serious ('*Putabat enim me iocari . . .*'), that he was poking fun at the Copenhagen astronomers. On the way home after dinner, Tycho convinced Pratensis of the existence of the new star by showing it to him shining brightly in Cassiopeia (Brahe, 1913–1929, vol. 3, pp. 93–94). Pratensis was as astonished as Tycho himself had been when he first beheld this star in the sky.

The experience of Tycho shows how difficult it is to transform the data of sense experience (a point of light that looks like a star) into a physical cause (a star appearing in the heavens where it should not be). The example of Galileo will show us the opposite effect, the strongly reinforcing influence of a commitment to the Copernican doctrines. Writers on the history of science often convey the impression that when Galileo pointed the newly invented astronomical telescope at the heavens in 1609 he discovered or "saw" mountains on the moon and satellites of Jupiter. Of course, he did nothing of the sort. A careful reading of Galileo's manuscript records or the published account of his discoveries that he presented in his *Starry Messenger* of 1610 shows that when Galileo examined the moon through the telescope he saw a larger number of spots than he had expected. Some of the spots were darker and very much bigger than others; they were called by Galileo 'the "large" or "ancient" spots', since these were the ones seen and reported by naked-eye observers during many centuries. They were to be distinguished from certain smaller and very numerous spots which had never been observed until the invention of the telescope, or—as Galileo said—'had never been seen by anyone before me'.[10] These new spots were the raw data of sense experience. Or, to put it another way, what Galileo actually *saw* through the telescope was a collection of spots of two sorts. It took some time until, as Galileo tells us, he transformed these sense data or visual images into a new concept: a lunar surface with mountains and valleys, the source and cause of what he had seen through the telescope. On this score there can be no doubt whatever, as Galileo himself made clear in his published account. Let him speak for himself (Galileo, 1957, p. 31):

> From observations of these spots repeated many times
> I have been led to the opinion and conviction that the
> surface of the moon is not smooth, uniform, and pre-
> cisely spherical as a great number of philosophers believe
> it (and the other heavenly bodies) to be, but is uneven,
> rough, and full of cavities and prominences, being not
> unlike the face of the earth, relieved by chains of moun-
> tains and deep valleys.

Then Galileo describes the actual observations he had made 'by which I was enabled to draw this conclusion'. We need not go into the details of these observations (ibid., pp. 32–34), but we may note that many of them suggested to Galileo's mind an analogy with terrestrial phenomena. For example, certain 'small blackish spots' had 'their blackened parts directed toward the sun', while on the side opposite the sun they appeared to be 'crowned with bright contours, like shining summits'. We see a similar phenomenon on earth at sunrise, Galileo remarks, 'when we behold the valleys not yet flooded with light though the mountains surrounding them are already ablaze with glowing splendor on the side opposite the sun'. Another 'astonishing' observation was a series of 'bright points' in the dark region of the moon well beyond the terminator. He found that these would gradually get larger and eventually join the 'rest of the lighted part [of the moon] which has now increased in size'. These, he concluded, were luminous mountain peaks rising so high from the surface of the moon that they are illuminated by the sun's light, even though their bases are in the region of shadow or in darkness. Again Galileo reminds his reader of a terrestrial analogy, since 'on the earth, before the rising of the sun, are not the highest peaks of the mountains illuminated by the sun's rays while the plains remain in shadow?'

The intellectual transformation of these lunar observations into conclusions that agree with what Galileo calls 'the old Pythagorean opinion that the moon is like another earth' was propelled by Galileo's commitment to the Copernican system. There must have been an enormous unconscious pressure to vindicate the Copernican position that the earth is merely another planet, that it is not different from the other planets and the moon. If the earth is not a unique body, it is not specially conditioned to be motionless and at the center of the universe. Galileo's commitment to Copernicanism thus caused him to transform the data of observation into an argument that the moon resembles the earth.

A somewhat similar process of transformation of the sense data of experience occurred in relation to what Galileo called 'the matter which in my opinion deserves to be considered the most important of all—the disclosure of four PLANETS never seen from the creation of the world up to our time' (Galileo, 1957, p. 50). Here Galileo is using the word "planet" in the original Greek sense of any wandering body in the heavens and is referring to his discovery of satellites of Jupiter, or secondary planets accompanying the primary planet Jupiter. What he actually "saw" was not a set of moons or satellites. He actually observed on 7 January 1610 'beside the planet . . . three starlets, small indeed, but very bright'. These points of light, looking like stars, despite their proximity to Jupiter, were the actual sense data. Galileo at first made only the simple and obvious transformation of the sight of these points of light and concluded that he had seen stars. As he says (Galileo, 1957, p. 51), 'I believed them to be among the host of fixed stars'. The only special aspect that aroused his curiosity, he goes on, was their 'appearing to lie in an exact straight line parallel to the ecliptic and . . . their being more splendid than others of their size'. So far was he from conceiving that these might be satellites of Jupiter that he tells us that he 'paid no attention to the distances between them and Jupiter, for at the outset I thought them to be fixed stars, as I have said'. His second observation occurred on the next night and showed 'three starlets . . . all to the west of Jupiter, close together, and at equal intervals from one another'. Even then, Galileo did not begin to guess that these were satellites. Rather, he tells us,

I began to concern myself with the question of how Jupiter could be east of all these stars when on the previous day it had been west of two of them. I commenced to wonder whether Jupiter was not moving eastward at that time, contrary to the computations of the astronomers, and had got in front of them by that motion. Hence it was with great interest that I awaited the next night.

After further observations, he eventually 'decided beyond all question that there existed in the heavens three stars wandering about Jupiter as do Venus and Mercury about the sun'. Before long he had found that there are 'four wanderers' that 'complete their revolutions about Jupiter'. It is not without interest that Galileo draws an analogy between the satellites or lesser lights moving around the greater light of Jupiter and the motion of Venus and Mercury about the brighter light of the sun. This analogy would indicate that Galileo's Copernicanism was directly related (according to his own testimony in so many words) to his transformation of the idea that there are *stars* moving *along with* Jupiter to the idea that there are *satellites* moving *around* Jupiter.[11]

The example of the satellites of Jupiter differs in one essential from the earlier experience with spots on the moon. Galileo's Copernicanism and anti-Aristotelianism obviously preconditioned his mind to the possibility that the moon would be earthlike. But there was nothing in his anti-Aristotelian bias or his pro-Copernican commitment to prepare him for the existence of a model of the Copernican system in miniature in the form of a satellite system around Jupiter. It might be argued that if the earth is not unique, then it should follow that the earth ought not to be the only planet with a satellite. This line of thought might possibly have been a part of Galileo's ultimate conception that there are satellites of Jupiter. But in fact Galileo does not mention the analogy with the earth's having a moon. In fact, there is an astonishingly great difference between a planet's having a single moon and the existence of a whole satellite system of four new "planets" encircling Jupiter. Even so firm a Copernican as Kepler was shattered by the news that Galileo had discovered four new planets or wandering stars, since he did not quite know how he could fit them into his scheme in which the separation between six planets was related to the existence of five and only five regular geometric solids.

Of course, there was an additional point to the new discovery, once it had been made, and that was that it answered the questions of the anti-Copernicans who argued that the earth could not move in its orbit (and remember that it does so at the enormous speed of about twenty miles per second) without losing its moon. Everyone admitted that Jupiter must move; well, if Jupiter could move in orbit and not lose four moons, surely there could be no objection to the earth's moving and not losing its single moon.

Before long, Galileo had made another remarkable discovery, namely that the sun has spots. These spots are the given, the data of sense observation. What is significant is how they were transformed or interpreted by the mind of Galileo. It is well known that Galileo showed these to be actual spots on the surface of the sun, and thus interpreted their motion as an indication that the sun rotates on its axis. Others, who were of a different scientific and philosophical point of view, tried to give another interpretation, holding that these were shadows cast on the sun, possibly by stars that 'revolve about it in the manner of Mercury or Venus'.[12] The two interpretations show how the point of view acts upon what is observed. An Aristotelian must believe that the sun itself is pure and unspotted, whereas an anti-Aristotelian like Galileo did not care whether the sun is spotted or unspotted, whether it is immutable or whether it undergoes changes from day to day. The sunspots are of interest in the present context in a historical sense, because it turns out that there had been in the Middle Ages a certain number of observations of sunspots, but these had tended to be interpreted as instances of the passage of a planet (Mercury or Venus) across the face of the sun, since the prevailing philosophy would not permit these observations to be transformed into the interpretive statement that the sun has spots on it (see Goldstein, 1969).

The doctrine of transformation tends to pinpoint the actual occasion on which the scientist's background, philosophical orientation, or scientific outlook interacts with sense data in order to provide the kind of base on which science advances. Thus the way is prepared for the next phase of investigation, which I shall not attempt to pursue further here. Obviously, it would be important to be able to identify, classify, and interpret those parts of the background which are operative in an important way in a number of examples. For instance, a first assignment would be to try to distinguish between the effect of the general background in philosophy and science and that of the particular personality of the scientist. It would be important to try to find the degree to which intellectual transformations are related to the background and are independent of the particular scientist. Only the barest beginnings have been made in this general area of the psychological background to discovery. In particular this was the subject of a very perceptive set of observations by N. R. Hanson (1958, pp. 4–43), and it has been explored by Leonard K. Nash (1963, ch. 1–2). Certainly, Gestalt psychology has much of importance to contribute here.[13] And there is no doubt that studies by experimental psychologists such as R. L. Gregory (1970) and art historians such as E. H. Gombrich (see, e.g., Gombrich, 1960; Gregory & Gombrich, 1973) will eventually do much to illuminate this topic.

4.8 *The uniqueness of scientific innovation: Freud on originality*

At the end of §4.6, it was pointed out that the concept of transformation leads to the supposition that there should be multiple simultaneous and quasi-independent scientific discoveries. And at once the question must be raised as to whether there is really any originality in science in any complete sense.

In discussing the subject of what he called 'apparent scientific originality', Sigmund Freud wrote that when a new idea appears in science, it is at first hailed as a discovery and then, 'as a rule, disputed as such'. But 'objective research soon afterwards reveals that after all it was in fact no novelty'. In fact, Freud says, it is usually found that the discovery has in fact not only been made previously, but repeatedly. The idea is proved to have had forerunners, to have been 'obscurely surmised or incompletely enunciated' (Freud, 1923, p. 261). Freud could well have been describing the common labors of the historian of ideas, and particularly the historian of scientific ideas.

What is particularly of interest in the present context, however, is what Freud then said about the 'subjective side of originality'. The scientist may ask himself 'what was the source of the ideas peculiar to himself which he has applied to his material'. In some cases, Freud said, the scientist 'will discover without much reflection the hints from which they were derived, the statements made by other people which he has picked out and modified and whose implications he has elaborated' (Freud, 1923, p. 261). Here, to all intents and purposes, Freud has stated the doctrine of transformation of scientific ideas.[1] Freud, however, went a step further than the above quotation might indicate. For he was equally concerned with those original ideas for which the scientist could himself find no source. In this case, Freud said, the scientist 'can only suppose that these thoughts and lines of approach were generated—he cannot tell how—in his own mental activity, and it is on them that he bases his claim to originality.' This claim to originality, however, according to Freud is diminished by 'careful psychological investigation'. In this way, he said, there may be revealed 'hidden and long-forgotten sources which gave the stimulus to the apparently original ideas', and the ostensible new creation is seen as a 're-vival of something forgotten applied to fresh material. There is nothing to regret in this; we had no right to expect that what was "original" could be untraceable and undetermined' (Freud, 1923, p. 261). Freud freely admitted that 'the originality of many of the new ideas employed by me in the interpretation of dreams and in psycho-analysis has evaporated in this way'.

Freud discussed the sole idea for which he could find no source in an essay on Josef Popper (1838–1921) or Popper-Lynkeus.[2] Here Freud called attention to the fact of his own awareness that the 'originality of many of the new ideas employed by me in the interpretation of dreams and in psycho-analysis' was somewhat different from the free creation ordinarily imagined, since each new idea could be traced back to some source. Freud, in fact, was as immodest in speaking about the sources of his ideas as he was about his own qualities as a scientist and discoverer. Thus he said concerning his work on dreams and psychoanalysis in general, 'I am ignorant of the source of only one of these ideas' (Freud, 1923, p. 261). The idea in question was that of dream-distortion, the phenomenon in which dreams assume such a 'strange, confused and senseless character', which Freud attributed to the fact that 'something was struggling for expression in them which was opposed by a resistance from other mental forces' (ibid., p. 262). Freud defined dream-censorship as the 'mental force in human beings' which keeps watch on the internal contradiction between impulses stirring in the mind of the dreamer and 'what might be called the dreamer's official ethical and aesthetic creed'. Dream censorship 'keeps watch on this internal contradiction and distorts the dream's primitive instinctual impulses in favour of conventional or of higher moral standards'.

In the present context, Freud's essay is significant because, as he said, it was precisely 'this essential part of my theory of dreams' which was 'discovered by Popper-Lynkeus independently'. Curiously, Freud did not further analyze this strange coincidence of discovery or invention. He merely concluded by stating his belief 'that what enabled *me* to discover the cause of dream-distortion was my moral courage', whereas in the case of Popper-Lynkeus, 'it was the purity, love of truth and moral serenity of his nature'.[3] It apparently never occurred to Freud to apply his own theory to suggest that both he and Popper-Lynkeus had either been reading the same works, which served as sources for their respective transformations, or had been reading different works which had originally derived from one and the same source. Nor did Freud consider that possibly both he and Popper-Lynkeus had based their transformations on elements found in common in the then-current general environment of ideas.

4.9   *Transformations and scientific revolutions*

It has not been my aim in these pages to produce an anatomy of scientific ideas, nor yet a complete taxonomy of the transformations of scientific ideas, although I may hope to have made a contribution to the achievement of both of those goals. But I could not discuss the Newtonian revolution in science without introducing these topics, since it would otherwise have seemed strange that a revolution could be other than a wholly new set of ideas: a turning over or turning out of the received concepts, principles, laws, and methods. Of course, the science of the *Principia* was new in that theretofore there had not existed a system so completely based on force and mass, on inertia and its manifestation in uniform and accelerated motions; and universal gravity as a quantified entity was new despite the many roots or steps of transformation that had enabled Newton to produce it (see Koyré, 1950*a*). And it is precisely in focusing attention on transformations that we may, I believe, see how a revolutionary system like Newton's could arise from elements that seem not to be new at all, but are only revealed in their stark novelty by the degree of transformation made by Newton. Without explicitly recognizing the degree of transformation in the ideas that make up so many of the essential features of the revolutionary Newtonian system, previous scholars understandably fell back on the imprecise word "synthesis".

Historians usually think of revolutions in terms of cataclysmic events. Thus scientific revolutions play a role in the development of the sciences akin to that of the French Revolution or the Russian Revolution in the history of human societies. The old way of life disappears, giving way to the new. In Thomas S. Kuhn's *Structure of Scientific Revolutions*, this kind of near-total and radical change is expressed as the rejection of a condition of "normal science" and the substitution of a new "normal science" in its place. With respect to the role of the *Principia* as the paradigm of the new normal science, Kuhn has undergone a notable change of mind between the first and second editions of his book, a matter that need not be discussed here.[1] Like most historians and historical analysts, Kuhn is concerned with what I think may be not unfairly called the gross features of scientific change: the substitution at large of one set of shared beliefs on the part of the scientific community for another. This may be called, following Georges Canguilhem, the macroscopic scale of the history of science.[2] Here, however, I am concerned with a different problem, not the totality of change introduced by the *Principia*, not the formation of a new paradigm for terrestrial and celestial mechanics, but the separate and separable stages of innovation whereby Newton altered so many of the accepted concepts and modes of analysis in the physical sciences. I am, in short, interested here in the detailed analysis of Newton's creative scientific thought, in the fine structure of the scientific revolution produced by the *Principia*, in the microscopic scale of the history of science. Or, to put matters in a different way, my concern is with the role of the individual in scientific change, even in scientific revolutions, as a means of understanding how science may undergo radical alterations of its systems of concepts, laws, and explanations. The analysis of revolutions into a series of transformations shows the continuity within the change, but does not thereby diminish the magnitude of the net change itself.

I referred earlier (in §4.2) to the problem of plausibility and reasonable belief: Does science in fact move ahead by great leaps and dramatic major revolutions? Many scientists and philosophers doubt the propriety of conceiving the growth of science in terms of revolutions. The idea of revolutions in science seems to physicists in particular to be a very inappropriate and inexact way to characterize the growth of their science. In a report on the state of physics issued a few years ago, under the aegis of the President's Committee on Science and Public Policy of the National Science Advisory Committee, Academy of Sciences (U.S.A.), the notion of such one-man revolutions was a specific target of attack.[3] More recently, Steven Weinberg has written that in 'the development of quantum field theory since 1930', the 'essential element of progress has been the realization, again and again, that a revolution is un-necessary'.[4]

A similar point of view was expressed some forty years ago, in an essay by Rutherford on the nature of scientific discovery (1938, pp. 73sq). It is 'not in the nature of things', he wrote, 'for any one man to make a sudden violent discovery':

> Science goes step by step, and every man depends on the work of his predecessors. When you hear of a sudden unexpected discovery—a bolt from the blue, as it were—you can always be sure that it has grown up by the influence of one man on another, and it is this mutual influence which makes the enormous possibility of scientific advance. Scientists are not dependent on the ideas of a single man, but on the combined wisdom of thousands of men, all thinking of the same problem, and each doing his little bit to add to the great structure of knowledge which is gradually being erected.

The notion of revolution by transformation provides a kinship between the historical reconstruction of scientific changes of the past and the actual development of science as we see it occurring here and now. In this sense, what Newton's successors called a "revolution" becomes a plausible event (or series of events) according to everything that we know about science as we have seen it develop in the present and the recent past, and in the Scientific Revolution.

As we trace Newton's transformations of the ideas, concepts, methods, laws, and principles by which the science of dynamics grew during the seventeenth century, we cannot help becoming aware of how much Newton owed to his illustrious predecessors and contemporaries. At the same time, however, we also gain an insight into the way in which he gave wholly new relevance and meaning to the work of fellow-scientists, one by one. Each such transformation reveals to us the operation of his creative genius. The grand generalization to a principle of universal gravitation was a final transformation to a conception that even Kepler never quite imagined. Of course, Newton was a genius, and in fact on the very highest level of genius in the exact sciences, comparable to a select group that (in my opinion) includes Archimedes, Laplace, Einstein, Rutherford, and Fermi. But to say that is no more helpful to our understanding than to speak of a "synthesis" in the formation of Newtonian science. We are (at least at present) helpless in analyzing Newton's genius,[5] and in attempting to trace either its nature or genesis and its nurture. But the study of the sequence of transformations that led him to universal gravity and the science of the *Principia* reveals the mode of action of his creative scientific imagination. Of course, the one essential step, the final transformation, the supreme creative effort that enabled Newton to display the universe and its functioning parts as a coherent system of rational mechanics held together by the inverse-square law—that still remains shrouded in mystery, although perhaps now removed from the category of the miraculous and placed on the partially comprehensible level of supreme human genius.

5

Newton and Kepler's laws: stages of transformation leading toward universal gravitation

5.1 Kepler's laws and Newtonian principles

There are two widely held opinions concerning the development of Newton's scientific ideas: that he found the law of universal "gravitation" in the 1660s and then refrained from publishing it for twenty years,[1] and that he found this law by "deducing" it from Kepler's "laws" (or, possibly, from only one of Kepler's laws).[2] The analysis presented in this chapter will show that according to any reasonable definition of universal "gravitation", Newton did not find this law until some time after November 1684, and before 1686, and then published it forthwith. It will be seen that Newton did not (and logically could not) "deduce" the law of universal gravitation from Kepler's laws. In any event, he was not consciously aware of the law of areas in a fruitful context of dynamics until somewhat later than the 1660s, possibly not until the time of a famous exchange of correspondence with Hooke in 1679–1680 (see §3.1). The clarification of the exact role of Kepler's three laws of planetary motion in Newton's thoughts about celestial motions will show Newton's successive steps and transformations leading up to the generalization of a universal force of gravity, and will reveal how the last step entailed a radical transformation of Kepler's laws.

This episode provides an example of Newton's creative mind at work, worth far more toward the understanding of whatever logic there may be in scientific discovery than a hundred precepts. Newton's use of Kepler's laws, and his eventual transformation of them, exhibits the delicate relation between the exploitation of mathematical systems or imaginative constructs and the search for "true" principles and laws of nature that characterizes the Newtonian science of the *Principia* and is the mark of the Newtonian style.

To help the reader follow the argument, a few of the results established by Newton in the *Principia* need to be stated clearly. Newton begins (sects. 2–3, bk. one) with an analysis of the physical significance or meaning of each of Kepler's three "laws" or "planetary hypotheses"[3] in the order in which they were presented (and discovered) by Kepler himself. First, the area law: that a "radius" (we would say "radius vector") from the center of the sun to any of the planets sweeps out equal areas in any equal times. Second, that the planets move in elliptical orbits, with the sun at one of the two foci. Third, the harmonic law: that the ratio a^3/T^2 is the same for all planets in the solar system, where a is the average distance of the planet from the sun (really the semiaxis major) and T is the period of revolution of that planet around the sun, reckoned with respect to the fixed stars.[4] Once Newton's analysis has shown the kind of physical system of the world represented by these three "laws" or "planetary hypotheses", it is apparent that this is not the true system of the world[5] but only a mathematical construct that is very different, and obviously different, from the real world. For the mathematical construct corresponds to what is essentially a one-body system and a center of force. We shall see how Newton alters this construct by first introducing a second, mutually interactive body in place of the mathematical center of force, so as to produce a system like that of the sun and a single massive planet.[6] Then, and only then, does he add yet other mutually interactive bodies so as to produce a system like that of the sun and its six planets. The transition to the dynamics of satellites of such planets requires a further separate step, as does the introduction of bodies with real sizes and shapes.

From the point of view of the dynamics of the *Principia*, Kepler's third law is doubly incorrect. Not only does it fail with respect to the perturbing effects of one planet upon another, which are most sensible with respect to the perturbations produced by Jupiter (which has by far the greatest mass of any planet in our solar system), but this law is far from true even for a system of planets with different masses, each of which mutually attracts and is attracted gravitationally only by the sun. As Karl Popper has put the matter, not only are 'Kepler's laws' found to be 'strictly invalid' (since they are 'only approximately valid') 'in Newton's theory', once we take cognizance of the 'mutual attraction between the planets'; but, additionally, the 'contradictions between the two theories' (Kepler's and Newton's) are 'more fundamental' than anything so 'obvious'. Even if 'we neglect the mutual attraction between the planets', he writes,

> Kepler's third law, considered from the point of view of Newton's dynamics, cannot be more than an approximation which is applicable to a very special case: to planets whose masses are equal or, if unequal, negligible as compared with the sun. Since it does not even approximately hold for two planets one of which is very light while the other is very heavy, it is clear that Kepler's third law contradicts Newton's theory in precisely the same sense as Galileo's [Popper, 1957, pp. 29sq; 1972, p. 200].

Newton, as we shall see below, showed that for any two-body system, say the sun (mass s) and a planet (mass p_n), there is a law that (for suitably chosen units) is $a^3{}_n \ / \ T^2{}_n = 1 + (p_n/s)$, or $a^3{}_n \ / \ T^2{}_n = p_n + s$. Here $n = 1,2,3, \ldots$ corresponds to the order in which a planet is found with respect to nearness to the sun. The values of $a^3{}_n \ / \ T^2{}_n$ for planets in our solar system, using Newton's data, will be $1 + 1/1067$ for Jupiter, $1 + 1/3021$ for Saturn, and $1 + 1/169,282$ for the earth. Thus the maximum difference between Kepler's third law and Newton's transformed version of Kepler's third law is about one part in a thousand or one-tenth of 1 percent.[7]

We may agree with Popper that Kepler's third "planetary hypothesis", $a^3/T^2 = $ constant, where we find 'the same constant for *all* planets of the solar system', can be valid only if all planets have the same mass (p_1), which is false, or if all masses are so small with respect to the mass of the sun that they may be taken as zero. The latter may be 'quite a good approximation from the point of view of Newton's theory', Popper concludes, but it 'is not only strictly speaking false, but unrealizable from the point of view of Newton's theory', since a 'body with zero mass would no longer obey Newton's laws of motion'. Thus, wholly apart from the actions of one planet upon another, 'Kepler's third law contradicts Newton's theory' which yields a quite different law (Popper, 1957, pp. 29–32; 1972, p. 200). This example shows the new high degree of exactness which Newton's *Principia* was capable of introducing into considerations of theoretical planetary astronomy.

5.2   The status of Kepler's laws in Newton's day

Before exhibiting the sequence of Newton's ideas, a few remarks about Kepler's laws may be in order. Today, many scholars make the unwarranted assumption that at the time of Newton, Kepler's three laws were well known and well established and that Newton's assignment was to deduce from them the law of gravitational force. This was not the case at all. The third (or harmonic) law was more generally accepted; after all, it could be simply confirmed by putting in the values of the periods and distances of the planets. As Newton said, in the *Principia*, 'This ratio [or rule], found by Kepler, is accepted by everyone'.[1] The law of elliptical orbits was commonly stated in books on astronomy in circulation in 1670, but it was not always used in practical problems. The law came under attack by Giovanni Domenico Cassini, director of the Paris Observatory, who proposed to substitute for the Keplerian elliptical planetary orbits a family of curves of his own invention, the ovals of Cassini (of which an individual member of the family was called a "Cassinoïde" or a "Cassinian"). This curve is defined as the locus of points, the product of whose distances from two fixed points is constant; it is a conceptual transformation by Cassini of Kepler's ellipse, which has a defining property of being the locus of points, the sum of whose distance from two fixed points (or foci) is constant.[2]

In some works on astronomy (such as the widely read book by Vincent Wing), the new Keplerian ellipse was brought into harmony with traditional epicyclic astronomy by constructing it neither as a section of a cone (in the style of Apollonius), nor as a locus with respect to two points (in the style of Kepler), but as the curve traced out by an epicycle with a period of rotation equal to the period of revolution of its center along the deferent.[3]

As for the law of areas, this was—to a large degree—ignored.[4] For example, in Thomas Streete's *Astronomia Carolina*, from which the young Newton made a note about the third law, the area law is conspicuously absent.[5] In its place there is a different law, based on the uniform rotation of a radius vector about the empty focus (or equant) of the ellipse. Reference has already been made to three forms of this substitute law that were current in Newton's day. The simplest of these substitutes for the law of areas had been invented by Kepler himself and then rejected by him before he had arrived at the "true" focal ellipse.[6] This law is a good approximation to the observed motions of the planets, but it is less accurate in the region of the octants than elsewhere in the orbit. Hence a correction factor was later brought in, by introducing an auxiliary circle, having the axis major of the ellipse as a diameter. As in Kepler's approximation, there is a radius vector rotating uniformly about the empty focus (or equant); but the intersection of this radius vector and the ellipse is no longer the planetary position, but merely a point used in the construction. Now a perpendicular is erected on the axis major, passing through this point on the ellipse and continuing until it meets the circle (as may be seen in fig. 5.1). Then a new radius vector is drawn from the empty focus to this point on the circle; its intersection with the ellipse determines the position of the planet in its elliptical orbit. This scheme appears more simply in fig. 5.1 than may be apparent from the verbal description; we may agree with Delambre that although this

Figure 5.1. Substitutes for the law of areas. In the simple construction, a radius vector is centered at the empty focus (*E*) of the ellipse and rotates uniformly. Its intersection with the ellipse (at *P* or *p*) determines the planetary position. In the more complex model, a circle is drawn having the major axis (*MN*) of the ellipse as diameter. A perpendicular (*RQ* or *rq*) is erected on this diameter (*MN*), passing through one of the previously determined points (*P* or *p*) and intersecting the circle (at *Q* or *q*). A new radius vector (*EQ* or *Eq*) is now drawn from the empty focus (E) to this point on the circle; its intersection (*P'* or *p'*) with the ellipse is the "corrected" planetary position. It will be observed that this "correction" shifts the orbital position determined in the simple model away from the nearer apsis (*P* to *P'*, away from aphelion; *p* to *p'*, away from perihelion).

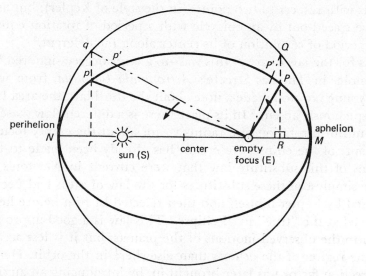

mode of calculation gives a good result, 'it is to go to too much trouble for a false hypothesis' (Delambre, 1821, vol. 2, p. 170). Nicolaus Mercator introduced yet another reckoning system, making use of a point on the axis major determined by the "divine section".[7] In Newton's day, the commonly used systems were the simple equant (centered on the empty focus), originally proposed by Kepler, and the simple equant plus auxiliary circle, associated with Ismaël Bullialdus (or Boulliau) and Seth Ward (astronomer and the Bishop of Salisbury, who later proposed Newton's candidacy for fellowship in the Royal Society). These methods of approximation were easier to apply than the area law, since they used a mean motion and a correction factor, both of which could be easily computed and tabulated.[8]

There is considerable misunderstanding in the literature of the history of science with respect to the area law itself. It must be kept in mind that in practice this law could be used only by the aid of some form of approximation. That is, if we know the area of some given sector of an ellipse reckoned from the solar focus, whose arc along the ellipse corresponds to the motion of a planet during a time T_1 from one position P_0 to another P_1 (see fig. 5.2), we cannot actually use the exact area law to find the position P_2 of the planet at the end of another time T_2. The area law tells us that the point

Figure 5.2. The use of the law of areas to determine a planet's future position depends on solving *quam proxime* a problem in geometry which has no exact solution: given points P_0, P_1 on an ellipse, to determine its point P_2 such that the areas of the related focal sectors (P_0SP_1 and P_1SP_2) are in a stated ratio, namely (by Kepler's area law), that of the times of passage from P_0 to P_1 and from P_1 to P_2. The sun is at a focus S of the ellipse.

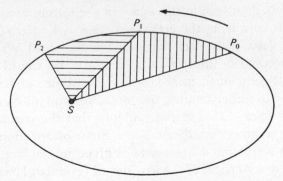

P_2 is located so that the area of the sector SP_1P_2 is to the area of the sector SP_0P_1 as T_2 is to T_1. But it is not possible, as Newton sought to prove in lem. 28, bk. one, of the *Principia*, to find such an area 'by means of equations finite in the number of their terms and their dimensions'; that is, the equation resulting from this problem is transcendental and leads to an infinite series.[9] Kepler himself was aware of the difficulty and appealed to the geometers of the world for help. In sect. 6 of the *Principia* Newton introduced a good approximation to what we know today as "Kepler's problem", or "Kepler's equation". He made use of the defining Cartesian equation of the cycloid, following in the footsteps of Christopher Wren.[10]

The approximations associated with the names of Seth Ward, Bullialdus, and Mercator were specifically intended as substitutes for Kepler's area law. The area law was not, however, the only rule devised by Kepler to account for the varying orbital speed of a planet. We have seen (§3.1) that he also had at one time believed that the speed of a planet is inversely proportional to the planet's distance from the sun. Wren, in fact, begins his essay, 'On solving Kepler's problem by means of a cycloid', by referring to the way that 'Kepler asserted from physical causes that each planet is carried about the sun in an elliptical orbit in such a way that the planet's [transverse] velocity is everywhere proportional to the planet's distance from the sun inversely'. Here, then, was another way of not having to use the law of areas.

As a matter of fact, although Wren does include the area law in his discussion of Kepler's problem, he refers to it only in passing and not in a way that would indicate that it had a true importance as a primary principle of celestial motion.[11] Wren was obviously more interested (at least in this essay) in the purely geometric problem than in advancing the actual practice of Keplero-Copernican astronomy. Hence, the fact that Newton had read Wren's essay[12] does not necessarily imply that, as a result, he would have been actively and consciously aware of the importance of Kepler's area law as a key to understanding the forces that might control planetary motions (see §5.5). Further evidence that the area law was not part of the common intellectual armament of astronomers in the 1670s, even Keplerian astronomers, is given by the fact that in the five hundred or so pages of the writings of Jeremiah Horrox (*Opera posthuma*, 1673) there is no explicit statement of this law.[13] Horrox was, no doubt, Kepler's chief disciple in England and his *Opera* are replete with praise of Kepler, expositions of his ideas, and improvements on his methods; the opening essay is characteristically entitled 'Jeremiae Horroxii; Astronomia Kepleriana, defensa & promota'. Hence, the absence of Kepler's law from Horrox's work is especially significant.

In considering the somewhat insecure position of the first two laws of Kepler in Newton's day, it must also be taken into account that Kepler's own proof of the area law was faulty, and that he had really worked out the elliptical orbit of only one planet (Mars) and "guessed" that the other orbits would also be ellipses. Hence it was an unusual and a very daring step to erect an astronomical system encompassing Kepler's three laws, as Newton did. Following the imaginative leap forward that Newton made, in showing the physical meaning and conditions of mathematical generality or applicability of each of Kepler's laws, this whole set of three laws gained a real status in exact science.[14] From that day onward, treatises on physical astronomy would begin—as the *Principia* does—with the three laws of Kepler, but it was something of a novelty to do so in 1687.[15] This aspect of Newton's work was publicly recognized when the first of the three "books" that make up the *Principia* was received at the Royal Society: it was perceptively and quite correctly described, probably by Halley, as 'a mathematical demonstration of the Copernican hypothesis, as proposed by Kepler'.

5.3   *Newton's early thoughts on orbital motion and Kepler's third law*

Some straightforward algebra shows that once the rule for centrifugal force $(f \propto v^2/r)$ is known for uniform circular motion (where f is the centrifugal force, v the "linear" speed along the circular path, and r the radius of the circle), then the law of force follows almost at once from Kepler's third planetary law.[1] The orbital speed is $v = 2\pi r/T$, where T is the period of revolution, so that $v^2 = 4\pi^2 r^2/T^2$ or $v^2 \propto r^2/T^2$. It requires very little mathematical insight for anyone who is familiar with Kepler's third law to see that with slight modification this last term r^2/T^2 can be easily turned into r^3/T^2 which is constant in the third law. This result would be all the more obvious if the uniform circular motion in question is that of the planets moving in orbits around the sun,[2] or of a system of planetary satellites, or even – by extension or transformation – the motion of the moon around the earth.[3] Hence in formally equivalent developments, $v^2 \propto r^2/T^2$ could become $v^2 \propto (1/r) \cdot (r^3/T^2) \propto 1/r$, since $r^3/T^2 = $ const., or $v^2 \propto r^2/T^2$ could become $v^2 \propto r^2/r^3 \propto 1/r$, since $T^2 \propto r^3$. It follows at once that $f \propto v^2/r \propto 1/r^2$. In a document of the late 1660s or early 1670s, Newton states the centrifugal law for 'different circles' in terms of 'endeavours <to recede> from the centers', which are 'as the diameters divided by the squares of the times of revolution, or as the diameters multiplied by <the squares of> the number of revolutions made in any one given time'. In this document (for which see the supplement to §5.3), Newton also states Kepler's third law for the planets in the form: '. . . the cubes of their distances from the sun are inversely as the squared numbers of their periods in a given time'. That is, if E is the endeavor of receding from the center, D the diameter, and T the period of revolution, Newton says that $E \propto D/T^2$ and that $T^2 \propto D^3$, so that it is at once obvious and elementary that $E \propto 1/D^2$. Small wonder that Newton could not consider it a significant feat of intellect to derive an inverse-square law for circular orbits once the centrifugal rule was known, as it was when published by Huygens, in his *Horologium oscillatorium* in 1673.

In Newton's old age, in late July or early August 1718, he wrote out an autobiographical statement in defense of a scenario in which he would have made his discoveries so early that there could be no doubt as to his priority. This was set down in the draft (superseded and canceled in the MS) of a letter to Des Maizeaux. This preliminary draft, which has been quoted again and again, declares his own procedure in discovery to have been (at least in outline) much like what has just been presented in algebraic terms:

> And the same year [probably 1666] I began to think
> of gravity extending to the orb of the Moon, & having
> found out how to estimate the force with which [a] globe
> revolving within a sphere presses the surface of the sphere:
> from Kepler's rule of the periodical times of the Planets
> being in a sesquialterate [3/2] proportion of their distance
> from the centers of their Orbs, I deduced that the forces
> which keep the Planets in their Orbs must be reciprocally
> as the squares of their distances from the centers about
> which they revolve.[4]

A number of observations may be made on this extraordinary

memorandum of the second decade of the eighteenth century. It contains four parts. First, he thought of gravity (heaviness toward the earth) 'extending' to the moon's orbit. Second, he had discovered 'how to estimate' centrifugal endeavor (not really centrifugal force) in uniform circular motion. Third, he was acquainted with 'Kepler's rule' (the third or harmonic law). Fourth, he had combined the law of centrifugal endeavor with Kepler's rule to find that the planetary 'force' is inversely proportional to the square of the distance from each planet to the center of its revolution,[5] evidently on the assumption that the orbits of the planets are (or may be considered to be) nearly circular.

This whole statement is typical of the scenarios composed by Newton, especially during the years from about 1715 to 1718, when he was deep in controversy about questions of method and of priority in discovery. In this case the first problem that faces the critical historian is that in the 1660s Newton was still thinking of the motion of orbiting planets or their satellites (and of our moon) in terms of a concept derived from his studies of Descartes, a 'conatus recedendi a centro' or an 'endeavor to recede from the center'. It is far from clear how he then conceived that such an endeavor might be balanced or counteracted by gravity, or the downward tendency of bodies toward the earth (or terrestrial weight) extended to the moon's orbit. It was only a decade and a half later that Newton was directed by Hooke (see §5.4) to consider such orbital motion as compounded of an accelerated motion toward the central body and an inertial (linear or tangential) motion. It was then, and only then, that the centrally directed force could meaningfully be considered as the centripetal force of terrestrial gravity extended to the moon. Further evidence that Newton was far indeed from his later concept of a simple centripetal force continually deflecting planets and moons from their inertial (rectilinear) paths into curved orbits was found by D. T. Whiteside. Whiteside (1964*b*, 1976) discovered that at this time, Newton was still thinking of planetary and lunar orbital motion in relation to Cartesian vortices, a concept that he sanctioned long after 1666. In other words, we must read Newton's statement about 'the forces which keep the Planets in their Orbs' with more than a single grain of salt.

Comparing Newton's statement of his alleged analysis of planetary motion of 1666 with the system written out in the *Principia* two decades later, we observe that in the mature work the planetary orbits begin as circles (prop. 4, bk. one) but rapidly become ellipses (prop. 11, bk. one). Further, by 1687 the centrifugal force has become a "centripetal force"—a "vis centripeta", so named by Newton as the contrary of centrifugal force, a name he retained in honor of Christiaan Huygens. What is of notable importance is that there is independent evidence that in the late 1660s or early 1670s Newton had used Kepler's third law and the equivalent of Huygens's v^2/r measure to show that the "centrifugal" endeavors of the planets are proportional to the inverse square of their distances, as he says in the 1718 Des Maizeaux memorandum.

With regard to the second point, that (after January 1665, at approximately the time he received his A.B.) Newton had found the measure of centrifugal endeavor, we are on surer ground. An outline of a demonstration, based upon his researches of twenty years earlier, was sent by him to Halley in July 1686 to be added at the end of the scholium following prop. 4, bk. one, of the *Principia*; where it has ever since been available in print for all to read. The cause of this addition was a letter from Halley, saying that Hooke wanted Newton to give him credit for having suggested the inverse-square law; Newton sent on this new paragraph to show how he could have found this law (for circular motion, at least) even before the publication of Huygens's *Horologium oscillatorium* of 1673. In the *Waste Book* Newton wrote out (early in 1665 or shortly afterwards) a first form of this demonstration, which has been published and analyzed by John Herivel.[6] We know from an entry in one of Newton's student notebooks that he learned of Kepler's third law from reading Thomas Streete's *Astronomia Carolina* (1661) (U.L.C. MS Add. 3996, fol. 29^r; cf. Whiteside, 1964a, p. 124).

In the autobiographical statement, furthermore, Newton says nothing whatsoever about the possibility that the forces which he alleged he had been considering in the mid-1660s might act mutually, that the moon might pull on the earth with a force equal to the alleged force of gravity extended out to the moon, that the planets might pull on the sun. And although Newton mentions here both 'gravity extending to the orb of the Moon' and 'the forces which keep the Planets in their Orbs', he does not discuss the grounds for supposing that these two "forces" might even be of the same kind (or "species").

In Newton's memorandum and in his early manuscript of the

1660s, there is as yet no explicit suggestion that there is a solar "force" exerted on the planets that is identical to the terrestrial force exerted on the moon (which he alleges that he supposed to be ordinary terrestrial gravity). The "endeavours to recede", in which he believed in the 1660s, are very different in concept from the simple centripetal forces continually deflecting planets and moons from their inertial (rectilinear) paths in which he came to believe some time later, in 1679–1680 or thereafter. Accordingly, there is no legitimate ground for saying that Newton had "known" of the law (or a law) of *universal gravitation* in 1665 and had "delayed" announcing it for twenty years. In fact, far from believing in a force that is "universal", Newton did not then even have an awareness of the possibility of a planetary action on the sun, a lunar action on the earth, much less an action of one planet on another.[7]

In the memorandum quoted above, Newton says that he 'began to think of gravity extending' to the moon. Presumably, he would have guessed–by analogy–that if the planetary "endeavours" to recede from the sun (in his mind they were not at that time "forces") vary inversely as the square of the distance, then it must be the same for the moon.[8] Hence, if gravity extends to the moon, and varies inversely as the square of the distance, the intensity of gravity at the moon's orbit should be $1/r^2$ times the intensity here on earth,[9] r being the earth's radius.

According to Newton's recall and reconstruction some fifty years or more after the event, the correctness of his reasoning in 1666 could be determined by an easy test. The memorandum continues: '[I] thereby compared the force requisite to keep the Moon in her Orb with the force of gravity at the surface of the earth, and found them answer pretty nearly'. From the age of Newton to our own day, there has been a considerable discussion of this alleged early "moon test". Before getting into the question of the numerical results, however, it is important to keep in mind that what would have been compared to 'the force of gravity at the surface of earth' was definitely not a 'force requisite to keep the Moon in her Orb' in the sense of a simple centripetal force acting on the moon and continually drawing the moon toward the earth away from a linear path. In the literature concerning Newton, stress has been given to the poor agreement between theory and the results of observation: not 'pretty nearly', as Newton would have us believe in 1718, but, rather, not quite good enough. Pemberton, who knew Newton well, said that the lack of agreement was caused by Newton's having chosen a poor value for the radius of the earth (Pemberton, 1728, preface; see supplement to §5.3). In the nineteenth century Glaisher publicized a suggestion by J. C. Adams that there was an additional problem, that Newton could then only guess that the earth (or any sphere that is either homogeneous or composed of a series of homogeneous concretric circular shells) acts gravitationally on an external body like an apple or a moon as if all of its mass were concentrated at its geometric center.[10] This latter factor was aggrandized by Florian Cajori (1928) into a cause for Newton's 'twenty years delay in announcing the law of universal gravitation'. There seems to be no doubt that Newton did not prove the theorem concerning the attraction of spheres until some twenty years later, but it is certainly wrong to speak of a 'delay' since Newton obviously was not as yet thinking in terms of a universal gravitating force, or even a centripetal force.

One additional bit of evidence that in 1666 Newton was far indeed from his later concept of universal gravitation is provided by William Whiston, Newton's successor as Lucasian professor at Cambridge. Whiston wrote about 'the Occasion of Sir *Isaac Newton's* . . . discovering his amazing Theory of Gravity'. He had heard Newton himself relate how he had made his discovery; the occasion, according to Whiston, was 'long ago, soon after my first Acquaintance with him, which was 1694'. Whiston's account (1749, pp. 35–38) reads:

> It was this. An Inclination came into Sir *Isaac*'s Mind to try, whether the same Power did not keep the Moon in her Orbit, notwithstanding her projectile Velocity, which he knew always tended to go along a strait Line the Tangent of that Orbit, which makes Stones and all heavy Bodies with us fall downward, and which we call *Gravity*? Taking this Postulatum, which had been thought of before, that such Power might decrease, in a duplicate Proportion of the Distances from the Earth's Center. Upon Sir *Isaac*'s first Trial, when he took a Degree of a great Circle on the Earth's Surface, whence a Degree at the Distance of the Moon was to be determined also, to be 60 measured Miles only, according to the gross Measures then in Use. He was, in some Degree, disappointed, and the

Power that restrained the Moon in her Orbit, measured
by the versed Sines of that Orbit, appeared not to be quite
the same that was to be expected, had it been the Power of
Gravity alone, by which the Moon was there influenced.
Upon this Disappointment, which made Sir *Isaac* suspect
that this Power was partly that of Gravity, and Partly that
of *Cartesius*'s Vortices, he threw aside the Paper of his
Calculation, and went to other Studies.

Pemberton adds some further details:

In this computation, being absent from books, he took
the common estimate in use among geographers and our
seamen, before *Norwood* had measured the earth, that 60
English miles were contained in one degree of latitude
on the surface of the earth. But as this is a very faulty supposition, each degree containing about $69\frac{1}{2}$ of our miles,
his computation did not answer expectation; whence he
concluded, that some other cause must at least join with
the action of the power of gravity on the moon. On this account he laid aside for that time any farther thoughts upon
this matter

Pemberton's 'some other cause' is not so specific as Whiston's
'*Cartesius*'s Vortices'. A few years ago, D. T. Whiteside found that
Newton at this time was indeed using Cartesian vortices in his
theory of the moon's motions.[11] If Newton gave any grounds for
the existence of lunar or planetary vortices, he obviously had not
yet reached the concept of universal gravitation.

The autobiographical memorandum discussed above would indicate that by 1666 Newton had found the Huygenian rule for
centrifugal force and had combined it with Kepler's third law to
get the inverse-square law of force. The proof of the Huygenian
rule, as given in the *Waste Book*, is close to a dated entry of 'Jan.
20th 1664' [O.S.], which would be 20 January 1665 in the New
Style; this may have been the reason why Newton later assigned
the date of 1665 to the processes of reasoning described in the
memorandum. But in that *Waste Book* entry, there is no advance
from the Huygenian rule–via Kepler's third law–to the law of the
inverse square. Nor is anything of that sort to be found elsewhere
in the *Waste Book*.[12]

But in an untitled essay, first printed by Rupert Hall in 1957,
Newton does attain the law of the inverse square for planets from

a combination of Kepler's third law and the rule for computing centrifugal endeavor. In this essay Newton writes of the planets' 'endeavours to recede from the sun', not of centrifugal (much less centripetal) "forces": '... in the primary planets, since the cubes of their distances from the Sun are reciprocally [or inversely] as the squared numbers of their revolutions in a given time, their endeavours to recede from the Sun will be reciprocally as the squares of their distances from the Sun'.[13] This paper, which may be dated at about 1667 or 1668 (see supplement to §5.3), does not include the moon test, but it does contain a computation of the magnitude of the moon's 'endeavour to recede' in relation to terrestrial gravity (see supplement to §5.3). Herivel (1965a, pp. 68sq) has analyzed the data in this document and others of about the same period to show that most probably 'the actual figure arrived at by Newton for the "fall" of the moon in her orbit was approximately 13.2 feet in one minute. This may be compared with the true figure of approximately 16 feet in one minute, corresponding to a fall of the same distance in one *second* at the surface of the Earth, a figure derived by Newton [in another manuscript]'. This discrepancy of some 17 percent accords with Whiston's report that Newton was 'in some degree disappointed' and Pemberton's remark that Newton's 'computation did not answer expectation'.

In the 1667/8 document just quoted, Newton (as has been mentioned) does not conceive of a planet or the moon falling under the action of a centripetal force directed respectively to the sun or to the earth. For the planets it is the 'conatus a sole recedendi' which 'reciproce erunt ut quadrata distantiarum a sole'; and for the moon it is the 'conatus Lunae recedendi a centro terrae'. Hence the 13.2 feet represent not so much a distance through which the moon will have fallen in one minute as 'the distance outward through which the Moon *would* have moved in one minute under the action of this *conatus*'.[14]

Newton's concept of *conatus*, or endeavor, is a transformation of the Cartesian concept of the same name, which he had encountered in reading Descartes's discussion of the motion of a stone being whirled in a sling.[15] The Cartesian 'conatus a centro' eventually became the Huygenian 'vis centrifuga', which Newton also used. Hence, in the case of the moon's motion, what he was comparing in 1667 or thereabouts is the '[conatus] Lunae et superficiei terrestris recedendi a centro terrae', the centrifugal endeavor of receding from the center of the earth calculated for the moon and for any object at or near the earth's surface. It was only much later that Newton abandoned the notion of centrifugal force and transformed it into a centripetal force, saying expressly that he had done so in honor of Huygens: 'Mr Huygens gave the name of vis centrifuga to the force by which revolving bodies recede from the centre of their motion. Mr Newton in honour of that author retained the name & called the contrary force vis centripeta.'[16]

It is an open question as to when Newton wholly abandoned the concept of centrifugal force. Did he (and for how long did he) hold to the view of Borelli that the stability of the moon in its orbit arose from a balance between two forces or tendencies: a centripetal and a centrifugal?[17] In the *Principia* he still on occasion wrote of centrifugal force,[18] but was well aware that what appears to be the action of a centrifugal force is only the result of an inertial motion along the tangent which thus tends to carry the body away from the center. No doubt, the use of the Huygenian 'vis centrifuga' and of the Cartesian 'conatus recedendi a centro' are only examples of the persistence of forceful phrases that remain long after the concepts for which they originally stood have been rejected. They stand as signs, to critically alert scholars, of the sources of ideas that have been replaced by being transformed, but whose persistent presence in a wholly new signification shows an element of continuity within change, a terminological invariance in a transformation. As we shall see below (§5.4), it was no doubt the influence of Hooke that shook Newton free from thinking in terms of centrifugal forces and set him on the path of centripetency to universal gravitation.

It is important to keep in mind in this connection that the concept of centrifugal force or "endeavor" inhibits the leap to universal gravitation. For the latter is a generalization of the apparent attraction that arises in special cases in which one body has a curved motion about another, whereas centrifugal force implies a repulsion, a tendency to move away from, rather than towards, the center of motion. Furthermore, the link between the centripetal force pulling the moon down toward the earth and the force of weight or gravity pulling nonorbiting bodies (such as falling stones or apples, projectiles, etc.) down toward the earth is established by analogy or similitude; whereas for centrifugal force, there is rather a suggestion of contrast or dissimilitude. That is, assuming the earth to be at rest, stones falling down would exhibit an attraction *toward* the earth's center, whereas the moon staying in its orbit would exhibit a force *away* from the earth's center. And what could the motion of the moon under the action of a centrifugal force have to do with the mutual attraction toward one another of two heavy bodies at rest? Here are additional grounds for concluding how far Newton was from the concept of universal gravitation in the 1660s and even in the 1670s.

Supplement to 5.3. An early computation of the moon's 'endeavour of receding' and a planetary inverse-square law
In an untitled MS document of the mid- or late 1660s (U.L.C. MS Add. 3958, sect. 5, fol. 87),[1] Newton computes the 'endeavour of receding from the center' ('conatus a centro' or 'conatus recedendi a centro') by determining how far out along a tangent a body would move in some given time if it had that same 'endeavour' in a linear direction along the tangent and there were no impediment. In short, Newton is measuring the 'endeavour of receding' (not yet "centrifugal force") by the acceleration, and the acceleration by the distance through which a body would move, according to Galileo's rule for uniform acceleration, freely along a straight line in a given time, 'the time of one revolution'. Newton then computes how far a body would descend if its 'endeavour of approaching toward the centre in virtue of its gravity' ('conatus accedendi ad centrum virtute gravitatis') were equal in magnitude to its 'endeavour of receding from the center' at the equator, as a result of the earth's daily rotation. In a period of one day, this would be 19 3/4 earth-radii, which Newton reckons at 69,087 miles, corresponding to 120 miles in an hour, 1/30 mile or 100/3 paces ('that is, 500/3 feet') in a minute, and hence 5/108 foot or 5/9 inch in a second.[2] But at the earth's surface the 'force of gravity' ('vis gravitatis') is of such a magnitude that it will drive a heavy body downward through about 16 feet in one second, which 'is about 350 times further in the same time than the endeavour from the center'.

Several observations may be made on the document as presented thus far. First, Newton assumes (as he will later on in his correspondence with Hooke) that even for so long a period of time the downward acceleration of gravity will be constant. Second, in this computation there is no question of balance of "centrifugal" and "centripetal" forces, but merely a comparison of the distances, so as to compare the effect of gravity and the 'endeavour of receding from the center' at the equator as a result of the earth's daily rotation. Newton does observe that the 'force of gravity is so many times greater that the earth in rotating may not cause bodies to recede and spring up into the air'.

As he makes this computation, he assumes that 19 3/4 (actually 19.7392) earth-radii correspond to 69,087 miles, or that the earth's equatorial radius has the magnitude of 69,087/19.7392 miles or 3499.9 miles. He reduces one revolution to hours, then to minutes, and then to seconds (see n. 2) by dividing by 24^2, 60^2, 60^2. The "mile" is reduced to paces (1000 paces = 1 mile) and feet (5 feet = 1 pace), on the basis of the "Italian mile" (5000 feet = 1 Italian mile) rather than the English statute mile of 5,280 feet.

In a 'corollary', Newton sets forth the general rule that 'in different circles the endeavours from the centers are as the diameters divided by the squares of the times of revolution, or as the diameters multiplied by the squares of the number of revolutions made in any given time'. He applies this D/T^2 rule to compare the moon's orbital 'endeavour to recede from the center' from the center of the earth with the 'endeavour to recede' at the earth's surface on the equator. The latter turns out to be 'about 12 1/2 times greater'. Hence, he concludes, 'the force of gravity is 4000 and more times greater than the endeavour of the moon to recede from the center of the earth'. In fact, the force of gravity, from Newton's computations, turns out to be 4375 (= 350 × 12 1/2) times greater than the moon's 'endeavour of receding'. It may be noted that if Newton had taken 4000 miles of 5280 feet/mile for the equatorial radius, the result would have been 3584 rather than 4375, very close to the "theoretical" value of 3600, that is, the value that might have been expected if Newton had assumed that the ratio of the moon's 'endeavour of receding' to the surface force of the earth's gravity is the inverse of the ratio of the square of the distance from the moon to the earth's center (60 earth-radii) to the square of the earth's radius. Of course, Newton *might* have made such a comparison in his mind without writing it down, which would have shown that the computed result of 4375 was off by 21.5 percent from the theoretical value of 3600, as he later alleged that he had done (in writing about these events, in 1718 or so, in the Des Maizeaux memorandum); but there is *no documentary evidence* that he had done so.

In the manuscript in which the foregoing calculations are described, Newton does derive an inverse-square law *for the planets*, by combining Kepler's third law for 'the primary planets' with their 'endeavours of receding' from the sun (in the manner described in the opening paragraph of §5.3). This occurs in some brief or summary paragraphs that are added onto the rather detailed discussion and exposition of the moon's 'endeavour of receding' and the 'endeavour of receding' at the earth's surface. Newton neither says expressly nor in any way implies in this document either that the earth's gravity may extend as far out as the moon's orbit or that the moon's 'endeavour of receding' is in accord with an inverse-square law of the distance. The only application that he makes of his calculations concerning the moon's orbital 'endeavour to recede' is an attempt to account for the fact that the moon always 'turns the same face toward the earth'.[3]

Later on, after Newton had learned how to analyze orbital motion in terms of the action of a centripetal force on a body with an initial component of inertial motion, and had written the *Principia*, he interpreted his early calculations *as if* they were essentially the "moon test" described in the scholium to prop. 4, bk. three, of the *Principia*. In the account that he gave of his supposed test of the inverse-square law, as reported by Whiston and by Pemberton, Newton, as we have seen, referred to the failure of the test, the poor agreement. He was 'disappointed' (according to Whiston) that 'the Power that restrained the Moon in her Orbit' was not 'the Power of Gravity alone', that (according to Pemberton) the 'computation did not answer expectation'.[4] But in the memorandum written in about 1718 for Des Maizeaux, Newton wanted it to be believed that the computed result of the mid-1660s agreed with the theoretical value, as he said, 'pretty nearly'. This phrase is Newton's English equivalent of the Latin word "quamproxime" which he used later on in *De motu* to characterize the results of the true moon test, and which is used frequently in the *Principia* for results that are not quite exact but very nearly so. Hence there is a striking contrast between Newton's oral reports to Whiston and Pemberton and his allegation in the Des Maizeaux memorandum.

In a letter to Halley (20 June 1686; Newton, 1959–1977, vol. 2, p. 436), when Newton was defending his independent claim to the inverse-square law, and insisting that he was not indebted to Hooke for having suggested it to him, he referred to 'one of my papers' in which 'the proportion of our gravity to the Moon's *conatus recedendi a centro Terrae* is calculated though not accurately enough'. In the document under analysis here, it is just this proportion that Newton has calculated and we would agree about the lack of sufficient exactness in the result.[5]

It is, in any case, to be noted that if Newton had made a proper moon test to see if the earth's gravity, diminishing according to the square of the distance, could account for the observed motion of the moon, he probably would have been surprised and puzzled if the results had provided an exact confirmation. For it was obvious to Newton that the moon does not move in a circular orbit and that the motion of the moon is more characterized by irregularities than by regularities. Furthermore, as D. T. Whiteside (1964*a*, esp. nn. 13, 54) has indicated, Newton in the 1660s believed firmly in the real existence of a separate centrifugal force or endeavor, which in the noncircular case could not be quantitatively equal to the centripetal inverse-square force. Hence a near agreement between the computations and theory would at that time have made him wonder why the moon's "vis centrifuga" should turn out to be equal and opposite to the theoretical inverse-square "vis centripeta".

5.4   *Newton and dynamical astronomy in the years before*
        1684: correspondence with Hooke in 1679–1680

Newton's interest in astronomy was a long-standing one, easily traceable to at least his student days at Cambridge in the 1660s. He read or studied such works as Streete's *Astronomia Carolina* and Wing's *Astronomia Britannica* (Whiteside, 1964*a*, esp. nn. 13, 54). Newton's copy of the latter work (London, 1669) survives, and it contains annotations in his hand concerning Huygens's observations of a moon of Saturn. On the end-papers one may still read Newton's hand-written calculations based upon Wing's tables.[1] There is adequate manuscript testimony in calculations and observations to document his continuing interest as a young scientist in astronomical problems and celestial phenomena. In 1672 Newton came publicly before the world of science with the publication (in the *Philosophical Transactions* of the Royal Society) of two communications: one was his analysis of sunlight, with his 'New Theory about Light and Colours', and the other the announcement of his newly invented 'Catadioptrical Telescope'.[2] In 1676, while he was still in his early thirties, his first published contribution to astronomy appeared. It was an explanation of the librations of the moon, published in Nicolaus Mercator's *Institutionum astronomicarum libri duo* (London, 1676, pp. 286sq), presented in a summary by Mercator of a letter from Newton, now lost. Newton's own mature presentation of the libration appears in prop. 17, bk. three, of the *Principia*.

Newton's attention was forcibly drawn to astronomical problems three years later, in 1679, when Robert Hooke (recently appointed secretary of the Royal Society) wrote to Newton, expressing the hope that Newton would renew his former 'philosophicall' exchanges with the Society. To start things off, Hooke invited Newton to comment on a 'hypothesis or opinion of mine . . . of compounding the celestiall motions of the planetts [out] of a direct motion by the tangent & an attractive motion towards the centrall body'.[3] In his reply, Newton declined to discuss Hooke's 'hypothesis'. Instead, he advanced a 'fansy' of his own: the effects of 'the Earth's diurnal motion' on the path of freely falling bodies (Newton to Hooke, 28 November 1679; Newton, 1959–1977, vol. 2, p. 301). Newton had erred, however, as Hooke was quick to discover. Newton had proposed a spiral path, the nature of which is not entirely clear from the diagrams usually accompanying the publication of Newton's letter to Hooke (28 November 1679); this matter was rectified only in 1967 by J. A. Lohne's study of 'The Increasing Corruption of Newton's Diagrams'. Hooke corrected Newton; he showed that the curve would not be (as Newton thought) 'a kind of spirall',[4] which after a few revolutions would bring the falling object to 'the Centre of the Earth'. The curve would be 'rather a kind [of] Elleptueid'. On 13 December Newton wrote to Hooke, 'I agree with you that . . . [the body] will not descend in a spiral to the very center but [will] circulate with an alternate ascent & descent . . .' (Newton, 1959–1977, vol. 2, p. 307). The importance of this problem is that finding the path of a body falling freely on a rotating earth is mathematically and physically equivalent to finding a planetary orbit. For the falling body will start out just as if it had been given a push, or had a component of inertial (tangential) motion, and then was continually attracted to a center (in this case, the earth's center).

Once Newton had acknowledged Hooke's correction, Hooke was emboldened to write to Newton of 'my supposition' concerning the attractive force that keeps the planets in their orbits; it was

'. . . that the Attraction always is in a duplicate proportion to the Distance from the Center Reciprocall, and Consequently that the Velocity will be in a sub-duplicate proportion to the Attraction and Consequently as Kepler Supposes Reciprocall to the Distance' (Hooke to Newton, 6 January 1679/80; Newton, 1959–1977, vol. 2, p. 309). Newton did not comment directly on this statement. His eventual opinion may be readily gathered from the fact that he proved that the velocity of a body (such as a planet or planetary satellite) moving in an elliptical orbit under the action of an inverse-square force is *not* 'as Kepler [and as Hooke] Supposes Reciprocall to the Distance', but is rather reciprocally as (or in inverse proportion to) the perpendicular distance from the center of force to the tangent to the planetary orbit.

Newton's result (as we have seen in §3.1) shows that this speed law of Kepler's may be used without sensible error only in the regions of the apsides. In the immediate neighborhood of the aphelion and perihelion, the direct distance from the sun to a planet is very nearly equal to the perpendicular distance from the sun to a line drawn through the planet's position, tangent to the orbit. Hooke would thus appear to have been unaware of the error in assuming a validity over the whole orbit of a rule that could properly be used only in two very limited regions near aphelion and perihelion. And evidently Hooke was wholly ignorant of the analytical inconsistency between the area law and this supposed speed law.[5]

When Hooke proposed to Newton his 'supposition' of an attraction varying as the inverse square of the distance, with the alleged consequence of a velocity inversely proportional to the distance, he stressed the importance of the problem, noting '. . . the finding out the proprietys of a Curve made by two such principles will be of great Concerne to Mankind, because the Invention of the Longitude by the Heavens is a necessary Consequence of it: for the composition of two such motions I conceive will make out that of the moon' (Hooke to Newton, 6 January 1679/80; Newton, 1959–1977, vol. 2, p. 309). Hooke read this letter (addressed to Newton) at a meeting of the Royal Society. A few days later Hooke wrote to Newton again, with a direct challenge that was to be of the utmost importance:

> It now remaines to know the proprietys of a curve Line
> (not circular nor concentricall) made by a centrall attrac-

tive power which makes the velocitys of Descent from the tangent Line or equall straight motion at all Distances in a Duplicate proportion to the Distances Reciprocally taken [i.e., inversely as the square of the distances]. I doubt not but that by your excellent method you will easily find out what that Curve must be, and its proprietys, and suggest a physicall Reason of this proportion [Hooke to Newton, 15 January 1679/80; Newton, 1959–1977, vol. 2, p. 313].

Since this statement by Hooke has often been cited by historians (and was used by Hooke himself) as evidence that Hooke discovered the inverse-square law of gravitation, it is of more than a little interest that Newton should have given so much prominence in the *Principia* to a proof that Hooke's velocity law is generally false. I have mentioned the fact that in all three editions of the *Principia* (1687, 1713, 1726), this result appears in prop. 16, bk. one.[6] Hooke made his claim to priority with respect to the inverse-square law after the MS of Newton's bk. one had been received by the Royal Society.[7] To shore up his own claim to original discovery, Newton sent Halley the new scholium to prop. 4, in which he gave his youthful demonstration of the law of acceleration (v^2/r) in uniform circular motion; presumably simple algebra would lead from v^2/r to the law of the inverse square – at least for circular orbits.[8] It would have been too late to alter the structure of the early propositions, or perhaps Newton did not then see how easy it would have been to confute Hooke's statement before prop. 16. But after the *Principia* had been published, he decided to transfer the corollaries to prop. 1 from their place in the original manuscript and in the first edition to a new location; they became corollaries to prop. 2, which had had no corollaries until this time. This enabled him to insert a wholly new set of corollaries to prop. 1, of which the first one was a statement of the law of velocity as Hooke should have written it:

> Corol. 1. The velocity of a body attracted towards an immoveable centre, in spaces void of resistance, is reciprocally as the perpendicular let fall from that centre on the right line that touches [i.e., is tangent to] the orbit [see Newton, 1972, vol. 1, pp. 90–93].

Here, in effect, was a crushing reply to Hooke's claims, placed prominently as corol. 1 to prop. 1, bk. one. If this interpretation is correct, Newton was implying that Hooke could not have a valid claim to the discovery of a law that he did not fully understand!

Newton did not ever reply directly to Hooke's challenge in 1680, as to finding planetary orbits by compounding an inertial rectilinear motion with a continual descent according to an inverse-square force. Some time later, he wrote briefly to Hooke on another matter, mentioning some experiments but not referring to the problem of celestial mechanics (Newton to Hooke, 3 December 1680; Newton, 1959–1977, vol. 2, p. 314).

Newton's reluctance to grapple directly and publicly with the problem set by Hooke is readily understandable. It was not in his character to develop major ideas in correspondence with others, and Hooke would have been a most unlikely candidate for receiving Newton's unpublished nascent ideas. Perhaps Newton was not at first convinced of the primary significance of Hooke's problem as the key to our understanding of the functioning of the solar system. Nor was he apparently moved by Hooke's assertion (in his letter to Newton of 6 January 1680) of the 'great Concerne to Mankind' in solving this problem. Furthermore, before Newton would be able to complete his solution of the problems of planetary motion in elliptical orbits or of the motion of planetary satellites, there were to be certain major conceptual and mathematical hurdles to be got over.[9] But even though Newton did not at once proceed to the law of universal gravitation, nor find the solution to the two-body problem (much less a three-body problem), he apparently did prove to his own satisfaction that under the action of a central force varying as the inverse square of the distance, a body moving with an initial component of inertial motion will have an elliptical orbit. Newton later admitted that this first major step toward his celestial mechanics based upon the law of universal gravitation was taken as a result of Hooke's stimulus in the years 1679–1680.[10]

The period during which Newton was corresponding with Hooke (24 November 1679–3 December 1680) was notable for the appearance of the great comet of 1680, which was visible from November 1680 through early March 1681. This comet attracted Newton's attention and he made observations of its changing position and speed and studied its apparent and true path (see Ruffner,

1966). Newton was in communication about its position and path with Flamsteed, at the Royal Observatory at Greenwich; this exchange marks the first contact between Newton and the newly founded observatory. Newton received information concerning this comet from Flamsteed through an intermediary, James Crompton.[11] In one of his letters in reply, addressed to Crompton for Flamsteed (February 1681), Newton rejected Flamsteed's view that comets follow curved orbits, being turned around by the sun's magnetic force. Although he said that he could 'easily allow an attractive power in the Sun whereby the Planets are kept in their courses about him from going away in tangent lines', he was 'lesse inclined to believe [as Flamsteed did] this attraction to be of a magnetick nature'; his chief objection to the theory of magnetic solar attraction was that the sun is 'a vehemently hot body & magnetick bodies when made red hot lose their vertue'. More significantly, Newton still seems to have believed in the then-current Keplerian view of the rectilinear paths of comets;[12] in this case, he was 'suspicious that the Comets of November & December which Mr. Flamsteed accounts one & the same Comet were two different ones', presumably appearing as a pair moving in opposite directions along nearly parallel paths (Newton, 1959–1977, vol. 2, p. 342). Newton had developed the mathematical theory of such rectilinear cometary motion in his university lectures on algebra for 1676, and he did so again in 1680, making use of a modification of Kepler's hypothesis that had been introduced by Wren.[13]

We do not know the exact date of Newton's conversion to the view that comets are a sort of planet, moving in curved orbits around the sun. This latter view does occur specifically in his tract *De motu*, which he wrote out after Halley's visit (presumably in August 1684) to ask him about the possibility that an inverse-square solar force could produce or maintain elliptical planetary orbits (for details see Cohen, 1971). Yet, as Newton wrote to Flamsteed on 19 September 1685, he had 'not yet computed[!] the orbit of a comet' and was about to do so; he noted that 'it seems very probable that those of November & December were the same comet' (Newton, 1959–, vol. 2, p. 419). Considering the dates, this apparently refers to what Whiteside has aptly called Newton's 'clumsy *ad hoc* graphical-cum-arithmetical technique', which was written for the first version of bk. three of the *Principia* (published posthumously as the *De mundi systemate liber*) (see Cohen, 1971, suppl. 6).

In a letter presumed to have been written in April 1681, Newton said: 'About the Comets path I have not yet made any computation though I think I have a direct method of doing it, whatever the line of its motion be'. And, in another letter, on 16 April 1681, he told Flamsteed that of all the difficulties in the curved orbit theory, 'this sways most with me: that to make the Comets of November & December but one is to make that one paradoxical. Did it go in such a bent line, other comets would do the like & yet no such thing was ever observed in them but rather the contrary' (Newton, 1959–1977, vol. 2, pp. 362, 364). Of course, it must be kept in mind that in the portion of the orbit where the comet turns around, the comet is close to the sun and not simply visible, and that the easily visible path is not very curved as compared, say, to a planetary orbit (even that of Saturn).

The theory of comets was a topic of major astronomical importance for Newton, especially during the 1680s, as may be seen in the mass of observational data and computations he amassed concerning comets he actually saw and those reported in the literature (Ruffner, 1966). In all editions of the *Principia*, the theory of comets (not just their motions, but their appearance and composition) occurs as the final scientific subject of the third and last book on the System of the World, accounting for between one third and one half of that book (37 out of 110 pages in the first edition, 59 out of 145 pages in the third edition). Eventually, comets came to be of special importance in exhibiting the action of solar gravitational force to great distances, beyond the limits of the visible solar system. Since he still believed comets to move in straight lines as late as 1681 and possibly a little later, he could hardly as yet have found a law of "universal" gravitation. But in the tract *De motu*, he simply assumed (in 1684) without discussion that some comets return to our part of the solar system and he indicated a way to 'know whether the same comet returns time and again' ('cognoscere an idem cometa ad nos saepius redeat'). Newton also recorded his observations of another comet, in August 1682; and in 1683 he received information about this comet from Arthur Storer, in Maryland – one of Newton's two American informants, the other being Thomas Brattle of Cambridge, Massachusetts.[14] His concern for several sorts of astronomical problems may be readily documented (by correspondence and other manuscripts) for the years 1679, 1680, and 1681.

5.5   *The Newtonian discovery of the dynamical significance of Kepler's area law: the concept of force*

In various unpublished documents, Newton admitted that in 1679–1680 Hooke had provided the occasion for his study of planetary dynamics, although he would not admit that Hooke had made any substantive contribution to his thinking. In one such statement, Newton describes Hooke's experiments on falling bodies and Hooke's comment that such bodies 'would not fall down to the center of the earth but rise up again & describe an Oval as the Planets do in their Orbs'. Newton continues:

> Whereupon I computed what would be the Orb described by the Planets. For I had found before by the sesquialterate proportion of the tempora periodica of the Planets to their distances from the Sun, that the forces which kept them in their Orbs about the Sun were as the squares of their mean distances from the Sun reciprocally: & I found now that whatsoever was the law of the forces which kept the Planets in their Orbs, the areas described by a Radius drawn from them to the Sun would be proportional to the times in which they were described. And [by the help of] these two Propositions I found that their Orbs would be such Ellipses as Kepler had described.[1]

In another memorandum, the one quoted earlier in §5.3, Newton did not mention Hooke and also placed the dates too early:

> At length [*changed from* And some year(s later)] in the winter between the years 1676 [*changed from* 1666] & 1677 I found the Proposition that by a centrifugal force reciprocally as the square of the distance a Planet must revolve in an Ellipsis about the center of the force placed in the lower umbilicus of the Ellipsis & with a radius drawn to that center describe areas proportional to the times [U.L.C. MS Add. 3968, sect. 41, fol. 85; see Cohen, 1971, p. 291].

Of course, this is bogus history, created by Newton in about 1718. If we replace the expression "centrifugal force" by "centripetal force", however, then we have a pretty accurate description of what Newton has done in the tract *De motu* (1684) and in the *Principia*, rather than in 1676–1677 as he alleged. There is no evidence (nor even a hint of any) that Newton might have achieved this result before his correspondence with Hooke in 1679–1680. If he then worked out the relation between the inverse-square law and elliptical orbits, as it is generally supposed today that he did, he might possibly have done so in terms of "centrifugal force" rather than "centripetal force"; but in the light of Hooke's actual suggestions and the text of *De motu*, it seems more likely that by this time he would have conceived of "centripetal force" and have used this concept in the demonstration. It is, accordingly, all the more interesting that in this last extract Newton has used the expression "centrifugal force", the concept he had been using in the 1660s, and not "centripetal force", the concept Newton actually uses in all the documents in which he shows that an elliptical orbit is produced by an inverse-square force. This was a rather obvious slip, possibly an unconscious transition from the expression in the previous sentence: 'What Mr Hugens has published since about centrifugal forces I suppose he had before me'.[2] In fact, in this particular sentence Newton had originally actually written "centripetal", thus applying to Huygens a Newtonian concept and a Newtonian name that nowhere appear in Huygens's presentation of this topic. We have seen that both the Huygenian concept and name had been transformed by Newton from "centrifugal" to "centripetal".[3] When did this transformation occur? Was Hooke's primary contribution to suggest to Newton that planetary motions should be compounded of a linear inertial component and the effects of a force directed *toward* the sun? It would seem so. This would be a far more important contribution to Newton's thought than a guess that the solar force varies inversely as the square of the distance; especially since, as mentioned above (§5.3), it would be fairly obvious after Huygens published the rule for centrifugal force in 1673 that such a force must vary as the inverse square for circular orbits. It would not be too much of a guess that the same force might somehow produce elliptical orbits too.

Science does not advance by guesses, but by demonstrations, although good guesses may lead to sound demonstrations. This was not so for Hooke. He did not have the double advantage of mathematics in disciplining his creative imagination and in giving it the powers necessary to enodate the tangle of Kepler's laws (see §3.1). In any event, as we shall see below, the discovery of the dynamical significance of each of Kepler's laws, and their transformation from kinematic or descriptive rules into dynamical or causal principles, was but a bare first step toward universal gravitation, although one of great significance.

If the shift (or transformation) from the concept of centrifugal to centripetal force was one basic element in Newton's development in 1679–1680,[4] another was the transformation resulting in the recognition of the dynamical significance of the area law. I have mentioned earlier that in Newton's day many astronomical treatises and manuals presented Keplerian astronomy without this law, using in its place one or another form of a substitute based on a mean motion with respect to the empty focus plus (most often) a correction factor. Newton, of course, had come upon the area law before 1679–1680. The law appears in Nicolaus Mercator's book of 1676, in which we have seen that there is a lengthy summary of Newton's own contribution to the analysis of the librations of the moon. But it is to be noted that in that book, once the law has been stated, it is then passed over in favor of Mercator's own form of substitute for it. Newton also had read Christopher Wren's solution of Kepler's problem, printed as a supplement to John Wallis's tract on the cycloid, in which the law is mentioned but not in any prominent or important way as an astronomical proposition or rule. In Newton's studies of that topic (in his "epistola prior" to Leibniz on 13 June 1676) he refers to this as 'Kepler's astronomical problem' ('Astronomicum illud Kepleri problema'), but in his solution by series, he does not mention or imply any particular astronomical application, treating this as a purely geometric problem of astronomical origin: to find a focal sector of an ellipse having a given ratio to another focal sector of the same ellipse. Nowhere in any of Newton's early writings – notes, extracts, correspondence, manuscript tracts – is there any explicit statement of Kepler's area law. In short, prior to the correspondence with Hooke in 1679–1680, the second law was not part of Newton's conscious armory of astronomical principles; it had never been active in his thinking about astronomy, as had been the case with respect to the third law (which had led him to the inverse-square law of force for circular planetary orbits).[5]

And then, perhaps suddenly, the significance of the area law would have burst upon his consciousness.[6] If, as I suppose, there is no reason to doubt that the discussion of Kepler's laws in the tract

De motu and in the beginning of the *Principia* follows the same line as his own discoveries, then we may see how important was Hooke's 'Theory', as he once called it, of compounding 'celestiall motions of the planetts' out of 'a direct motion by the tangent' and 'an attractive motion towards the central body' (Hooke to Newton, 24 November 1679; Newton, 1959–1977, vol. 2, p. 297; see §5.4). For in prop. 1 of both *De motu* and the *Principia*, Newton does just that. For clarity, let me recapitulate the argument, as described in §4.4, in relation to the transition from individual force impulses to a continually acting force. First, he considers the inertial (linear) motion of a body in the absence of any external forces; and he shows that with regard to any point in space not in the line of motion, a line drawn from the body to that point will sweep out equal areas in any equal times. What an extraordinary revelation! For the first time it was seen that the law of inertia is intimately bound up with the law of areas. Newton then "compounds" this inertial motion with a blow, a sudden impact, an "impulsive" force (as it later came to be known), directed toward that point, and he proves geometrically that area is still conserved. In the new inertial motion following the first blow, a second blow is given – again in the direction of the point (or center) – and again area is conserved. These blows follow one another at regular intervals and produce a polygonal path, whose sides, together with the lines from the central point to the extremities of the sides, determine a set of equal-area triangles. Then, in the limit, as the time between successive blows becomes indefinitely small, or as 'the number of triangles [is] . . . increased and . . . the width of the triangles diminished indefinitely', the 'ultimate perimeter' will be a curve and 'the centripetal force by which the body is perpetually drawn back from the tangent of this curve will act continually' (see §4.4). Prop. 2 proves the converse, that the area law implies inertial motion in a central force field.[7] Thus the transformed area law provides, in Newton's hands, a necessary and sufficient condition for a centripetal force. Armed with this law (presumably post-1679), and an ingenious measure of force (prop. 6, bk. one, of the *Principia*), Newton was able to show that planetary motion in elliptical orbits (and cometary motion in elliptical or parabolic orbits) is a consequence of the combination of inertial (linear or tangential) motion and a particular kind of centripetal force, one that varies inversely as the square of the distance.

As we have seen, Newton recorded that as a result of his correspondence with Hooke, and in fact goaded by Hooke,

> I found now that whatsoever was the law of the forces
> which kept the Planets in their Orbs, the areas described
> by a Radius drawn from them to the Sun would be
> proportional to the times in which they were described.
> And by the help of these two Propositions I found
> that their Orbs would be such Ellipses as Kepler had
> described.[8]

There can be no doubt of the importance attached by Newton to
the law of areas.[9] In the version of the story recorded by Conduitt,
it is said that when Hooke 'writt to him that the Curve would be
an Ellipsis & that the body would move according to Kepler's notion [sic]', this gave Newton 'an occasion to examine the thing
thoroughly & for the foundation of the Calculus he intended laid
down this proposition, that the areas described in equal times were
equal, which though assumed by Kepler was not by him demonstrated, of which demonstration the first glory is due to Sir Isaac.'[10]
Newton himself did not make quite so overt a claim for the area
law, but he did do so in the matter of the elliptical orbits. In the
Principia he gave explicit credit to Kepler for having discovered
the third or harmonic law, but did not even mention Kepler's
name in association with either the area law or the law of elliptical
orbits.[11]

Now, in addition to a transformation from centrifugal to centripetal force, and a recognition of the significance of the area law,
there is a further factor to be given its due account: Newton's willingness to consider the action of forces of attraction that may produce elliptical orbits at millions of miles of distance. I have mentioned earlier (§3.8, end) that R. S. Westfall (1971, p. 377) has
presented a highly original analysis of the 'redirection of his [New-
ton's] philosophy of nature undertaken . . . about 1679', whereby
Newton went beyond 'the orthodox mechanical philosophy' which
'insisted that physical reality consists solely of material particles in
motion, characterized by size, shape, and solidity alone' and 'now
added forces of attraction and repulsion, considered as properties
of such particles, to the catalogue of nature's ontology'. Westfall
would associate this change in Newton's philosophy of nature with
a decline in his belief in the aether, possibly as a result of pendulum experiments performed about 1679.

There can be little doubt that Newton came to consider centripetal forces as if they were entities in their own right, and to concentrate directly on their action to produce observed motions (that is, to initiate a motion, to terminate a motion, to change a motion by altering the direction or magnitude or both – in short, to produce an acceleration). This required, on Newton's part, major transformations of his ideas of force that included a shift away from the concept of a body's "conatus" or "endeavour" to that of an external force acting on a body, and then the admission of the possibility that bodies can exert forces on other bodies at great distances (that is, at distances that are great with respect to the dimensions of the bodies themselves) – a possibility that Newton considered in both its mathematical and physical implications. Additionally there had to be a change from thinking in terms of a body's centrifugal force to conceiving that in curved motion there is a centripetal force acting on a body. We have seen that it was the failure to develop this kind of concept of force that prevented Huygens from having any real part in the discovery of the law of gravity.

What gave Newton the power to forge ahead with the reconstruction of dynamics and the formulation of a system of the world based on celestial dynamics, what enabled him to develop the concept of universal gravity and to explore its consequences and to discover the gravitational law, was indeed a novel approach to the problem of forces. But I believe that the key to Newton's creative thought in celestial dynamics was not that he considered forces to be real, primary, or essential properties of bodies or of gross matter, but rather that he could explore the conditions and properties of such forces *as if* they were real, and without needing to be able to find a satisfactory answer (or any answer at all) to questions about the independent reality of such forces. The Newtonian style, developed around 1679–1680, enabled him to avoid and to postpone any problems concerning the reconciliation of the concept of centripetal or attractive forces with the received mechanical philosophy until after phase three, when he had found evidence that the force of universal gravity 'really exists'.[12] And then, as we saw in §3.8, Newton became concerned to find out – in a sequel to phase three – how to make such a force a consequence or a secondary effect of some cause that was philosophically acceptable. That is, Newton was convinced that there is some kind of "action" that draws or impels one body toward another, as if one body were attracting another over great distances, hundreds of millions of miles in the case of the sun's action on planets and yet farther in the case of comets. But Newton did not believe that a force could act over such great distances unless there was "something" in the space between to produce what would appear to be such an "attraction". This level of enquiry and speculation was not primarily relevant to the working out of the mathematical consequences of the imaginative systems or constructs that Newton proposed (phase one), or to the comparison of the mathematical results to the world of physical nature (phase two).

It is a moot question whether, from 1679–1680 to 1687 and afterwards, Newton was willing to deal with forces acting at a distance because of the strength of his inner conviction and hope (in vain as it turned out) that such forces would eventually prove reducible to some effect of matter and motion (as in the impulses of aether particles). But it is a plain fact that he *did* undertake an analysis of planetary motion according to the Hookean mode of conceiving a centrally directed force to be acting on a planet that had a component of linear inertial motion, and he *was* able to do so even though he still had not fully given up his adherence to the concept of aethereal vortices and though he apparently did not really believe in centripetal forces acting at a distance. Whatever his beliefs, the Newtonian style enabled him to explore the properties of this kind of force and eventually to discover, as a result of what I have called phase three of his enquiry, that universal gravity is useful and even necessary, and that it 'really exists' (as he later declared in the concluding general scholium of the *Principia* in 1713) and acts according to the laws he had set forth. We have seen that for Newton such a conclusion was not the end of the exercise, as it would have been had he been able to accept the existence of universal gravity without further question. The very "reality" of the existence of universal gravity – as an effect and not as a primary cause – demanded that gravitational action be "explained" in terms of the received philosophy of nature, be reduced to the principles of the "mechanical philosophy", according to which all phenomena must be ultimately accounted for by the principles of "matter and motion". The Newtonian style, and the Newtonian philosophy of science based upon it, boldly declared the reasonableness of accepting the Newtonian system of the world – and the Newtonian celestial mechanics which was its fundament – without such an explanation having been found, but Newton himself evidently believed that it would not be too long until this kind of explanation would be forthcoming. The "reality" of universal gravity did not mean for Newton an independent existence apart from a strict "reduction" to "mechanical principles".

It is not my purpose here to enquire into the reality and independent existence of short-range forces, associated with particles of matter, save in relation to universal gravity.[13] This type of force may have arisen as a concept in Newton's mind in the context of his alchemical studies or his inquiries into the structure and properties of matter; or it may have found strong reinforcement as a result of his alchemical and chemical investigations.[14] What is of concern here is not the analysis of the ways in which Newton studied the properties of particulate forces but only the manner in which he conceived that particulate forces may be related to the forces of gross bodies.[15] In the statements made by Newton about this relation, he argues that his success in the *Principia* concerning macro-forces leads him to hope that eventually there should be an equally successful science of matter based on micro-forces. Thus, in the published preface to the first edition, he expresses the wish 'that we could derive the other phenomena of nature from mechanical principles by the same kind of reasoning' that had proved to be so successful in celestial mechanics, in dealing with gross bodies. 'For', he concluded, 'many reasons lead me to suspect somewhat that all of them may possibly depend on certain forces by which the particles of bodies . . . either are impelled toward one another and cohere in regular figures, or are repelled from one another and recede'. What is impressive about such statements by Newton, however, is their diffidence, as if he were expressing hopes or beliefs about particulate forces – that is, his aims or expectations – rather than making assertions about accomplished facts or guaranteed truths.[16] There was a good reason for adopting this position, since Newton had no real experimental or phenomenological evidence for the existence of such short-range particulate forces, nor had he ever constructed a successful mathematical system encompassing such forces in the Newtonian style. Newton's writings on the structure of matter are characterized by a general absence of mathematics in any true sense.[17]

There is thus a fundamental difference between Newton's statements about universal gravity and about particulate forces. He introduced considerations of the latter in the suppressed draft preface intended for the first edition of the *Principia*, using such phrases as: 'For I suspect that all these things depend upon certain forces'; 'I therefore propose the inquiry whether or not there be many forces of this kind, never yet perceived' (Hall & Hall, 1962, pp. 302sqq). In a suppressed conclusion, he wrote in a similar vein: 'I suspect that the latter depend upon the lesser forces, as yet unobserved, of insensible particles'; 'I do not clearly see why lesser ones should not act . . . by similar forces'; 'It is probable that the particles . . . cohere by a mutual attraction'; '. . . is, I suspect, to be attributed to the mutual attraction of particles'; 'I have briefly set these matters out, not in order to make a rash assertion that there are attractive and repulsive forces'; 'For I think that burning bodies are . . .'; 'Where I suspect that . . .'; 'if only it be possible to prove that forces of this kind do exist'; 'I am far from affirming that my views are correct' (ibid., pp. 320sqq). In a 'Hypoth. 2', written for a version of the *Opticks* in the early 1690s, he admitted: 'The truth of this Hypothesis I assert not because I cannot prove it, but it very probable . . .' (Cohen, 1966, p. 180; Westfall, 1971, pp. 371–382).

What a contrast this is to the *Principia*, in which he insists without any qualification that universal gravity 'really exists' ('revera existat') and that 'gravity' . . . acts according to the laws which we have set forth', whereas we have seen that in the published preface he says only that he has been led 'to suspect somewhat' ('multa me movent, ut nonnihil suspicer') that the 'other phenomena of nature . . . may possibly depend on certain forces by which the particles of bodies . . . either are impelled toward one another and cohere . . . or are repelled from one another'. That is, he asserts the existence of universal gravity even though he cannot understand it and cannot find a cause for its mode of action, but with respect to the particulate forces (which, he says, are like gravity in that their 'causes [are] not yet known') he only voices a suspicion. The *evidence* for universal gravity was firm and, to his way of thinking, unassailable; but the evidence for particulate forces was very shaky.[18]

In the physics of the *Principia*, the forces associated with gross bodies have aspects that perplex the simple view of Newtonian physics, as expounded in elementary textbooks and in many histories of science. We have seen, for example, that Newton wrote about a "vis insita", an inner or inherent or essential "force" that, as its name suggests, may have been "implanted" in matter; this is the same "force" that Newton said was a "vis inertiae" or a force of inactivity (*Principia*, def. 3). Newton contrasted this "inherent force" (or "force of inertia") with "impressed force"; the latter, he said, 'consists in' the 'action alone', i.e., the 'action exerted on a body to change its state either of resting or of moving uniformly straight forward'; and it 'does not remain in a body after the action has ceased' (Newton's explanation of defs. 3 and 4, *Principia*). He noted that 'a body perseveres in every new state by the force of inertia alone'. Newton then referred to three 'origins' of 'impressed force': 'percussion, pressure, or centripetal force' (explanation of def. 4). That is, changes in state can result from an instantaneous contact force, from a continuously applied contact force, or from a force that acts at a distance so as to push or pull a body toward a center.

Although it may seem confusing to have used the same word "force" (*vis*) for two such different concepts or physical entities as the external cause of a body's change of state and the internal cause of a body's keeping itself in whatever state it has attained,[19] I believe that there was never any real confusion in Newton's own mind as to the very different modes of action of these two types of forces. There is no example in the *Principia* or in any other writing of Newton's (in print or in MS) that I have ever encountered in which he makes the conceptual error of assuming that these two "forces" are of the same kind or act in the same manner.[20] Newton (as we have seen in §4.5) was merely using a then-common and traditional expression, "vis insita", which he coupled with his own concept of "inertia" (transformed into a "vis inertiae")[21] to indicate his belief that the inertial property of matter needed some explanatory cause to account for it.[22]

Not only does the Newtonian philosophy make use of these varieties of "forces" related to the behavior of gross bodies; there are also forces of attraction and cohesion and of repulsion between particles. These are developed in various essays and particularly in the *Opticks*, and there is a reference to their probable or possible properties in the preface to the first edition of the *Principia*. In the queries of the *Opticks* Newton also introduces the notion of "active" forces and "passive" forces. It is thus clear that a thorough analysis of Newton's physics and his philosophy of nature must

entail a study of force in Newton's physics, as R. S. Westfall's mono-
graph on this topic is entitled.

For the immediate purposes of the study of Newton and Kepler's laws during the period 1679–1686 (from the time of the correspondence with Hooke to the completion of the *Principia*), however, no such analysis is needed.[23] All that is required is to be aware
that Newton made use of external forces which alter a body's state
of motion and of rest and internal "forces" which do not produce
any such alteration but merely maintain a body in whatever state
it is already in or has obtained by the action of forces of the first
kind; thus it is by internal "forces" that bodies resist attempts to
change their state of motion or of rest. It is also necessary to be
aware that Newton (as in the demonstrations concerning the area
law) had a method of conceiving the action of a continuous force
to be the limit of the action of a sequence of blows, as the time between blows is indefinitely diminished. And it must be kept in
mind that the Newtonian style, independently of his views on the
ontogeny of forces, enabled him to deal with forces as if they were
real entities, and to achieve the law of universal gravitation, without consideration of the reality of such forces until he had found
that they so fully explained nature that the question of their actual
existence and possible mode of action could no longer be bypassed.

5.6   *From Kepler's laws to universal gravitation*

In August 1684 Halley made his famous visit to Newton,
in order to ask him what paths the planets would describe if continually attracted toward the sun with a force varying inversely as
the square of the distance (see Cohen, 1971). This problem had
been discussed in London by Wren, Halley, and Hooke; Hooke
claimed to have had a solution but never produced it, even though
Wren offered him a prize for doing so. When Halley put the question to Newton in Cambridge, Newton 'immediately' (so the story
is recorded by Conduitt) replied that the path would be an ellipse.
How did he know this? 'Why, saith he, I have calculated it.' But he
could not find his calculations and so, after Halley had gone, he
once again worked up his thoughts on the subject and wrote out
the demonstrations in a tract which Edward Paget took to Halley
in London. Halley was so impressed that he returned to Cambridge
to get Newton to record, if not make public, his results. Newton
thereupon sent him a version of a tract now generally known as

De motu (possibly the same tract, though not necessarily the same version, previously sent via Paget), which was duly recorded, and may still be read in vol. 6 of the Register of the Royal Society.[1] Now stimulated by the work he had undertaken, and encouraged by Halley and the Royal Society, Newton entered yet more deeply into the subjects of astronomy and celestial mechanics, and eventually wrote and published his *Principia*. We may gain some insight into the rise of the fundamental physical and astronomical principles of Isaac Newton by a comparison and contrast of some aspects of this first tract *De motu* and the final *Principia*, taking some account in passing of certain other intermediary documents.[2]

In *De motu* (theor. 1) as in the *Principia* (prop. 1), Newton begins by showing that the area law follows from the assumption that a body with an initial inertial motion is acted on by a force continually directed toward the point about which the areas are reckoned. This proof is general in that there is no reference to any particular condition of force or to any specified curve; Newton merely shows how the area law is related to inertial motion in any central force field. Additionally, as we have seen (§4.4), Newton shows that the area law holds in the special case of a null force field for a body with uniform rectilinear (i.e., inertial) motion. In the *Principia* (but not in *De motu*), Newton then proves the converse (prop. 2), that the area law implies a force continually directed to the point about which the equal areas are reckoned; and he there also introduces an area-law-cum-central-force theorem (prop. 3) about the motion of a body (*L*) around another body (*T*) that may be moving.[3]

In prop. 1 Newton postulates an 'immovable centre of force';[4] in prop. 2 the center may be 'either immovable, or moving forward with a uniform rectilinear motion'. Then, in both *De motu* and the *Principia*, Newton introduces (*De motu*, theor. 2; *Principia*, prop. 4) a general theorem plus corollaries, concerning the measure of centripetal force in uniform circular motions.[5] In both works (*De motu*, corol. 5, theor. 2; *Principia*, corol. 6, prop. 4), Newton observes that if the squares of the periodic times are as the cubes of the radii, the centripetal forces will be inversely as the squares of the radii. In *De motu* (schol. to theor. 2; translated by D. T. Whiteside in Newton, 1967–, vol. 6, pp. 31sqq), Newton states that this relation holds 'for the major planets circulating about the Sun and for the minor ones circulating about Jupiter and Saturn'. A similar statement occurs in the *Principia* (schol. to prop. 4), in which Newton says that the 'case' of this corollary 'obtains in the celestial bodies'. In the *Principia* there is an additional parenthetical statement, not found in *De motu*: 'As our colleagues, Wren, Hooke, and Halley, have severally observed'. In *De motu*, it is said that the above relation (that the 'squares of the periodic times [of the celestial bodies] are as the cubes of the distances from a common centre about which they revolve') is 'now accepted by astronomers'. In the *Principia*, Newton's statement is more concise: since this relation holds for celestial bodies, he intends 'to treat more at length of those things that relate to centripetal force decreasing as the squares of distances from centres'.[6] In the *Principia*, but not in *De motu*, there follows (prop. 5) a neat way to find the center of force, given an orbit and the velocity at any two points.

Then (*De motu*, theor. 3; *Principia*, prop. 6) Newton introduces a highly original measure of centripetal force, the foundation on which the analysis of elliptical orbits (and much else) is built.[7] This leads him to a further discussion of motion in a circle. In *De motu* (prob. 1), the centripetal force is directed to a point in the circumference; in the *Principia* (prop. 7), the force is 'directed to any point'. There is no equivalent in *De motu* to props. 8 and 9 of the *Principia*, in which Newton considers motion in a semicircle, where the center of the force is 'so remote' that all lines drawn from that center to any point on the semicircle 'may be taken for parallels', and explores the law of centripetal force in motion in a spiral.

Next (*De motu*, prob. 2; *Principia*, prop. 10), elliptic motion is introduced, with the force directed to the center of the ellipse. And then Newton is led to the principal fruit of the exercise set by Halley: the derivation of the 'law of centripetal force directed to a focus of an ellipse' (*De motu*, prob. 3; *Principia*, prop. 11). It is shown that this force is 'inversely as the square of the distance.' As was the case for the area law and the harmonic law, Newton has transformed Kepler's kinematic or observational rule into a causal principle about the forces in planetary motion. In *De motu*, this leads Newton to the following conclusion:

> Scholium: Therefore the major planets revolve in ellipses having a focus in the centre of the sun, and radii drawn [from the planets] to the sun describe areas proportional to the time, entirely [*omnino*] as Kepler supposed.

This scholium, or its equivalent, is not to be found in the *Principia*; the reason, as we shall see, is that in the latter work Newton has gone beyond the level of *De motu* and, in fact, proves that the planets do *not* actually move in this fashion.

If the truth be known, Newton has not actually proved the above scholium in *De motu*. He has only proved a limited form of it, namely, the case of an immovable sun, equivalent to a one-body system of a single planet moving about a fixed (or noninteractive) center of force. In the world of astronomy, in the realm of phenomena and observation, as Newton was to show in the *Principia*, this is only a simple mathematical construct or imagined system, although it is closely approximated by the motion of the planets in their orbits, and it must be modified in two major ways in order to be like the system of the world. First, in every two-body system of sun and planet, each of the two bodies will move in an elliptical orbit with respect to the center of mass of the system. Second, the motion of each planet suffers perturbations from the gravitating force of all the other celestial objects, most notably the outer planets. The first of these modifications comes from the Newtonian concept that the attraction between the sun and the earth is mutual, that the earth attracts and moves the sun just as the sun attracts and moves the earth. This result arises as a special instance of the general law that action and reaction are always equal in magnitude but opposed in sense or direction: this is Newton's third law of motion, whose first appearance in the *Principia* is one of the notable aspects of that treatise. In *De motu* only the first law of motion is stated explicitly. The second law is present by implication, in its impulsive form, in the proof of the relation of the area law to a central force. But the reader would be hard put to find the third law, even in embryo, since Newton is essentially dealing with a one-body problem in a central force field, and the question of action and reaction arises only when two or more bodies interact. The transformation that yielded the Newtonian mutual force of sun and planets from the Hookean concept of a solar centripetal force acting on a planet – a transformation produced by the application of the general Newtonian law of action and reaction (itself a generalization by transformation of a rule for impact) – was a step of the greatest significance in the history of the exact sciences. It led, by an almost inevitable logic, to universal gravitation. Of course, as mentioned earlier, there also had to be a modification caused by the transformation of the attracting and attracted bodies from essentially mass points into bodies with physical dimensions and shapes.

The first modification of the simple construct of a one-body system arises from the recognition that in any sun-planet system, it does not suffice to consider the sun as a fixed body, incapable of any motion or of attraction by the planet. The third law causes a shift from a one-body to a two-body system. At this stage, the solar universe would consist of a series of somewhat independent two-body systems, each of which has the sun as one of the two bodies, the other being a planet. There follows a second modification which arises as a consequence of the first one. For if each planet attracts the sun and is therefore a center of attraction as well as an attracted body, then each planet ought also to attract and be attracted by every other planet. This leads to the concept of interplanetary perturbing forces and a system of three or more interacting bodies. Simple logic, imaginatively applied by Newton, makes an additional transformation of the concept of planetary force. The earlier transformation had led Newton from the idea of a solar force acting on a planet to that of a mutual force, the planet pulling on the sun to the same extent that the sun pulls on the planet. The second transformation leads from mutual solar-planetary forces to mutual interplanetary forces. As is well known, Newton showed (primarily using Kepler's third law) that one and the same force acts between the sun and each planet, so that if any planet were instantaneously shifted to another's orbit, it would there have the speed and period of revolution of the planet that is normally there (see *Principia*, bk. one, props. 14 and 15; also Cohen, 1967*b*). By this transformation, Kepler's third law, no longer a generalization about empirically determined planetary distances and periods, became a key to unlocking the secrets of a universal force. As a next stage, it is shown that this same force is exerted by the earth on the moon and by the moon on the earth (and is mutually exerted between Jupiter and the Jovian satellites); and that this force is gravity, the cause of weight in terrestrial objects. From this point on, Newton refers to this force as "gravity" rather than the 'centripetal . . . force by which planets are kept in their orbits', since (as he says in the scholium following prop. 5, bk. three) it 'is now established that this force is gravity'. In his progression toward the proposition that 'gravity is a property of all planets universally' (corol. 1, prop. 5), he assumed both that 'Venus, Mercury, and the rest are bodies of the same kind as Jupiter and Saturn' and that 'every attraction is mutual, by the third law of motion'. In a final transformation, Newton generalized this concept of force to a *universal gravitation*, a force that acts mutually on and between any two samples of matter anywhere in the universe. As we shall see at the end of this section, the third law (action = reaction) was crucial for Newton in attaining this final generalization.

This sequence of successive transformations and generalizations, found in the *Principia*, is absent from *De motu*. Furthermore, in the *Principia* Newton goes on from elliptical orbits to consider bodies moving in hyperbolas (prop. 12) and parabolas (prop. 13); in each case he finds that the force directed to the focus varies inversely as the square of the distance. Thus in the *Principia* Newton deals with the problem in general, whereas in *De motu* he keeps more closely to the limitations set by Halley's problem of a planet moving in an ellipse. In the inverse problem (prob. 4, *De Motu*; prop. 17, *Principia*), Newton starts out with an inverse-square force of a given magnitude, and a given initial rectilinear or projected speed (both the starting point and direction are specified, as is the magnitude). In *De motu* Newton states the problem: 'required to find the ellipse which a body shall describe' under these conditions. In the *Principia*, however, what is required is to find the orbit in general, the curved line of the projected body's path. The resulting possible trajectory is shown to be an ellipse, a parabola, or a hyperbola, depending on the initial conditions of velocity. This end result does not appear in *De motu*, but only as a kind of afterthought or addendum to a problem expressly stated for ellipses. It is introduced thus: 'This argument holds when the figure is an ellipse. It can, of course, happen that the body moves in a parabola or a hyperbola . . .' In *De motu*, in a scholium following theorem 4 on elliptic orbits, Newton outlines a plain method for ascertaining the transverse (or major) axes of the planetary orbits from their periodic times, and for determining the actual orbits from specific observations, using one orbit to correct another. But be it noted that Newton still appears to assume here that planetary orbits may be exact ellipses, with the sun fixed at a focus; and there is no hint of planetary perturbations.

In book 1 of the *Principia*, Newton proves that for orbital motion in an ellipse (prop. 11), in a hyperbola (prop. 12), or in a parabola (prop. 13), there is a force (directed to a focus) that varies inversely as the square of the distance. A corollary to prop. 13 then states the converse, that if there is such an inverse-square force, the orbit will be one of the conic sections. In the first edition, this statement appears without proof, which could have given the impression that Newton was unable to prove it. (But surely no one should assume that Newton believed a proof had already been given in props. 11–13, that the proof of "A implies B" proves also that "B implies A"). As D. T. Whiteside points out (Newton, 1967–, vol. 6, pp. 148–149 n.), Johann Bernoulli 'fairly criticised' Newton for merely presupposing the truth of this statement 'sans le démontrer' (7 Oct. 1710, in a letter to J. Hermann, of which an extract was published in 1713 in the *Mémoires de mathématique et de physique* of the Paris Académie des Sciences for 1710). Newton himself had become aware of the need for a proof even earlier and on 11 Oct. 1709 directed Roger Cotes, then preparing the text for a second edition, to add a pair of sentences indicating the steps to show that the conic sections are the only curves possible under the given conditions. In the second edition, Newton thus begins with the proved result that under the action of an inverse-square force there *can be* orbital motion in a conic section and then indicates how to prove that *no other* orbits are possible. What is curious is that Newton considered this general result to be so obvious or trivial that he relegated it to the status of a corollary rather than a numbered proposition and originally published it without proof. Even more curious is the fact that in giving a particular example to illustrate the great prop. 41, bk. 1, of the *Principia* (to find the orbit under any given conditions of centripetal force), he chose (corol. 3) to discuss a force varying as the inverse cube of the distance and not of an inverse-square force.

In a copy of the tract written out by Newton's amanuensis, Humphrey Newton (who later made the fair copy of the *Principia* for the printer), there are two insertions of significant astronomical interest.[8] In one of them Newton invokes a poorly defined 'Planetarum commune centrum gravitatis . . .' (which Whiteside suggests is the 'instantaneous centre' of 'interacting planetary force'; Newton, 1967–, vol. 6, p. 78, n. 10), and which has been subjected to criticism by Curtis Wilson (1970), who points out that Newton has no grounds for assuming that this center either lies within the body of the sun or very near it. Newton concludes that, as a result of the deviation of this center from the sun (the sun's center?), 'the planets neither move exactly in ellipses nor revolve twice in the same orbit'. And this leads him to the following result:

> So that there are as many orbits to a planet as it has revolutions, as in the motion of the Moon, and the orbit of any one planet depends on the combined motion of all the planets, not to mention the action of all these on each other. But to consider simultaneously all these causes of motion and to define these motions by exact laws allowing of convenient calculation exceeds, unless I am mistaken, the force of the entire human intellect [Herivel, 1965a, p. 301].

He confesses that only by 'ignoring those minutiae' will 'the simple orbit and the mean among all errors . . . be the ellipse of which I have already treated'.

Quite evidently, Newton has now learned that in the world of experience, in the actual solar system, the imagined system or construct which led to pure elliptical orbits will not suffice for an absolutely accurate and true representation of the world. Considerations of a physical system or construct of greater complexity, based on a two-body system, with external planets producing perturbations, have shown him the modifications that must be made in Kepler's simple "laws". Of course, this addendum might possibly have been only a clarification which Newton thought was lacking in the previous version, in which case it would not be a new and radical step by Newton toward universal gravitation. I favor the view that this represents a real intellectual leap forward, a fundamental transformation of his first limited conception. For (as Herivel has noted) some of the 'substantial additions' in this version 'represent important conceptual advances in Newton's dynamical thought'.[9] In the present context, the most significant phrases in this addendum may be those relating to a focal point of the planetary orbits other than the sun's fixed center and the possible action of the planets upon one another.

In another addendum, Newton introduces the motion of the moon – a topic absent from the original version of *De motu*. Newton is not concerned here with the "inequalities" in the moon's motions, but with the "falling" of the moon, or the moon test, a major link in the chain to universal gravitation. This is the demonstration that terrestrial gravity, if projected out to the moon by a diminishing factor of the inverse-square law, provides exactly the centripetal force for the moon's acceleration, a proof worked up in full in prop. 4, bk. three, of the *Principia* (see §5.3). In the *Principia* Newton compares these forces in terms of the distance fallen in one second; a theoretical value of 15 Paris feet, 1 inch, 1 4/9 lines[10] compared to a 'terrestrially' observed value of 15 Paris feet, 1 inch, 1 7/9 lines, from which Newton concluded in more positive terms of exactness of agreement than he did in *De motu*, where he described his result as 'quamproxime' or 'pretty nearly' or 'very nearly'. (The two numerical values actually agree to within about 0.03 percent.[11]) In *De motu* Newton does not give any numerical results, but says only, 'my calculations reveal that the centripetal force by which our moon is held in her monthly motion about the earth is to the force of gravity at the surface of the earth very nearly (*quamproxime*) as the reciprocal of the square of the distance from the center of the earth.'[12]

We have no documents that show us how or why Newton came to perceive that planets act gravitationally on one another. But there can be no doubt that in the new paragraph he states unambiguously that there exists a gravitational action of one planet on another, referring specifically and directly to 'eorum omnium actiones in se invicem' ('the actions of all of them on one another'). The effect of this mutual gravitational interaction is that Kepler's three laws of planetary motion are not strictly or exactly true in the world of physics and are true only for a mathematical construct, an imagined realm in which mass points (which do not react with one another) revolve in orbit about either a mathematical center of force or a central attracting body fixed in a system. This aspect of Kepler's laws, one of the features of my presentation of the revolutionary Newtonian celestial dynamics, implies that these "laws" are laws only in the realm of mathematics or of the imagination and that in the realm of physics they are planetary "hypotheses" (as Newton indicated when he called them "hypotheses" in the beginning of book 3 of the *Principia* on the system of the world). Since I have explored this topic at some length elsewhere in this book, I need not say more concerning it at this point. But it is of great interest to observe that when Newton added the new paragraph about planetary gravitational interactions, implying that the planets do *not* move in orbit 'exactly as Kepler supposed', he did not then cancel the scholium to prob. 3, in which it is said that they do so move. Since the text of *De motu* in which the new paragraph appears (U.L.C. MS Add. 3965 §7, fol.) is not written in Newton's hand, but is 'basically Humphrey Newton's secretarial transcript of the corrected state of Newton's primary autograph' (Newton, 1967–, vol. 6, p. 74, n.1), it is likely that Newton composed the new paragraph (or even dictated it) for insertion at the end of the scholium to theor. 4 without carefully rereading the whole text of *De motu*. He would thus not have been aware that the new paragraph contradicted the prior scholium to prob. 3, which accordingly would have required major revision.

I have assumed that the logic whereby Newton arrived at the concept of mutual gravitational perturbations of planetary orbits must have depended heavily on the third law of motion: that to every action there is also an equal and opposite reaction. But I must repeat that we have no direct evidence to support this point of view, nor do we have any document whatever in which there is an antecedent version of Newton's conclusion concerning 'eorum omnium actiones in se invicem'. Newton most probably wrote this paragraph in December 1684, within a month after he had sent a "fair copy" to London to the Royal Society. At the latest, the date would be the first month or so of 1685, since by spring 1685 he was well on his way to the first draft of what was to become book 1 of the *Principia*, of which some pages are still in existence in the text which Newton deposited in the University Library as his professorial lectures (Cohen 1971, pp. 89–91, 311–19; Newton 1967–, vol. 6, pp. 92sqq.). A companion to that early book 1 was a book 2 on the system of the world, which Newton later suppressed in favor of the book 3 of the *Principia*, in which the subject is developed in a more mathematical style. In this first version Newton explains clearly the steps that led to the concept of planetary gravitational interactions or perturbations. I see no reason not to believe that this was the very same set of steps that had led him to this concept some few months earlier, at the time when he wrote out the new paragraph for the scholium to theor. 4 of *De motu*, referring to 'eorum omnium actiones in se invicem'.

In this first version of Newton's system of the world, there is no room for doubt concerning the progression of ideas and the central role of the third law of motion. In his own words:

20. *The agreement between the analogies.*

And since the action of centripetal force upon the attracted body, at equal distances, is proportional to the matter in this body, it is reasonable, too, that it is also proportional to the matter in the attracting body. For the action is mutual, and causes the bodies by mutual endeavor (by law 3) to approach each other, and accordingly ought to be similar to itself [i.e., the same] in both bodies. One body can be considered as attracting and the other as attracted, but this distinction is more mathematical than natural. The attraction is really that of either of the two bodies towards the other, and thus is of the same kind in each of the bodies.

21. And their coincidence.

And hence it is that the attractive force is found in both bodies. The sun attracts Jupiter and the other planets, Jupiter attracts its satellites; and similarly the satellites act on one another and on Jupiter, and all the planets on one another. And although the actions of each of a pair of planets on the other can be distinguished from each other and can be considered as two actions by which each attracts the other, yet inasmuch as they are between the same two bodies they are not two but a simple operation between two termini. By the contraction of one rope between them, two bodies can be drawn to each other. The cause of the action is two-fold, namely, the disposition of each of the two bodies; the action is likewise two-fold insofar as it is upon two bodies; but insofar as it is between two bodies it is single and one. There is not, for example, one operation by which the sun attracts Jupiter and another operation by which Jupiter attracts the sun, but one operation by which the sun and Jupiter endeavor to approach each other. By the action with which the sun attracts Jupiter, Jupiter and the sun endeavor to approach each other (by law 3), and by the action with which Jupiter attracts the sun, Jupiter and the sun also endeavor to approach each other. Moreover, the sun is not attracted by a two-fold action towards Jupiter, nor is Jupiter attracted by a two-fold action towards the sun, but there is a single action between them by which both approach each other. [Trans. from Newton's MS by I. B. Cohen and Anne Whitman.]

Newton was fully aware, as he said next, that 'according to this law all bodies must attract each other', a conclusion he proudly affirmed, explaining why the magnitude of this force between pairs of terrestrial bodies would be so small as to be unobservable. 'It is possible', he added, 'to observe these forces only in the huge bodies of the planets', that is, in the primary and secondary planets, or in the planets and their satellites.

In book 3 of the *Principia*, Newton treats this topic in essentially the same sequence (prop. 5 and corol. 1, 3; prop. 6; prop. 7). The first transformation is to extend terrestrial gravity or the weight-force to the moon and to show, by the moon-test, that this force varies inversely as the square of the distance. Next a transformation identifies this terrestrial force with a solar force acting on planets and a planetary force acting on planets' satellites; all these forces are now to be called "gravity". The application of the third law of motion transforms the solar force acting on planets into a mutual force between sun and planets and similarly transforms the planetary force into a mutual force between planets and their satellites. It follows, as a further transformation, that planets attract (or interact gravitationally with) one another as do satellites and also that all satellites and all planets attract one another. The final transformation is the law of universal gravity. This logical sequence does not, of course, minimize the creative role of Newton's unique genius.

It has been mentioned earlier (§5.4) that some time after 1681, Newton had become converted to nonlinear orbits of comets that curve around the sun. In the earliest version of *De motu* (schol. following prob. 4), Newton uses the problem on finding the elliptical trajectory (given an inverse-square centripetal force, a point of origin of a body's projection, and the magnitude and direction of that projected motion) in order to 'define the orbits of comets, and thereby their periods of revolution, and then to ascertain by a comparison of their orbital magnitudes, eccentricities, aphelia, inclinations to the ecliptic plane and their nodes whether the same comet returns with some frequency to us.' In the *Principia* Newton develops in bk. three a theory of cometary motions in which he satisfactorily approximates a parabolic path to the observed part of the comet's orbital motion (i.e., in its approach to the sun, and recess therefrom: into and through the solar system). This method was successfully applied to both periodic comets (with elliptical orbits) and nonperiodic ones (with parabolic orbits), neglecting the planetary perturbations that cause a departure from orbits that would be pure conic sections. This final theory was long in maturing, and Newton continued to try to use the rectilinear path as an adjunct to computing until almost the completion of the *Principia*.[13] The first edition of his *Principia* (1687) concludes with a lengthy presentation of his final method of using observed places of the comet of 1680 to determine the elements of its orbit, using the parabolic approximation.[14]

Newton's recognition that comets are 'a kind of planet'[15] introduced an additional transformation of the gravitational force by making it even more general than a mutual interaction between the earth and terrestrial objects, between sun and planets, between planets and their satellites, and between one planet and another.[16] Including a sun-comet force among gravitational effects not only enlarged the scope of action of gravity, but showed that this force can extend far beyond the confines of the visible solar system, that the range of gravity as an effective force is many times farther from the sun than the orbit of Saturn, since the force of the sun's gravity acts on comets at great distances so as to turn them around and bring them back into the neighborhood of the sun.

This analysis of the steps toward universal gravitation shows the importance of Newton's third law, the equality of action and reaction. It was by this law, as Newton said in the revision of *De motu*, that he was led to the view that each planet attracts the sun, and that planets with satellites are attracted by the satellites. We have seen that it follows at once that if planets are centers of attraction, as well as subjects of attraction, they should attract one another. The same law of action and reaction implies that any sample of matter near the earth (or near another planet or the moon) will attract the earth (or the other planet or the moon) as well as be attracted. By essentially the same argument that led from Newton's third law of action and reaction to interplanetary attraction, it follows that all samples of matter ought to attract one another. This is the final transformation of the Hookean concept of a planetary centripetal force into the great Newtonian generalization about universal gravity. It was not a generalization based on a sequence of inductions in the classic Baconian manner so much as a result of a sequence of imaginative transformations arising from the successive application of the Newtonian law of action and reaction. This example illustrates the general thesis that the best way to learn about Newton's "method" is to analyze the way in which he solved particular scientific problems rather than to attempt to analyze and to interpret his 'Rules of Philosophizing' about nature.[17]

5.7   *The role of mass in Newtonian celestial mechanics*

In Newton's celestial mechanics, mass enters in two ways: as a measure of a body's resistance to being accelerated or undergoing a change of state when acted on by a gravitational force,[1] and as a measure of the force aroused by a body placed in a gravitational field.[2] The first of the concepts of mass arises in the three laws of motion and their applications in problems of dynamics, the second during the stages toward the law of universal gravitation. Newton was apparently not at first aware of the conceptual problem introduced by having these two differing concepts of mass. After completing *De motu*, however (apparently some time in 1685), he recognized the need for an experiment to establish this equivalence (or, more strictly, the proportionality) of these two kinds of mass.

In a fragment containing *Definitiones*, *Leges motus*, and *Lemmata*, which is entitled *De motu corporum in mediis regulariter cedentibus* ('On the Motion of Bodies in uniformly yielding media'; U.L.C. MS Add. 3965, fols. 23–26), he describes such an experiment, along with what may be his first formal and general statement of a truly universal law of action and reaction. This incomplete work, begun after *De motu* and before the *Principia*,[3] contains a def. 11, in which Newton introduces 'a quantity of body' (which he will later, in the *Principia*, transform into 'quantity of matter' or 'mass'). He now says of this quantity: '[It] is reckoned from the bulk of corporeal matter, which is generally proportional to its gravity. The oscillations of two equal pendulums with bodies of the same weight are to be counted, and the bulk of matter in each of them will be reciprocally as the number of oscillations made in the same time.'[4] The language used by Newton in this passage seems to indicate that as yet he has not actually performed the experiment. But shortly thereafter, he wrote out a revised and augmented series of *De motu corporum definitiones* (Herivel, 1965a, pp. 315–320) in which (def. 7) he actually lists substances on which he has made this experiment: gold, silver, lead, glass, sand, common salt, water, wood, and wheat (p. 317). In bk. three of the *Principia*, and in the prior discarded version (see supplement to §3.5, n. 1), both written shortly after the *De motu corporum definitiones*, Newton describes how he had actually performed such experiments, using these very same nine substances (*Principia*, bk. three, prop. 6; *System of the World*, sect. 9).

In the act of transforming or generalizing terrestrial weight-force (gravity) into a universal gravity or universal force of gravitation, both of these varieties of mass enter into the quantitative formulation. For there is set forth a magnitude equality of the force aroused by any body in a gravitational field (km_g) and the acceleration (A) produced on that same body's mass (m_iA) by the force, where m_g is the gravitational mass and m_i the inertial mass. The proportionality factor k depends on the distance (r) from the second (or "central") body in the inverse-square proportion, and – by mutuality – must also include the mass (M) of that other body (strictly speaking, M_g). Thus for km_g we may write $G(Mm_g/r^2)$, where G is what we now call the universal gravitational constant. Accordingly we have $G(Mm_g/r^2) = m_iA$; by the "equivalence" or proportionality of m_g and m_i, this becomes $A = G(M/r^2)$.

In the case of a planet with a satellite, where $A = v^2/r$, this law, that $A = G(M/r^2)$, leads at once to $M = \dfrac{v^2r}{G}$ and gives a measure of the planet's mass. In the case of freely falling terrestrial bodies, at any given place on earth, A becomes the familiar acceleration (g) of free fall, which is seen to be $g = G(M/r^2)$ and to depend only on the mass and radius of the earth, and to be independent of the mass (m) of the falling body, as Galileo's "tower" experiment had shown. Or, as Newton put it (*Principia*, prop. 6, bk. three), the Newtonian pendulum experiments had shown with great accuracy the result long observed that, but for the small factor of air resistance, all sorts of heavy bodies fall to the earth from equal heights in equal times. In this analysis, there is need of Newton's finding that a homogeneous sphere (or a sphere made up of homogeneous spherical shells) will gravitate as if all of its mass were concentrated at its geometric center. Newton, of course, did not know how to measure the value of G, and generally compared masses of different bodies, such as planets, and did not ever compute their absolute values.

In the discussion of def. 8, at the very beginning of the *Principia*, Newton anticipated this later result (essentially $A = G(M/r^2)$) to explain that at any one place all bodies fall freely with the same acceleration. Then he showed that if we move farther and farther away from the earth, the acceleration would be diminished. In this case, he said, the weight (which is 'always . . . as the product of the body [or mass] times the accelerative gravity') 'would be equally diminished'. Newton did not employ such expressions as "gravitational mass", "inertial mass", or "equivalence"; what he demonstrated was the *proportionality* of *weight* to *mass*. Nor, of course, did Newton write any equations such as those used above to give the main outlines of his argument.

5.8   *Kepler's laws, the motion of the moon, the Principia, and the Newtonian Scientific Revolution*

Newton's mature astronomy may be seen in the application he made of his new principles of celestial dynamics to Kepler's three laws, or "planetary hypotheses". We have seen that he first develops a mathematical system or imagined construct (essentially a one-body system with a center of force, which need not necessarily be at rest) in which Kepler's laws are true. Then he shows what modifications must be made for a two-body system, in which both bodies mutually attract and move one another, and finally introduces a many-bodied system, in which all bodies attract one another and perturb one another's motions. This is his procedure in bk. one, in developing the general principles, and again in bk. three, in applying these general principles to the solar system. Thus, in prop. 13, bk. three, he says that:

> Since the weights of the planets toward the sun are reciprocally as the squares of the distances from the center of the sun, it follows (from bk. 1, prop. 1 and 11 and prop. 13, corol. 1) that if the sun were at rest and the remaining planets did not act upon one another, their orbits would be elliptical, having the sun in their common focus, and they would describe areas proportional to the times. The mutual actions of the planets upon one another, however, are very small (so that they can be ignored) and perturb the motions of the planets in ellipses about the mobile sun less (by bk. 1, prop. 66) than if those motions were being performed about the sun at rest.

He then does observe, however, that 'the action of Jupiter upon Saturn is not to be ignored' (see §3.5); there is 'so sensible a perturbation of the orbit of Saturn at every conjunction of this planet with Jupiter that astronomers cannot ignore it'. On the other hand,

> the perturbation of the orbit of Jupiter is far less than that of Saturn's. The perturbations of the remaining orbits are still less by far, except that the orbit of the earth is sensibly perturbed by the moon. The common center of gravity of the earth and moon traverses an ellipse about the sun, one in which the sun is located at a focus, and this center of gravity, by a radius drawn to the sun, describes areas proportional to the times; the earth, during this time, revolves around this common center with a monthly motion.

As Newton was perfectly aware, the reason why Kepler's laws hold so well is that almost all of the mass of the solar system is in the sun, so that (prop. 12, bk. three) the effect of the gravitating force of all the planets does not cause the sun ever to recede 'far from the common center of gravity of all the planets'.[1]

In sect. 11 of bk. one of the *Principia*,[2] Newton modifies the original simple construct of a one-body system, for which Kepler's laws are valid. Here Newton shows what modifications must be made in the transition from a one-body to a two-body system. For instance, in prop. 60 of sect. 11, he shows that in a two-body system of bodies having masses S and P, the primitive or simple harmonic law $a^3/T^2 = k$ becomes more complex, because the planetary masses must be introduced into the equation.[3] Prop. 66 of this sect. 11, with its many corollaries, lays a theoretical foundation for the simplified three-body problem corresponding to the motion of the moon (see §3.5). At the end of sect. 11 he takes account of the fact that there is no single fixed force-center (the only possible condition for the truth of the area law) for all the bodies in a multibody system such as the solar system, and hence (corol. to prop. 68) planetary orbits 'will approach nearer' to an ellipse, 'and the descriptions of areas will be more nearly uniform', if the center be taken as the center of 'gravity' (actually, of mass[4]) of the system comprising the central body (the sun) plus all the planets interior to the one whose orbit is under consideration. And this leads to the final prop. 69 of this sect. 11, in which Newton shows that the 'absolute force' of each attracting planet in a system of planets is as its 'body' or mass.[5]

While developing these consequences of the gravitational law, Newton called upon Flamsteed for information concerning Jupiter's satellites and concerning the possible corrections to the Keplerian motion of Saturn, due to a perturbation of Saturn by Jupiter.[6] Their exchange of letters shows that by December 1684, shortly after writing the first tract *De motu*, Newton was already concerned about interplanetary perturbations and was thus well on the road toward the concept of universal gravitation. The information about Jupiter's satellites was especially important for showing how exactly the third law of Kepler could be verified, so that he could conclude that the sun's perturbing force does not noticeably alter the satellites' motions. Hence, as Curtis Wilson (1970, p. 165) has observed, Newton 'will later conclude . . . that the weights of Jupiter and its satellites toward the sun are proportional to their respective quantities of matter [or masses]'. Wilson also calls attention to other observations and conclusions of Flamsteed's, whereby 'the position of Jupiter's shadow and hence its heliocentric longitude' become known 'at the same time as its geocentric longitude is given by observation'. Thus Flamsteed enabled Newton to have a 'direct confirmation' of the area law 'independently of any hypothesis as to the planet's orbit' (ibid.).

In the *Principia*, in 'The System of the World' or bk. three, we have seen that Newton starts out with the system of Jupiter's satellites before turning to the more difficult problems of the earth and moon or the planets' motions about the sun. And it is notable that in the set of "phaenomena" given at the beginning of bk. three (those "phaenomena" which provide the empirical foundation on which Newton's system of the world rests), there is given the area law for the moon and for the planets, and for the satellites of Jupiter (and of Saturn), together with the third or harmonic law for these same bodies, but *not* the law of elliptical orbits. Planetary ellipses appear for the first time in bk. three in prop. 13, 'The planets move in ellipses which have their common focus in the centre of the sun . . .'[7]

If the roles of Kepler's laws in the *Principia* are to be described accurately, the following propositions will hence emerge. (1) Newton discloses the physical significance of each of Kepler's three laws, and thus reveals (2) that Kepler's laws are true of a one-body system, and that (3) they cannot be accurate descriptions of our solar system. (4) He then shows what transformations of these laws must be made in considering their application to a two-body system, such as the sun and the earth or the sun and any other planet, and finally (5) introduces yet additional transformations for a system of more than two interactive bodies, such as the earth–sun–moon system, or the sun–Jupiter–Saturn system. On the scale of either the highest possible or imaginable observational accuracy, or of mathematical exactitude, Kepler's laws are thus strictly false; they accord with phenomena to a moderately high degree of approximation, but they are "true" only when transformed in the manner indicated by Newton. I have said that I believe that this aspect of Kepler's "laws" explains why, in bk. three of the *Principia*, they were originally designated "hypotheses". Newton was evidently saying that these "laws" as stated by Kepler are not really "true" of the actual solar system but are accurate enough to be taken as working principles. After the first edition had been published, Newton changed his mind about the earlier designation of "hypotheses" (see §3.6, n. 5) and decided to call these "planetary hypotheses" of Kepler's by a new name, "phaenomena". He evidently now meant that these statements were only "phenomenologically" true: true, that is, within the limits of calculations based on the data of observation. And it was the burden of bk. three on the system of the world, which follows these "phaenomena", to find out just how great are the departures of the solar system from these idealized laws which are strictly "true" only of a simpler model or construct.

In bk. three of the *Principia* Newton developed at some length the concept of the three-body problem in the special case of sun, earth, and moon. He then allowed a revised statement of his solution of the problem of lunar motion to be published in 1702, as a part of David Gregory's textbook on astronomy (see Newton, 1975). These results were then corrected and revised and in large measure introduced into the second edition of the *Principia* (1713), where they occur in a new scholium following prop. 35 of bk. three. There he alleged that his 'computations of the lunar motions' showed 'that by the theory of gravity the motions of the moon could be calculated from their [physical] causes'. In part it was true, since he had derived at least one of the known inequalities from gravitational considerations and had similarly introduced some new ones. (He was really successful only in accounting for the variation and the nodal motion.) But in order to get the accuracy of which he boasted, he would – at least in part – have had to "fudge" the data (to use the traditional expression)[8] and to have recourse to geometric models rather than to pure gravitational theory, notably the original Horroxian model and its Flamsteedian "improvement"; and so it was by the partial use of deferents and epicycles and not by gravitational theory alone that he obtained his results (see Whiteside, 1976).

The astronomical triumphs of the *Principia* included the explanation that tides are caused by the gravitational pull of sun and moon, yet another example of the universality of the gravitational force. Yet it must be kept in mind that, astonishing and significant as his work on tides was and is, the gravitational explanation of tides is incomplete, save for tidal phenomena on the high seas, since it does not take into account the physical conditions that determine the particular response of any given body of water to the gravitational attractions of sun and moon. For example, the gravitational attractions of sun and moon do not determine the number of tides in any given port or harbor. Newton himself was aware that local circumstances might be of significance and he adumbrated one sort of explanation based on local circumstances to account for the unusual tides at Batsha on the Tonkin Gulf (see bk. three, end of discussion of prop. 24, 'The ebb and flow of the sea arise from the actions of the sun and moon'). His mighty intellect is also revealed in his association of the precession of the earth's axis with a physical cause, the nonsymmetrical pull of the moon on the earth's supposed equatorial bulge.

The theory of the moon is especially noteworthy because of the hoped-for practical importance of the subject for navigation: the determination of longitude at sea (see Forbes, 1971). One of the reviewers of the second edition of the *Principia* called attention to this very matter, in his analysis of the novelties introduced in the revision. It is paradoxical, however, that the theory of the moon was not advanced by Newton himself to the high level of which he boasted. His successors (as Eric Forbes, 1971, and Craig Waff, 1975, 1976 have recently shown; also Chandler, 1975) were left with the bulk of the job still to do. We have seen that at one stage it even seemed likely that the Newtonian law of attraction might have to be modified.

It is legitimate to consider the Newtonian theory of the moon to have been a failure, since it is not, in its greatest part, gravitational at all, despite Newton's protestations. A true "Newtonian" lunar theory, that is, a lunar theory truly based on Newtonian gravitation, did not exist until it was created long after Newton's death by Clairaut, d'Alembert, and Euler. But in another sense, the theory of the moon was an outstanding triumph. For it was a partial demonstration – at least as far as it went – that even as complex a phenomenon as the moon's motion might yield to a physical theory, to gravitational theory (as indeed it would eventually do). Thus Newton inaugurated a displacement from trial-and-error practice or artisanship to a new practicality based upon higher mathematics: the applications of pure theoretical science. In the case of the moon's motion, Newton's program (only in small part achieved by him) implied an eventual radical shift from the use of arbitrary geometric computing schemes to computations based on the demonstrable forces producing the actually observed motions in the universe, that is, a knowledge of the "true" causes of phenomena.[9] We may see this revolutionary aspect of Newtonian science, as it was displayed in the age of Newton, by examining a popular textbook of practical astronomy: Nicolas-Louis de La Caille's oft-reprinted and -translated *Elements of Astronomy* (English edition, London, 1750). As the full title indicates, this was an astronomy 'Deduced from Observations; and Demonstrated upon the Mathematical Principles of the Newtonian Philosophy: with Practical Rules whereby the Principal Phenomena are determined'. In the Newtonian and the post-Newtonian age, it was no longer possible to separate practical astronomy from the theoretical (physical and mathematical) principles on which it depends. The proof of the validity of Newtonian exact science, and of the Newtonian style displayed in the *Principia*, was its successful application in practical astronomy. I believe this to have been a primary message of Newton's *Principia*: an understanding of physical phenomena (in this case, astronomical phenomena), based on the forces producing them, and elucidated by mathematics, and systematized as 'The Mathematical Principles of Natural Philosophy'. The Newtonian revolution in science was not confined to a set of results that radically restructured the system of physical explanations of the phenomena of our world, but embodied a new way of obtaining those results – by means of a new system of concepts and laws, but chiefly by a new method in which mathematics was applied to imagined constructs which were then altered and given new properties so as to conform ever more closely to the world of experiment and observation. It did not matter so much that Newton had not actually solved all the problems, that he had only taken the first steps in some important examples such as the analysis of the motion of the moon; what counted for men like Clairaut and Lagrange and Laplace was that Newton had charted a way to solve physical problems and had set a course which the exact sciences have been following ever since.

Supplement: History of the concept of transformation: a personal account

The concept of the transformation of scientific ideas took its present form during the spring and summer of 1965, while I was contemplating the factors in scientific revolutions, a feature of the Wiles Lectures, on which this book is based. This concept was put to the test in the spring of 1966, while I wrote the first of the lectures, on the history of inertia (of which a portion is embodied in Ch. 4), a topic which notably displays the complexities of ways in which a scientist uses the ideas (and the names of ideas) of a predecessor or contemporary. As the reader may see for himself in this sample, the very facts of this history stridently declare that in each stage of the development of the concept and law of inertia, a scientist altered and adapted something that he had encountered in his reading and studying of the writings of a contemporary or predecessor.

The Wiles Lectures were delivered at Queen's University, Belfast, in May 1966. In the following autumn the doctrine of transformation was expounded in a privately distributed edition of the first two of the lectures, bearing the general title, *Isaac Newton: The Creative Scientific Mind at Work*.[1] A revised and expanded version of the chapter on the history of inertia was presented for discussion at a symposium held in Prague on 25–29 September 1967, on the occasion of the three-hundredth anniversary of the death of Joannes Marcus Marci of Cronland, of which the general theme was 'La révolution scientifique du 17e siècle et les sciences mathématiques et physiques'. My communication, entitled 'Dynamics: the Key to the "New Science" of the Seventeenth Century', was published in *Acta Historiae Rerum Naturalium necnon Technicarum*, special issue 3 (Prague, 1967), pp. 79–114; the introductory section of this communication was devoted to ' "Transformations" in the History of Scientific Ideas', a theme developed in relation to inertia and Newton's second law. My point of view was stated (p. 81) as follows:

> My own focus of interest in the scientific revolution seen in the new dynamics can best be summarized by three words: continuity, transformation, innovation. The use of such an expression as "vis impressa" by Galileo and by Newton is an instance of the continuity of scientific thought from the XIVth to the XVIIth century. But, unlike "intensio et remissio", which was used by Newton in almost exactly the same way as by Suiseth (the "Calculator") and Oresme in the XIVth century, "vis impressa" was used in the XVIIth century in a sense far different from what we find in late scholastic physics; furthermore, rather distinct and sharply contrasting meanings were given to this "vis impressa" by men such as Galileo and Newton. Thus, it is not merely the case that such revolutionary figures as Galileo and Newton continued to use older expressions with new meanings, rather than inventing a vocabulary of their own. Such a statement ignores the main element of continuity in the development of new concepts, since it overly stresses the form of expression at the expense of the changing content. For as I read the history of scientific thought, some of the boldest new ideas tend to appear as transformations of previously existing concepts, definitions, and laws. It is for this reason that the degree of innovation in science may be made precise by analyzing the particular qualities of the transformation. To carry this mathematical analogy yet one step further forward, let me say that viewing innovation as transformation enables the historian to seek for the invariants, those aspects of the concept or law that may be unaffected by the transformation: be it the name, the form, the field of application, the relation to other concepts or laws, or even the kind of justification that science permits for the concept or law, or the nature of the theory in which the concept or law may be embodied.

This meeting had the admirable effect of bringing together scholars from both sides of the Iron Curtain for fruitful discussions.

My own ideas profited greatly from the comments and discussion, chiefly the remarks made by Eric J. Aiton (of Great Britain) and J. B. Pogrebysski (of the USSR).[2]

The concept of transformation of scientific ideas was developed further in a communication at the International Congress of the History of Science (Moscow, 1971),[3] and in two articles, one on 'History and the Philosopher of Science' (Cohen 1974a), and the other on 'Newton's Theory vs. Kepler's Theory and Galileo's Theory' (Cohen, 1974d). I also used this concept in my graduate seminars at Harvard to illuminate the main issues in the Scientific Revolution and in the history of the exact sciences. This point of view has been used by students and colleagues, among them Yehuda Elkana, Ramunas Kondratas, and Frank Sulloway.

Naturally curious about the origins and history of this concept, and committed as I was to the point of view that even the most "original" ideas tend to be transformations of older ones, I sought for possible sources or prior uses of "transformation" as a guiding historical principle. I recalled that I had once read, somewhere in the writings of Ernst Mach, that all creative innovations are essentially transformations. Then, while rereading Alexandre Koyré's essay on the 'Newtonian Synthesis', I found that he referred to the Scientific Revolution as 'one of the deepest, if not the deepest, mutations and transformations accomplished – or suffered – by the human mind since the invention of the cosmos by the Greeks, two thousand years before'. This general reference to "transformations" is followed a few pages later by a more specific one, more akin to the sense I have adopted, in which Koyré says:

> The transformation of the concept of motion by substituting for the empirical concept the hypostasized mathematical one is inevitable if we have to subject motion to number in order to deal with it mathematically, to build up a mathematical physics. But this is not enough. Conversely, mathematics itself has to be transformed (and to have achieved this transformation is the undying merit of Newton) [Koyré, 1950b; 1965, p. 10].

Although Koyré does not use the expression "transformation" in a systematic way throughout his writings, the concept itself is implicit in much of his research.[4]

Another writer who, on at least one occasion, referred to such a transformation is Pierre Duhem. In his *Structure of Physical The-*

ory, Duhem asked: 'Is not his [i.e., Newton's] theory of gravitation derived entirely from the laws which were revealed to Kepler by observation, laws which problematic reasoning transforms and whose consequences induction generalizes?'[5] I do not believe that Duhem was ever particularly concerned with developing a philosophy of scientific change based on the notion of "transformation", but, again and again, in his *Aim and Structure of Physical Theory* (1906, 1914) he refers to the development of the thought of an individual scientist by means of a process that is all but the same as that which I have called "transformation of ideas". Duhem also refers to the "transformation of experience" made by an individual scientist.

The concept of transformation occurs in a significant way in Duhem's discussion of Ampère, who claimed that he had produced a *Mathematical Theory of Electrodynamic Phenomena, derived only from experience*. The final phrase of the title of his book, 'uniquement déduite de l'expérience', is taken directly from the concluding general scholium of Newton's *Principia*, where Newton says that he is not feigning hypotheses and that 'whatever is neither a phenomenon nor derived from phenomena is a hypothesis'. Duhem points out that Ampère's treatise does not follow the method prescribed in the title, since the 'facts of experience taken in their primitive rawness cannot serve mathematical reasoning'. In order 'to feed this reasoning', Duhem said (1954, Ch. 6, sect. 5, p. 196), such facts 'have to be transformed and put into a symbolic form. This transformation Ampère did make them undergo'. Duhem is referring here to a kind of transformation of experience by which a scientist, in the first instance, must 'translate experimental facts symbolically before introducing them into his reasoning'. The necessity of doing this, in Ampère's case (according to Duhem), 'renders the purely inductive path Ampère drew impracticable'; this path was 'also forbidden to him because each of the observed laws is not exact but merely approximate' (ibid., p. 197).

It is quite different for Ernst Mach. The concept of transformation (and the word itself) occurs again and again in his *Popular Scientific Lectures*. Comparing the design of new scientific instruments to the inventions made in practical technology, he writes about the way 'some very unobtrusive idea' can be 'productive of so much fruit and of such extensive transformations in physical technics'.[6] Using a biological analogy, he invoked the manner in which 'one thought is transformed into a different thought, as in all likelihood one animal species is gradually transformed into a new species'. Mach's gradualism (pre-mutation theory) was coupled with the idea of variation leading to survival: 'Many ideas arise simultaneously. They fight the battle for existence not otherwise than do the Ichthyosaurus, the Brahman, and the horse.'[7] A whole essay is devoted by Mach to the application of Darwin's ideas on the evolution of species to the development of science. It is called 'On Transformation and Adaptation in Scientific Thought'.[8] His main thesis was that 'if Darwin reasoned rightly, the general imprint of evolution and transformation must be noticeable in ideas also'. He concludes that the 'transformation of ideas thus appears as a part of the general evolution of life'. He uses the example of the motion of bodies to illustrate the 'mental transformative process in detail' (Mach, 1898, pp. 224sqq).

When I was glancing through Mach's essays with a wholly different purpose in mind, at the time of preparing the final version of the present work, I could not help being struck by the fact that Mach had announced so clearly and dramatically the historical doctrine of the transformation of scientific ideas that I had thought to be my own. It then came to mind that I had read all of these essays long before putting together, for the Wiles Lectures, my own thoughts concerning the ways in which scientific ideas develop. Although I did not specifically recall Mach's use of "transformation", I did remember vividly another essay in the same volume ('The Economical Nature of Physical Inquiry'), and I was aware that I had referred again and again to one of Mach's similes for the search for truth appearing in this volume of essays. The poets and painters, Mach said, have immortalized the scene of a high garden wall that separates a youth and a maiden who sigh and wait, neither one dreaming 'how near the other is'. But the history of science does not show such scenes. 'The inquirer seeks the truth', according to Mach, but 'I do not know if the truth seeks the inquirer'. In fact, 'Truth suffers herself to be courted, but she has evidently no desire to be won. She flirts at times disgracefully' but 'she is determined to be merited, and has nought but contempt for the man who will win her too quickly' (Mach, 1898, p. 45).

While I was reflecting on this all-but-hidden source of my own idea (or, at least, of an idea that I had adapted and transformed), I recalled also that Freud had written incisively on the hidden sources of our apparent intellectual originality. In fact, I myself had written about this notion of Freud's in an essay dealing with the general subject of unorthodoxies in science (Cohen, 1952, p. 506, n. 7). The occasion of Freud's essay was a chapter of a book by Havelock Ellis (1919), who had exhibited Garth Wilkinson as a precursor of Freud and of Freudian psychoanalysis. Freud and Ellis were poles apart,[9] and there can be no doubt that Ellis's essay or chapter was correctly interpreted by Freud (1920, p. 263) as 'a fresh turn taken by resistance and as a repudiation of analysis', since Ellis's main point was 'to show that the writings of the creator of analysis should be judged not as a piece of scientific work but as an artistic production'.

Garth Wilkinson, a physician as well as a Swedenborgian mystic and poet, published what Ellis described as 'a volume of mystic doggerel verse written by what he considered "a new method", the method of "Impression" '. Wilkinson would choose or write down a theme; then, he wrote, 'the first impression upon the mind which succeeds the act of writing the title is the beginning of the evolution of that theme, no matter how strange or alien the word or phrase may seem'. The first word that comes to mind, according to Wilkinson, would be 'the response to the mind's desire for the unfolding of the subject'.[10] Ellis was no doubt correct in describing Garth Wilkinson's method as 'a kind of exalted *laissez-faire*, a command to the deepest unconscious instincts to express themselves'. For in this process, 'reason and will ... are left aside', and the faculties of the mind are 'directed to ends they know not of'. Ellis concluded by observing that although Wilkinson was a physician, he never used his method 'for scientific or medical ends', but only in his religious and literary pursuits. Nevertheless, Ellis insisted, 'it is easy to see that essentially it is the method of psycho-analysis applied to oneself', which Ellis saw as 'further evidence as to how much Freud's method is an artist's method'.[11]

In his reply, Freud drew attention to a somewhat similar passage in the correspondence of Friedrich Schiller with Körner, in which in 1788 'the great poet and thinker recommends anyone who desires to be productive to adopt the method of free association'.[12] Then, writing about himself in the third person, Freud observed that the new technique alleged by Ellis to have been devised by Garth Wilkinson 'had already occurred to the minds of many others'. He insisted that the systematic application of this method in psychoanalysis 'is not evidence so much of Freud's artistic nature as of his conviction, amounting almost to a prejudice, that all mental events are completely determined'. Freud then effectively denied that either Schiller or Wilkinson 'had in fact any influence on the choice of psycho-analytic technique'.

But after having ruled out the above-mentioned two literary men as possible founts of the new technique, Freud brought his readers' attention to yet another source, which he described as 'a short essay covering only four and a half pages, by Ludwig Börne', and entitled 'The Art of Becoming an Original Writer in Three Days'. Börne's essay ends with these precepts:

> And here follows the practical application that was
> promised. Take a few sheets of paper and for three days
> on end write down, without fabrication or hypocrisy,
> everything that comes into your head. Write down what
> you think of yourself, of your wife, of the Turkish War,
> of Goethe, ... of the Last Judgment, of your ... superiors
> – and when three days have passed you will be quite out
> of your senses with astonishment at the new and unheard-
> of thoughts you have had. This is the art of becoming
> an original writer in three days.[13]

This anticipation of an aspect of Freudian technique had been brought to Freud's attention by Sandor Ferenczi.[14] Freud was always intensely interested in the history of his own ideas, and especially in any precursorship of them. Thus he proudly entitled his essay in reply to Ellis 'The Prehistory of the Technique of Analysis', and published it in the official *Internationale Zeitschrift für Psychoanalyse* in 1920 for all psychoanalysts to read. Freud says that when he read Börne's essay carefully, he found in it a number of significant points relating to 'the prehistory of the psychoanalytic use of free associations'.

What was most important, Freud said, was that when he was fourteen years of age, 'he had been given Börne's works as a present, that he still possessed the book now, fifty years later, and that it was the only one that had survived from his boyhood.' Börne, furthermore, 'had been the first author into whose writings he had penetrated deeply'. While Freud did not remember this particular essay, some of the others in the volume 'kept on recurring to his mind for no obvious reason over a long period of years'. Quite obviously, this had been a very significant book, whose conscious and unconscious effects on his thinking must have been profound and of long duration.

What particularly struck Freud on rereading 'The Art of Becoming an Original Writer in Three Days' was to find that Börne, in the advice that he gave to a prospective original writer, had expressed certain opinions which Freud himself 'had always cherished and vindicated'. One of these was the notion of a 'censorship' which is 'exercised by public opinion over our intellectual productions', and which Börne found to be even more oppressive than the censorship of government. Freud pointed out (1920, p. 265) that 'there is a reference here to a "censorship" which reappears in psycho-analysis as the dream-censorship'.

Freud's concluding sentence may be considered an anticipation of the doctrine of transformation of ideas in the same sense that Börne may have anticipated some of Freud's own ideas and technique; or, to translate these statements into the language of this book, it is possible that this aspect of the development of Freud's concepts, which I discussed in 1952, may have been the source of the transformation (however unconscious) which has produced my present point of view. For Freud concluded his essay with this powerful insight: 'Thus it seems not impossible that this hint may have brought to light the fragment of cryptomnesia which in so many cases may be suspected to lie behind apparent originality.' This concept of cryptomnesia (defined as the 'appearance in consciousness of memory images which are not recognized as such but appear as original creations' [15]), and its illustration by Freud, may be taken as a primary statement of the concept of transformation of ideas.

In another essay, Freud wrote of how pleased he was when he came upon one of his theories 'in the writings of one of the great thinkers of ancient Greece'. This led him to conclude (1937, p. 245): 'I am very ready to give up the prestige of originality for the sake of such a confirmation, especially as I can never be certain, in view of the wide extent of my reading in early years, whether what I took for a new creation might not be an effect of cryptomnesia.' How could we help but honor the historical honesty of so mighty an intellect! [16]

I shall end this history with a brief notice of the use of the concept of transformation by Michel Foucault. Although Foucault's works have been translated into English, they have as yet not really been assimilated into the tradition of Anglo-American scholarly thinking (see White, 1973). In his *L'archéologie du savoir* (1969), translated as *The Archaeology of Knowledge* (1972), there are many references to transformations of all sorts, and a whole chapter is devoted to 'Change and Transformation'. Transformation is stated at the outset to be one of the concepts of historical analysis (along with discontinuity, rupture, threshold, limit, and series) which Foucault proposes to study in relation to what 'we call the history of ideas, or of thought, or of science, or of knowledge'.[17] In the introduction Foucault (1972, p. 4) states his concern to find the 'ruptures' underlying the apparently 'great continuities of thought', ruptures which he likens to Gaston Bachelard's 'epistemological acts and thresholds'. And this leads him to stress 'the *displacements* and *transformations* of concepts'; he points to the analyses of Georges Canguilhem 'as models'.[18]

Foucault does not particularly isolate the concept of transformation of ideas but usually presents it as part of an impressionistic series:

> statements that are no longer accepted or discussed, and
> which consequently no longer define either a body of truth
> or a domain of validity, but in relation to which relations
> of filiation, genesis, transformation, continuity and
> historical discontinuity can be established . . .

> relations of resemblance, proximity, distance, difference,
> transformation . . .

> one can define a regularity (an order, correlations,
> positions and functionings, transformations) . . . [Foucault,
> 1972, pp. 58, 44, 38].

Nevertheless, the many references to the 'very large number of transformations that affect both . . . concepts and their relations'[19] show how important this concept is for Foucault. And occasionally, he does refer to 'conceptual transformations' such as 'the new definition of genus' (1972, p. 147), or 'the transformation of one positivity into another' that marked 'the transition from Natural History . . . to biology . . . at the time of Cuvier'.[20] While Foucault invokes the idea of transformation of scientific ideas, his main focus of interest is not the specific creative acts of transformation made by individual scientists. Rather, his aim is to contrast an 'archaeological . . . concern to establish thresholds, ruptures, and transformations with the true work of historians, which is to reveal continuities'.[21] Whereas I have sought in transformations of ideas the continuities behind the ruptures or scientific revolutions, Foucault has concentrated on transformations as the actual moments of such change or rupture.[22]

NOTES

General note: The extracts from Newton's *Principia* and *System of the World* are either from new translations (in progress) by I. B. Cohen and Anne Whitman or are generally revisions of existing translations.
§1.1

1. The Newtonian revolution in science

1 In recent years, much of the discussion concerning scientific revolutions has centered on Kuhn (1962). For comments on Kuhn's views, see Lakatos & Musgrave (1970). For a modified statement of Kuhn's views, see his paper, 'Second thoughts on paradigms', in Suppe (1974). The propriety of using the term "revolution" to describe scientific change is denied in Toulmin (1972), vol. 1, pp. 96–130, esp. pp. 117sq.

2 This expression is used by historians in a generally uncritical manner that does not necessarily imply adherence to any particular concept of revolution or even to a specific and clearly formulated doctrine of historical change. On the history of this concept and name, see Ch. 2 and Cohen (1977e).

3 Pierre Duhem was largely responsible for the view that many of the discoveries traditionally attributed to Galileo had been anticipated by late medieval thinkers. Duhem's thesis concerning the medieval origin of modern science has been put forth in a new way in Crombie (1953).

4 Still valuable are such older works as Ornstein (1928), the only comprehensive work ever produced on this subject, and Brown (1934); to be supplemented by such recent works as Hahn (1971), Middleton (1971), and Purver (1967).

5 The official name of the Royal Society is: The Royal Society of London for Improving Natural Knowledge.

6 On the history of scientific journals, see Thornton & Tully (1971), and Knight (1975), esp. ch. 4.

7 For Bacon's views on utility, see his *Novum organum*, bk. 1, aph. 73, aph. 124; bk. 2, aph. 3. Descartes's statements concerning the ways in which science can make us 'the masters and possessors, as it were, of nature' (chiefly 'the conservation of health' and 'the invention of . . . artifacts which would allow us the effortless enjoyment of the fruits of the earth . . .') are to be found in pt. 6 of his *Discours de la méthode* (chiefly the end of the second and the beginning of the third paragraphs, and toward the conclusion).

8 Two now-classic statements concerning social influences on seventeenth-century science are given in Hessen (1931) and Merton (1938). A stunning example of how to investigate 'the place of science within the conceptual framework of economic, social, political and religious ideas' has been given for the period in Webster (1975), and cf. the long and thoughtful review of Webster's work by Quentin Skinner, *Times Literary Supplement* (2 July 1976), no. 3877, pp. 810–812.

9 On Newton and ship design, see Cohen (1974*b*); on the longitude at sea, see Newton (1975), introduction, pt. 5; concerning the telescope, see Newton (1958), sect. 2, §§3–5 (§17 contains a description of another instrument, a reflecting octant for use in practical navigation, which was found among Newton's papers and which he never saw fit to make public).

10 Descartes, in his *Discours de la méthode* (1637), says explicitly that he does not consider himself to be a man of more than average mental capacity; hence if he has done anything extraordinary, the reason must be that he had a sound method (see Descartes, 1956, p. 2).

11 Galileo's own account occurs in the fifth paragraph of the text of his *Sidereus nuncius* (1610); Galileo (1890–1909), vol. 3, pt. 1, pp. 60sq; trans. by Drake in Galileo (1957), p. 29.

12 Newton (1672), p. 3075; reprinted in facs. in Newton (1958), p. 47. See Newton (1959–1977), vol. 1, p. 92.

13 The exception was Domingo de Soto (*d.* 1560), who in a commentary on Aristotle's *Physics* (1545) 'was the first to apply the expression "uniformly difform" to the motion of falling bodies, thereby indicating that they accelerate uniformly when they fall and thus adumbrating Galileo's law of falling bodies'. Quoted from William A. Wallace's (1975) account of Soto in the *Dictionary of Scientific Biography*. See Beltrán de Heredia (1961); Wallace (1968); Clagett (1959), pp. 257, 555sq, 658.

14 *Principia*, bk. one, prop. 4, corol. 6, corol. 7. For other examples of Newton's considerations of mathematical relations that do not occur in nature, see §3.3.

15 In the scholium following prop. 78 (bk. one) of the *Principia*, Newton refers to these two as 'major cases of attractions' and finds it 'worthy of note' that under both conditions, the attractive force of a spherical body follows the same law as that of the particles which compose it. On this topic see §3.1, n. 5.

16 In reaction, Alexandre Koyré came to the opposite conclusion: that, far from relying on experiment (and being the founder of the modern experimental method), Galileo was not primarily an experimenter. Furthermore, Koyré even asserted that many of Galileo's most celebrated experiments could not have been performed, at least not in the manner described. See Koyré (1943), (1950*a*), and (1960*c*); these are collected in Koyré (1968). On Galileo's thought-experiments, see Shea (1972), pp. 63–65, 156, 157sq. It is generally recognized today that Koyré's point of view was extreme and needs some modification. Some of the "unperformable" experiments cited by Koyré have since then been performed and yield the very results described by Galileo; see Settle (1961), (1967), and MacLachlan (1973). Drake has recently found that experiment very likely played a significant role in Galileo's discoveries of the principles of motion.

§1.2

1 Leibniz and Newton both had a share in this revolution (see n. 2 to §2.2). But it must be kept in mind that Newton made a large number of discoveries or inventions in mathematics, among them the general binomial expansion of $(a + b)^n$, the fundamental theorem that finding the area under a curve and finding the tangent to a curve are inverse operations, the methods of both the differential and integral calculus, the classification of cubic curves, various properties of infinite series, both the Taylor and Maclaurin expansions, modes of calculation and methods of numerical analysis (including the methods of successive iteration, interpolation, etc.), plus other aspects of geometry, analysis, and algebra. On these topics see Whiteside's introduction to Newton (1964–1967), and Whiteside's introductions and running commentary to his edition of Newton's *Mathematical Papers* (Newton, 1967–).

2 Newton's public positive contributions to chemistry are conveniently summed up in Partington (1961), ch. 13.

3 The contents of these queries are conveniently summarized in Duane H. D. Roller's analytical table of contents in the Dover edition of the *Opticks* (Newton, 1952, pp. lxxix–cxvi) and in Cohen (1956), pp. 164–171, 174–177. On the development of the Queries, see Koyré (1960c).

4 On this tract see Newton (1958), pp. 241–248, 256–258; also (1959–1977), vol. 3, pp. 205–214.

5 All extracts from Newton's *Principia* are given in the text of a new translation, now in progress, by I. B. Cohen and Anne M. Whitman, or are revisions of Andrew Motte's translation.

6 From Newton's unpublished *Conclusio*, trans. by A. R. & M. B. Hall (1962), p. 333.

7 See Roller's analytical table of contents (Newton, 1952, pp. lxxix–cxvi).

8 Young (1855), vol. 1, pp. 161, 183sq; see Peacock (1855), pp. 150–153. But it should not be thought that every number given in the published *Opticks* represents an exact measurement or the result of a computation based on such direct measurement.

9 For example, Roberts & Thomas (1934) is subtitled 'A study of one of the earliest examples of scientific method' and is part of a series with the general title, Classics of Scientific Method.

10 The strictly methodological portions of the *Opticks* are to be found in the final paragraph of qu. 28 and in the concluding pages of the lengthy qu. 31 with which the second English edition of 1717/1718 concludes; both had appeared in earlier versions in the Latin edition of 1706.

11 Quoted from Dover ed. (Newton, 1952), pp. 369sq. This query was first published in the Latin edition (1706) as qu. 20 and then appeared in revised form in English in the second English edition (1717/1718).

12 Quoted from Dover ed. (Newton, 1952), pp. 404sq. This query also appeared first in the Latin edition (as qu. 23) and then in revised form in the second English edition.

13 The word "scenario" is used because Newton wrote up his initial experiments with prisms in an apparently autobiographical manner, but his MSS hint that he was trying to impose a Baconian and experimental-inductivist scenario upon what must have been the sequence of his prior beliefs, experiences, and conclusions. On this subject see Lohne (1965), (1968); Sabra (1967), pp. 245–250.

14 See §3.6, n. 5. These "phenomena" were called "hypotheses" in the first edition of the *Principia*; see Koyré (1955*b*), Cohen (1966).

15 This is made evident in a table I have prepared for a commentary on the *Principia* (in progress), in which I have tabulated the occurrence of every explicit reference to a definition or law, as well as to a preceding proposition or lemma, and – in the case of bk. three – the rules, phenomena, and hypotheses.

16 The question of "analysis" and "synthesis" may cause real confusion in relation to Newton's scientific work. This pair of words (of Greek origin) and their counterparts (of Latin origin) "resolution" and "composition" are used by Newton in a general scientific sense and specifically, in qu. 31 of the *Opticks*, to describe how 'by this way of Analysis we may proceed from Compounds to Ingredients, and from Motions to the Forces producing them' and 'in general, from Effects to their Causes, and from particular Causes to more general ones, till the Argument end in the most general'. Then 'the Synthesis consists in assuming the Causes discovered, and established as Principles, and by them explaining the Phaenomena proceeding from them, and proving the Explanations'. Newton also refers to the 'Two Methods of doing things' of 'Mathematicians . . . which they call Composition & Resolution'.

Long ago Dugald Stewart showed that "analysis" and "synthesis" have different meanings in mathematics and in physics and that Newton therefore speaks with some imprecision in apparently relating the modes of investigation in physics or natural philosophy and in mathematics. Stewart even shows how in some cases "analysis" and "synthesis" may have opposing senses in the two realms. See Stewart's *Elements of the Philosophy of the Human Mind*, ch. 4 ('Logic of induction'), sect. 3 ('Of the import of the words analysis and synthesis in the language of modern philosophy'), subsect. 2 ('Critical remarks on the vague use, among modern writers, of the terms *analysis* and *synthesis*'); Stewart (1877), vol. 3, pp. 272sqq.

For a recent example of an attempt to make the method of analysis and synthesis, as expounded in qu. 31 of the *Opticks*, apply to the *Principia*, see Guerlac (1973). For Newton's published statement on analysis and synthesis in mathematics, and in the *Principia*, see n. 21 *infra*.

17 Turbayne (1962), pp. 46, 49. On the geometric form of presentation in essentially nonmathematical books, see §3.11.

18 The *Principia* was first written as two "books" (*De motu corporum*). Newton then expanded the end of the first "book" into a second "book" (on motion in resisting fluids, pendulum motion, wave motion, etc.), calling *these* two "books" *De motu corporum*. The subject matter of the original second "book" was wholly recast and became the third "book" of the *Principia* (called Liber tertius, *De mundi systemate*). After Newton's death, the text of the original bk. two was published in Latin and in an English version called respectively *De mundi systemate liber* (London, 1728) and *A Treatise of the System of the World* (London, 1728; London, 1731). It is in the English version of this latter work that the famous phrase, 'in a mathematical way', appears. See my introduction to Newton (1975), p. xix.

19 Cuvier (1812), 'Discours préliminaire', p. 3. 'Sans doute les astronomes ont marché plus vite que les naturalistes, et l'époque où se trouve

aujourd'hui la théorie de la terre, ressemble un peu à celle où quelques philosophes croyoient le ciel de pierres de taille, et la lune grande comme le Péloponèse: mais après les Anaxagoras, il est venu des Copernic et des Kepler, qui ont frayé la route à Newton; et pourquoi l'histoire naturelle n'auroit-elle pas aussi un jour son Newton?'

According to John T. Edsall (personal communication), Otto Warburg, in discussing the problem of biological oxidations around 1930, said: 'Heute, wie vor fünfzig Jahren, gilt das van't Hoffsche Wort: Der Newton der Chemie is noch nicht gekommen.' No doubt Warburg was referring to the general introduction ('An die Leser') to vol. 1 of the *Zeitschrift für Physikalische Chemie* (Leipzig, 1887), p. 2, where the state of chemistry is compared with the condition of astronomy in 'Kopernikus' und Kepler's Zeit', and the need is expressed for a 'Newton der Chemie'. This preface was signed jointly by Van't Hoff and Ostwald.

20 A major exception is the general scholium, at the end of sect. 6, bk. two (which was at the end of sect. 7 in the first edition), and the scholium at the end of sect. 7 (first published in the second edition). The latter scholium describes Newton's investigations of the resistances of fluids by experiments on bodies falling in air and in water. The general scholium is devoted to Newton's experiments on the resistances of fluids, in which he studied the oscillations of pendulums under various conditions and compared the motion of pendulums in air, water, and mercury.

21 In his later life, Newton attempted to superimpose on the history of the *Principia* a chronology in which he would have developed and used the new fluxional calculus in an algorithmic form so as to have discovered the main propositions by analysis, and then have recast them in the form of Greek geometry according to the method of synthesis. Thus he wrote: 'By the help of the new Analysis Mr. *Newton* found out most of the Propositions in his *Principia Philosophiae*: but because the Ancients for making things certain admitted nothing into Geometry before it was demonstrated synthetically, he demonstrated the Propositions synthetically, that the Systeme of the Heavens might be founded upon good Geometry. And this makes it now difficult for unskilful Men to see the Analysis by which those Propositions were found out'; quoted from Newton (1715), p. 206; cf. Cohen (1971), p. 295. There is no documentary evidence whatever to support this scenario, while abundant evidence favors the view that Newton's mode of discovery follows more or less the form of presentation in the published *Principia*.

§1.3

1 On Hales see Guerlac (1972); Cohen (1976*b*); F. Darwin (1917), pp. 115–139.

2 A contemporaneous English translation of Van Helmont's account of his experiment is given in Partington (1961), p. 223.

3 *De motu cordis*, ch. 9; quoted from Harvey (1928). Cf. Kilgour (1954) and especially Pagel (1967), pp. 73sqq.

4 Hence the quantitative method used by Harvey is at least as revolutionary as his conclusions about the circulation, and possibly even more so. From a seventeenth-century point of view, such subjects as theoretical statics, kinematics, and dynamics were exact mathematical sciences which became physical sciences only when applied to physics.

5 Galileo (1974), p. 147; (1890–1909), vol. 8, p. 190. Strictly speaking, Galileo

did not ever express his physical laws as the algebraic proportions
$s \propto t^2$ or $v \propto t$. As a matter of fact it is misleading even to write out his
results in the form $s_1{:}s_2 = t_1{}^2{:}t_2{}^2$, much less $(S_1/S_2) = (t_1/t_2)$. On this point
see n. 8 *infra*. In what follows, I shall (as a kind of shorthand) refer to
relations that Galileo found as $s \propto t^2$ or $v \propto t$, but without any intended
implication that these are Galileo's formulations of such laws.

6 'On naturally accelerated motion', third day, prop. 2; Galileo (1974),
p. 166; (1890–1909), vol. 8, p. 209.

7 In corol. 1 to prop. 2, Galileo shows that although the total distance
traversed is proportional to the square of the time, the distances traversed
in each successive equal interval of time are as the odd numbers starting
from unity–a result that follows from number theory, since the sequence
1, 4, 9, 16, 25, . . . leads to the sequence $1(= 1 - 0), 3(= 4 - 1), 5(= 9 - 4),$
$7(= 16 - 9), 9(= 25 - 16), . . .$

8 Galileo did not restrict himself to such number relations. Thus (third day,
prop. 2 on accelerated motion: 1974, p. 166; 1890–1909, vol. 8, p. 209):
'If a moveable descends from rest in uniformly accelerated motion,
the spaces run through in any times whatever are to each other as the
duplicate ratio of their times'. Galileo's proportion is thus $\mathrm{space}_1{:}\mathrm{space}_2 =$
$(\mathrm{time}_1{:}\mathrm{time}_2)^2$; he does not use the functional relation $s \propto t^2$. On this
point see the comments by Drake in the introduction to Galileo
(1974), pp. xxi–xxiv. But in discussing quantities of the same kind
(e.g., line segments), Galileo would use verbal equivalents of equations,
such as "*HB* est excessus *NE* super *BL*" ($HB = NE - BL$).

9 Quoted from *Il saggiatore* ('The assayer'), sect. 6, trans. in Crombie
(1969), vol. 2, p. 151; Galileo (1890–1909), vol. 6, p. 232. This statement
is omitted in Drake's version, Galileo (1957). I do not wish to enter here
into discussions of Galileo's possible Platonism, for which see Koyré (1943)
and a rebuttal by Clavelin (1974). Geymonat (1965), pp. 198sq, warns
against reading this particular remark of Galileo's out of context.

10 Newton's *Principia* is apt to be described, on the basis of superficial
examination, as a treatise in the style of Greek geometry. Although the
external form displays a geometrical style reminiscent of Euclid, a
closer examination shows that Newton's method is not at all like
that of the classic Greek geometers; rather, proposition by proposition
and lemma by lemma, he usually proceeds by establishing geometrical
conditions and their corresponding ratios and then at once introducing
some carefully defined limiting process. In sect. 1 of bk. one, Newton sets
forth general principles of limits (which he calls the method of first
and last ratios), so that he may apply some degree of rigor to problems
using nascent or evanescent quantities (or ratios of such quantities) in the
rest of the treatise. Furthermore, even in the matter of ratios and
proportions, Newton is a "modern"; he does not follow the Greek style,
in the sense that Galileo does and that Kepler tends to do. That is,
he writes "mixed" proportions, implying a direct functional relationship, for instance stating that a force may be directly or inversely
proportional to some condition of a distance. Traditionally one would have
had to say that one force is to another as a condition of some distance
is to that same condition of another distance. Finally, one of the
distinctive features of the *Principia*, as noted by Halley in his review,
was the extensive and innovative use of the method of infinite series,

which shows the degree to which the *Principia* is definitely not a treatise in Greek geometry. On this topic see Cohen (1974*c*), pp. 65sqq, and Whiteside (1970*b*).

11 Kepler's system of nested spheres is delineated in his *Mysterium cosmographicum* (1596; rev. ed. 1621). An annotated English version of this work has been completed by Eric J. Aiton and Alistair M. Duncan (Kepler, in press).

In a letter of 1595, Kepler said: 'The world of motion must be considered as made up of rectilinear [regular solids]. Of these, however, there are five. Hence if they are to be regarded as the boundaries or partitions . . . they can separate no more than six objects. Therefore six movable bodies revolve around the sun. Here is the reason for the number of planets'; quoted from Kepler (1965), p. 63. This example shows how considerations of shape and geometry were not necessarily free of numerical aspects.

12 In the *Narratio prima* (or *First Account*), trans. in Rosen (1971), p. 147, Rheticus said, 'What is more agreeable to God's handiwork than that this first and most perfect work should be summed up in this first and most perfect number?' For a history of this problem, see Cohen (1977*d*).

13 Quoted from Kepler's *Mysterium cosmographicum* (1596) in Kepler (1937–), vol. 1, p. 9; trans. in Rufus (1931), p. 10. Cf. Koyré (1973), p. 128.

14 Sizi found other grounds for his assertion: the seven primary metals in alchemy, the time when the embryo starts to form in a mother's womb (seven hours after coitus), the date at which a human fetus is sufficiently alive to survive if born prematurely (seven months after conception). See Drake (1958) and Ronchi's introduction to Clelia Pighetti's translation of Sizi (1964).

15 Kepler (1965), p. 14. Since the earth has one satellite and Jupiter has four, a geometric sequence would yield two for Mars, eight for Saturn, and none for Mercury and Venus. The number 6, which Kepler suggests as an alternative for the eight satellites attributed to Saturn, is more difficult to account for. We may well understand, accordingly, why some scholars have made a "silent correction" of this 6, so as to have it be 7 or 5. This number 6 does fit the arithmetic progression 2,4,6, but in this case the earth would have no satellite, which would actually negate the basis of assigning two to Mars. Furthermore, Kepler also suggested that Venus and Mercury might have one satellite each. These two numbers would break the sequence, but there could be no other choice if each planet is to have no more satellites than the immediately superior planet.

16 For details see Cohen (1977*b*), (1977*d*).

17 For other examples of numerology in nineteenth- and twentieth-century science, see Cohen (1977*d*). An outstanding example is the so-called Bode's law (or the Titius-Bode law), which gives reasonably good values for the planetary distances (up to Uranus), including a place for the asteroids. It too fails for the first term. See, further, Nieto (1972).

§1.4

1 Quoted from *Mysterium cosmographicum* (1596) in Kepler (1937–), vol. 1, p. 9; trans. in Rufus (1931), p. 9. Cf. Koyré (1973), p. 138. In fact, Kepler went on to say that he 'was induced to try and discover them [i.e.,

these three things] because of the wonderful resemblance between motionless objects, namely, the sun, the fixed stars and intermediate space, and God the Father, God the Son, and God the Holy Ghost; this analogy I shall develop further in my cosmography'.

2 *Mysterium cosmographicum*, ch. 20. To see how this law works, note that the periods of Mercury and Venus are respectively 88^d and 224$\frac{2}{3}$^{d}; hence half the increase in period is $\frac{1}{2}$(224$\frac{2}{3}$^{d} − 88^d) = 68$\frac{1}{3}$^{d}. Kepler's rule is that 88:(88 + 68$\frac{1}{3}$) = dist. of Mercury:dist of Venus. The results, as given by Dreyer (1906), p. 379, are Jupiter:Saturn 0.574 (0.572), Mars:Jupiter 0.274 (0.290), earth:Mars 0.694 (0.658), Venus:earth 0.762 (0.719), Mercury:Venus 0.563 (0.500); the number in parentheses in each case is the Copernican value. In an equation, Kepler's rule would read $(T_n + T_{n-1})/2T_{n-1} = A_n/A_{n-1}$; cf. Koyré (1973), pp. 153sq.

3 See Kepler's *Harmonice mundi*, in Kepler (1937–), vol. 6, p. 302. Kepler is astonishingly silent as to how he came upon this law. Koyré (1973), p. 455, n. 27, discusses some conjectures on this topic made by J. B. Delambre and R. Small (he gives his own opinion on p. 339). See, further, Gingerich (1977).

4 This difference between the third law and the first two is seen in Newton's treatment of them. He admits that the third law, 'found by Kepler, is accepted by everyone' (hypoth. 6, ed. 1, phen. 4, eds. 2 and 3, *Principia*, bk. three); but he does not give Kepler credit in the *Principia* for either the area law or the law of elliptical orbits, and at least once he claimed that Kepler had only guessed planetary orbits 'to be elliptical'.

5 Galileo's Platonic "discovery" did not embody a "causal" explanation in the sense of assigning a physical cause to the supposed accelerated motions of the planets toward the sun: for example, by supposing forces that might be operative in the celestial system to produce such accelerations.

6 Cf. Koyré (1960*b*), reprinted in Koyré (1965), where (p. 218, n. 3) Koyré discusses how A. E. Taylor believed (erroneously, as it turned out) that he had found the source of this supposed cosmological doctrine of Plato.

7 Cf. Cohen (1967*c*). Newton also pointed out other faults in Galileo's assumptions; see Whiteside's commentary in Newton (1967–), vol. 6, pp. 56sq, n. 73.

8 Galileo was aware that in projectile motion there is an acceleration in the same direction as gravity or weight; whereas there is no acceleration or deceleration (save for the slight retardation caused by air resistance) at right angles to that downward direction. This is not, however, a real anticipation (however limited) of the second law, since Galileo does not specify clearly that the "impeto" is an external force acting on a body in order to produce an acceleration. And the same is true of Galileo's analysis of motion along an inclined plane, where both the "impeto" of gravity and the acceleration are diminished in the ratio of the sine of the angle of elevation. Drake has discussed Galileo's concept of cause in his introduction to Galileo (1974), pp. xxvii–xxviii; see, further, Drake (1977).

9 For a convenient summary of the medieval physics of motion see Grant (1971), ch. 4. For texts and translations see Clagett (1959) and Grant (1974), sects. 40–51.

10 For Galileo's description of this series of experiments, see Galileo (1974), pp. 169sq; (1890–1909), vol. 8, pp. 212sq.

11 Whereas Kepler begins with the nature of the force, Newton concludes in the inquiry into the nature of a force with certain properties that have come to light during the antecedent investigations: as that it diminishes with the square of the distance, extends to great distances, and is proportional to the mass of bodies, etc.

12 Concluding general scholium to the *Principia*: see, further, §3.2.

13 *Almagest*, bk. 9, sect. 1.1. In his *Planetary Hypotheses*, Ptolemy developed a physical system or a physical model of astronomy in addition to the mathematical computing models of astronomy described in the *Almagest*. Cf. Hartner (1964), supplemented by Goldstein (1967).

§1.5

1 A doctoral dissertation on Kepler's optics was completed in 1970 by Stephen Straker (Indiana University).

2 Galileo was aware that if a moving body were to continue in motion along a horizontal path (tangent to the earth), it would in effect be getting farther and farther away from the earth's center or surface: rising up, as it were, *sponte sua*.

3 Galileo himself fell into this trap in his arguments for the Copernican system. He developed a theory in which the tides are produced by a combination of the motions of the earth. Hence, he believed (and argued), God must have created the universe with the earth rotating on its axis and revolving in an orbit, just as Copernicus said. Pope Urban VIII argued against the "conclusiveness" of this proof of the Copernican system on the grounds that it would limit God's omnipotence. Galileo had only shown that his version of the Copernican system would imply tidal phenomena similar to those we observe; he had not demonstrated the converse. His Copernican system was a sufficient condition to explain the tides, but it was not a necessary condition. On Galileo's theory of the tides, see Aiton (1954) and Burstyn (1962); also Aiton (1963) and Burstyn (1963).

4 Of course, as we shall see below in Ch. 4, a system, construct, or model could successively gain additional features that would bring it so nearly into harmony with experience that it would seem to be a description of reality.

5 Koyré (1973), p. 166, says that 'the very title of Kepler's work proclaims, rather than foretells, a revolution'.

6 Both Kepler and Newton held that the demise of the concept of crystalline spheres required a theory of planetary motions based on forces.

7 On Borelli, see Koyré (1952a). There is no adequate study of the celestial system of Descartes or of Bullialdus.

8 In Galileo's presentation of Plato's cosmological scheme (see §1.4, nn. 5 and 6, as well as Galileo, 1953, pp. 29sq; 1890–1909, vol. 4, pp. 53sq; also 1974, pp. 232–234, and 1890–1909, vol. 8, pp. 283sq), he seems to have assumed that when a planet was started off in its orbit with the proper speed, it would then move in its orbit without needing the action of any force.

9 Unsuccessful and faulty as a system in general, Kepler's dynamics did, however, serve to establish the first two Keplerian laws of planetary motion. See Koyré (1973), pp. 185–244; Krafft (1973).

10 *Mysterium cosmographicum* (1596), quoted in Duhem (1969), p. 101; Kepler (1937–), vol. 1, p. 16.

2. Revolution in science and the Newtonian revolution as historical concepts

§2.1

1 In mathematics and the physical sciences, this word still is used in its original sense: as in "solid of revolution" (a solid formed by the turning of a plane figure about an axis through 360°) and in the "revolution of a planet" (the motion of a planet in its orbit through 360°).
2 On the history of the concept and name of "revolution", see Cohen (1977e).
3 See Clairaut (1743) and §2.2, and the statement by Fontenelle in §2.2.
4 Then there occurs a series of short paragraphs (by $O = $ d'Alembert) on revolution as a term of geometry and of astronomy, plus a brief statement concerning 'Révolutions de la terre'; and then a technological study, occupying almost two pages, on horology (by Joh. Romilly).

§2.2

1 Fontenelle (1790), vol. 6, p. 43. Fontenelle also used the term "revolution" several times in relation to the Marquis de l'Hôpital's textbook on the calculus (in the *éloge* of l'Hôpital and in the *éloge* of Rolle); see Cohen (1976a), pp. 267–269, and Cohen (1977e).
2 In the preface to the 1727 *Eléments de la géométrie de l'infini*, Fontenelle said of the calculus: 'Newton trouva le premier ce merveilleux calcul, Leibnitz le publia le premier. Que Leibnitz soit inventeur aussi bien que Newton, c'est une question dont nous avons rapporté l'histoire en 1716, et nous ne la répéterons pas ici'.
3 It is to be observed that d'Alembert refers to both Newton's *Opticks* and *Principia*, but that his opening description applies more particularly to the *Principia*.
4 For an analysis of Lalande's views on revolutions in science, see Cohen (1977e).
5 Condorcet used the term "révolution" in the *éloges* of Duhamel du Monceau (1783), Albrecht von Haller (1778), d'Alembert (1783), and Euler (1783). In the first three he coupled the word "époque" with "révolution".
6 For details, see Cohen (1976a), (1977e).
7 Quoted in Berthelot (1890), p. 48.

§2.3

1 I use the word "science" here in distinction to "mathematics"; see the second paragraph of §2.2.

3. The Newtonian revolution and the Newtonian style

§3.1

1 I qualify "observation" by the adjectives "critical" and "precise" so as to make it clear that I do not have in mind merely "what everyone knows who has eyes in his head". I refer specifically to observations of planetary

positions, rates of clocks, various measured magnitudes, and so on.

2 Newton proves that a body moving with uniform rectilinear motion
will sweep out equal areas in any equal times; it is supposed that the point
is not in the line of motion. The proof appears in prop. 1, bk. one, of the
Principia; see §5.5.

3 See Routh (1896–1902), vol. 2, p. 44, sect. 99. The attraction of a uniform
solid sphere at an internal point at a distance r from the center can be
computed in two parts. The shell from the distance r to the outer
surface exerts no gravitational force on a particle within the shell (prop.
70, bk. one). The inner core-sphere of radius r acts as if all its mass
were concentrated at its center. Hence the force is proportional to $1/r^2$.
But since the force is also proportional to the mass (which in a uniform sphere is proportional to the volume), the force must be proportional
to $(1/r^2) \cdot (r^3)$ or the distance r (see *Principia*, prop. 72, bk. one).

4 By the time Newton seriously addressed himself to the problem of
elliptical orbits (in 1679 and thereafter), there was in use another kind
of planetary speed law, which for practical purposes had replaced the area
law; see §5.2. Astronomers associated this law and its modifications with
the names of Bullialdus, Seth Ward, and N. Mercator. See, further,
Whiteside (1964*b*); Wilson (1970); Maeyama (1971). On Kepler and the
law that the speed of a planet is inversely proportional to its distance
from the sun, see Aiton (1969).

5 Hooke to Newton, 6 Jan. 1679/80; Newton (1959–1977), vol. 2, p. 309;
see §5.4.

6 See also prop. 16, bk. one, of the *Principia*; and corol. 1, prop. 1, bk one
(in the second and third editions of the *Principia*). See, further, §5.4.

7 In the scholium following the definitions in the beginning of the *Principia*;
see, further, Cohen (1970) and (1974*c*), p. 69.

8 U.L.C. MS Add. 3958, sect. 3, fols. 48–63, first published by Hall & Hall,
eds. (1962), pp. 15–64; published with a commentary by Whiteside in
Newton (1967–), vol. 1, pp. 400–448, under the title 'The October
1666 tract on fluxions' (see *Waste Book*, U.L.C. MS Add. 4004, fol. 51;
and Newton, 1967–, vol. 1, pp. 392–399, for a prior version).

9 It is with regard to the "mechanical" curves (those which cannot be
written simply as algebraic equations, and which are defined as the locus
of a point moving according to certain specifications) that Newton
introduces the principles usually associated with the physics of motion
into considerations of pure mathematics. For the documents in which
Newton establishes the principles of the calculus for "mechanical" curves
by using the principles of motion, together with an illuminating commentary by D. T. Whiteside, see Newton (1967–) vol. 1, pp. 369sqq
('How to draw Tangents to Mechanicall lines', 30[?] Oct. 1665); pp. 377sqq
('How to Draw Tangents to Mechanicall Lines', 8 Nov. 1665). The latter
has a clear opening statement of the composition of velocities. The
parallelogram rule is to be found in the *Waste Book* as of autumn 1664,
while the concept of uniform (or inertial) motion occurs (fol. 12) as of
January 1665. Other documents are given on pp. 382sqq ('To find
the velocitys of bodys by the lines they describe', 13 Nov. 1665); pp.
390sqq (a restatement of how to 'resolve . . . Problems', 14 May 1666,
containing not only the parallelogram rule, but a proposition inspired
by Galileo's proposition – *Two New Sciences*, 3d day, foll. corol. 3, prop. 6

on accelerated motion [Galileo, 1974] – that bodies "falling" along any chords from a given point on a circle to the circumference will reach the circumference in the same time); pp. 392sqq (a recasting of 16 May 1666). These papers are followed by what has been called (following Newton's own appellation) 'The October 1666 tract on fluxions', ibid., pp. 400sqq ('To resolve Problems by Motion . . .'). See also Newton (1967–) vol. 2, pp. 194sqq ('De Solutione Problematum per Motum', *c.* 1668, a revised version in Latin of the English paper 'To resolve Problems by motion' of 16 May 1666; vol. 1, pp. 392sqq).

10 *Waste Book* (U.L.C. MS Add. 4004), fols. 50–51; transcribed and edited, with a commentary, in Newton (1967–), vol. 1, pp. 377–392. For another expression of the properties of curves by the use of kinematics, see ibid., p. 382, 'To find the velocitys of bodys by the lines they describe'. Newton apparently recognized within a week that this was an improper application of the vector parallelogram to drawing tangents.

11 Since the *Waste Book* contains, among other things, tracts on geometry and tracts on kinematics and dynamics, a first glance may not be sufficient for a scholar to tell whether a given page may have pure mathematics or the physics of motion as its subject. Portions of the *Waste Book* have been published in Newton (1967–) and in Herivel (1965a).

12 Newton (1737), p. 26; cf. Newton (1967–), vol. 3, p. 71. Whiteside (pp. 17, 71) sees in Newton's use of velocities a possible influence of Barrow. Whiteside observes (p. 71, n. 82) that Newton will elaborate, as problems 1 and 2, the method of finding, 'in the geometrical model of a line-segment traversed continuously in time, . . . the 'celerity' or 'fluxional' speed of a variable quantity as its derivative and, conversely, the determination of that 'fluent' quantity as the integral of the fluxional speed, where in either case 'time' is the independent variable'.

13 Newton (1737), p. 27; Newton (1967–), vol. 3, p. 73; the "dotted" fluxions represent a Newtonian notation which later became standard, but which he 'did not introduce . . . till late 1691' (1967–, vol. 3, pp. 72–73, n. 86).

14 Harris (1704), s.v. "Fluxions". This example is actually taken from the *Principia*, bk. 2, lem. 2, case 1, but is there presented in terms of *genita* and *momenta*. The rectangle of sides x and y is first diminished to one of sides $x - (\dot{x}/2)$ and $y - (\dot{y}/2)$ and the product or the diminished area will be $xy - y(\dot{x}/2) - x(\dot{y}/2) + (\dot{x}\dot{y})/4$. The rectangle is then increased by these same 'half Moments or Fluxions' and the new area will be $xy + y(\dot{x}/2) + x(\dot{y}/2) + (\dot{x}\dot{y}/4)$. Subtracting one from the other gives $x\dot{y} + \dot{x}y$, the fluxion of xy. The subtly fallacious part of Newton's argument was first detected by D. T. Whiteside: see Newton (1967–), vol. 4, p. 523, n. 6.

15 See Newton (1715); the many drafts in Newton's hand of this review, published anonymously, are to be found in U.L.C. MS Add. 3968. A draft of the commission's report (in his hand) exists; this report was published as if it were an impartial report of the Royal Society's commission of inquiry.

16 That is, these are the subjects of bks. one and three of the final *Principia*; see n. 18 to §1.2. The subjects of book two (as published in the *Principia*) embrace motion in various types of resisting mediums, vibrations of pendulums in resisting and nonresisting mediums, the physics

of deformable bodies, wave motion, and the transmission of sound, and kindred topics. On the several books of the *Principia* and their successive stages of composition, see Cohen (1970), Introduction; on the contents of bk. two, see Truesdell (1970).

17 Cf. n. 10 to §1.3. Newton himself was delighted by the statement made by the Marquis de l'Hôpital, in the preface to his 1696 *Analyse des infiniment petits* (the first textbook on the new infinitesimal calculus), that Newton's 'excellent Livre intitulé *Philosophiae Naturalis principia Mathematica . . .* est presque tout de ce calcul'.

18 One aspect of Newton's approach to the mathematics of natural philosophy which I have not explored is the place of his work in the tradition of the mathematization of space which Alexandre Koyré has shown to be of such importance for the new science of motion. The physics of inertia, he has shown, depends upon three presuppositions: '(a) the possibility of isolating a given body from all its physical environment, (b) the conception of space which identifies it with the homogeneous, infinite space of Euclidean geometry, and (c) a conception of movement – and of rest – which considers them as *states* and places them on the same ontological level of being'; quoted from Koyré (1968), p. 4. The changing concept of space in relation to the concept of a dynamical trajectory is to be found in Koyré (1939), pp. 107–136, 318–341. Readers of Koyré's writings on these topics will be aware of the degree to which his views have been the starting point from which my own have developed.

19 It does not follow, however, that this harmony between Newton's kinematic approach to mathematics and his analysis of the physics of motion should have produced significant innovations on the purely mathematical level in the *Principia*.

§3.2

1 Preliminary draft of Newton (1715), U.L.C. MS Add. 3968.

2 On this topic, see Koyré (1965), pp. 139–148. The question of occult qualities in Newtonian science was introduced by Leibniz and by N. Hartsoeker. Replies were made by Cotes in the preface he wrote for the second edition of the *Principia* and by Newton in the *Recensio libri* (1715), pp. 222 sq (reprinted by Koyré) and in qu. 23 of the *Optice* (1706), translated into English and revised as qu. 31 of the second English edition of the *Opticks* (1717/1718).

3 Here, and throughout this book, I have used the expression "one-body system", even though strictly speaking a single body cannot of and by itself constitute a "system". But I know of no other way to indicate simply the kinship between such a "one-body system" and a two-body system, three-body system, many-body system. Furthermore, the Newtonian "one-body system" is a "system" to the extent that it is composed of two entities, even though these are not homologous, as in the case of a system of two bodies: these are a single body (or mass point) and a center of force.

4 In this it differs from Descartes's hypotheses or "comparisons", such as those introduced in the beginning of his *Dioptrique*; see §3.7.

5 We shall see below (in Ch. 5) that the mutuality of the solar-planetary force does not occur in the earliest version of the tract *De motu*, but only in a later revision, or afterthought. Hence this step must have been

taken after the first version, later than November 1684 (see Cohen 1971,
ch. 3, pt. 2).

6 Or, what is essentially the same, a system of a body or point mass
moving around a fixed body; or, a set of such moving bodies that form
a system but do not interact with (or act upon) one another.

7 Newton hazarded a guess that there might be electrical forces that would
not require friction to excite them, and he was aware that the earth is
a magnetic body.

8 One must be careful not to exaggerate Newton's success in the matter
of the tides, as happens all too often in secondary accounts. He could explain the periodicity factors in both ebb and flow and the cycle from
spring tide to neap tide to spring tide. But of course he was powerless to
predict the local times of high or low tides, the heights of the tides,
and so on. He was also unable to account for the responses of bodies of
water in terms of their geographical and physical conditions, and
thus (for example) could not account for the possibility of one tide or
two tides per day. But he did try to explain the occurrence of single
daily tides in the Tonkin Gulf, by supposing hypothetical conditions
under which there could be both "destructive" and "constructive"
interference phenomena (see *Principia*, bk. three, end of prop. 24; Cohen,
1940). What Newton did accomplish, however, was to show the action
of the sun's and moon's gravitating forces in the production of tidal
phenomena.

9 As shall be seen below, on at least two occasions (in the mid-eighteenth
century [Clairaut] and in the early nineteenth century [Le Verrier]),
mathematical astronomers raised doubts about the absolute accuracy of
the inverse-square rule of the law of gravitation, but they did not
doubt that there was a force of universal gravity.

10 I do not wish to enter here into the debate whether the Newtonian
theory has been shown to be a special case of a more general Einsteinian
theory or whether relativity has displaced classical Newtonian dynamics.

11 For Newton this would no doubt have included a solution of the
lunar problem by the methods of gravitational celestial mechanics to
replace his use of geometric models for calculation, but it would certainly
not have included the many aspects of the dynamics of rigid bodies
and deformable bodies that were developed by post-Newtonian mathematical physicists such as Euler.

§3.3

1 These two laws of force, essentially $F \propto d(mV)$ and $F \propto (d/dt)(mV)$, and the
Newtonian transition from one to the other, are discussed in §4.4. Basically
there is no distinction between these two laws if there is a uniformly
flowing mathematical "time", so that $dt = $ constant; that is, $d(mV)$
and $d(mV)/dt$ differ only by a constant of proportionality, dt. And it is
the same with respect to a force F, its "impulse" $F \cdot dt$, and its spatial
argument $\frac{1}{2} F \cdot dt^2$; the context determines which of these Newton
has in mind.

Newton never wrote the second law in an equation or proportion using
letters or differentials (or even fluxions). But he did state his proportions
in the "mixed" form of the moderns. Thus, unlike Galileo (who traditionally would write that the ratio of one distance to another distance

is as the square of the ratio of the first time to the second), Newton would write that the distance is proportional to the square of the time. Thus the second law actually reads that the 'change of motion is proportional to the motive force impressed'. But it appears also in more general form (as at the end of the proof of prop. 39, bk. one): 'and the force is directly as the increment of the velocity and inversely as the time'. For an apparently unique example of fluxions applied to the second law, see Newton (1967–), vol. 7, p. 128: 'velocitatis fluxio est ut corporis gravitas' ('the fluxion of the velocity is as the weight of the body').

In addition to these two modes of action (instantaneous and continuous), Newton introduced three forms of 'impressed force', according to its 'diverse origins': 'percussion, pressure, or centripetal force' (def. 4).

2 The problem of a planetary "force", which could be caused by a physical sun-centered vortex, is different from that of a force mutually acting between sun and planets since that mutuality could not be accounted for by a simple vortex (see, further, §3.4).

3 J. T. Desaguliers was one such (see Cohen, 1956, pp. 249–251). Newton himself referred to magnetism in the *Principia*, and used electricity and magnetism in qu. 22 of the *Opticks* to show an aethereal medium can be so "rare" as to offer an "inconsiderable" resistance to the motion of bodies, and yet be potent enough to produce gravitational and optical effects.

4 Newton himself does not use the word "limit" in the title of sect. 1, which is entitled 'On the method of first and last ratios'. But in the text of sect. 1, the word "limit" does occur, as in corol. 4, lem. 3, where Newton refers to the 'curvilinear limits of rectilinear figures'; and in the scholium at the end of sect. 1, he refers to 'the limits of . . . sums and ratios' and 'the sums and ratios of limits', etc.

5 The tenth of the eleven lemmas in sect. 1 is concerned with the 'spaces that a body describes when urged by a finite force, whether that force is continually increased or continually diminished'; Newton proves that 'at the very beginning of a motion' these spaces are as the squares of the times. Here the force is conceived in a purely mathematical way, divorced in its context from physical application or exemplification.

6 That is, the arguments and proofs are mathematical and do not depend upon "proof by experiment" at any stage of the argument.

7 As in the case of the area law (props. 1–2), there is no statement made by Newton that the elliptical orbits may occur in nature. In bk. three, in the opening set of "phenomena" (among the "hypotheses" in the first edition), the area law is given, but *not* the elliptical orbits, which are first introduced in bk. three in prop. 13; see §5.6.

8 On Newton's solution as 'essentially' the same as Christopher Wren's (1650), see Whiteside's n. 134 in (Newton (1967–), vol. 6, p. 310, and also n. 128, p. 308. Kepler's problem is to invert the equation $T = \vartheta \mp e \sin \vartheta$ so as to yield ϑ 'explicitly as an algebraic function of T'.

9 These are the two that are of real significance in the gravitational world of nature: $f \propto 1/r^2$ for the centripetal force outside of a uniform spherical shell or a solid sphere that is either uniform or made up of a set of such uniform spherical shells, and $f \propto r$ for the case of a body within a uniform solid sphere at a distance r from the center.

10 This is a recurring phrase in the *Principia*. For the actual quadratures

(or integrations) which Newton assumes, see Newton (1967–), vol. 3, pp. 210–292, esp. pp. 236–254.

11 On this proposition, see Brougham & Routh (1855), pp. 80–87, based on Whewell (1832), pp. 61sqq, and Whiteside's commentary in Newton (1967–), vol. 6, pp. 345sqq, esp. par. 2 of the note on p. 349.

12 That Newton is actually dealing with a mathematical system, or with motion in mathematical rather than experiential or physical space, may be seen by the fact that the "line segments" or "displacements", as in prop. 1 bk. one, are (as D. T. Whiteside has shown) not finite (as they superficially appear to be), but infinitesimal, and even in some cases infinitesimals of the second order.

13 I use the expressions "mathematical construct" or "mathematical system" or "imagined system" rather than "model", since in today's usage a "model" is a different type of entity, designed in order to explain a set of phenomenological rules or other results of experiment or observation. On models, see the admirable succinct survey in Hesse (1967); also Hesse (1966) and Leatherdale (1974).

14 Even so astute a scholar as the late Alexandre Koyré could not understand how Newton might have believed that by 'using familiar language' such as the word attraction, he would 'be more easily understood by mathematical readers' (see Koyré, 1965, pp. 150–154).

15 In sect. 2 of the *System of the World* (see n. 1 to suppl. to §3.5), Newton said, 'It is our purpose to examine its quantity and properties [that is, the quantity and properties of the force that keeps orbiting bodies in their curved paths] and mathematically to investigate its effects in the moving of bodies . . .'; this extract is given in full in suppl. to §3.5, above n. 2, where its significance is discussed.

16 The analysis of the reaction of Newton's contemporaries to the word "attraction" is a notable feature of Koyré (1965).

17 In this phrase, surely, Newton is giving us more than a clue as to his personal belief in the cause of gravity being a shower of some kind of aethereal particles.

18 Quoted from the end of the lengthy discussion following def. 8; def. 5 deals with centripetal force, and defs. 6–8 with the measures of centripetal force.

19 He is obviously referring to the quotation given in n. 15 above.

20 There have been many such mathematical systems in bks. one and two; it is the final and most complex one which fits the conditions of observation.

21 Dalton did not recognize, in other words, that the law $f \propto 1/r$ is a necessary and sufficient condition for Boyle's law *if and only if* there is a force of repulsion between particles (or if and only if the static model based on repulsion is both sound or workable and holds in nature). Newton, furthermore, was aware that he had only considered that such forces act between immediately adjacent particles. On Dalton, see Roscoe & Harden (1896), p. 13, quoting a lecture of 27 January 1810, reading in part: 'Newton had demonstrated clearly, in the 23rd Prop. of Book 2 of the *Principia*, that an elastic fluid is constituted of small particles or atoms of matter, which repel each other by a force increasing in proportion as their distance diminishes.'

22 Later on, in qu. 31 of the *Opticks*, Newton would suggest that the attractive force that is operative when metals are dissolved in acid, and which 'can

reach but to a small distance from them', can turn into repulsion at greater distances. On this occasion, he used a simile from mathematics, that, 'as in Algebra, where affirmative Quantities vanish and cease, there negative ones begin; so in Mechanicks, where attraction ceases, there a repulsive virtue ought to succeed'.

§3.4

1 This general scholium was written for the second edition (1713).

2 Of course, vorticists (like Huygens) could not conceive of mutual gravitation, since the central body in a vortex has a purely passive role and cannot affect any body being pushed into it by the vortical motion. Thus Huygens believed in a scheme of vortices that would cause the earth to be pulled or pushed in toward the center of the vortex, and not necessarily toward a physical body, the sun, placed there (see the quotation from Huygens, above n. 5). But Huygens's vortices were not strictly those of Descartes; for the differences and similarities, see Koyré (1965), ch. 3, app. A, and especially Aiton (1972). Mutual gravitation thus would rule out the vortex as a cause. Newton's theory of the tides (caused by the pull of sun and moon on the water in the seas) also could not be explained by a vortex. Hence many aspects of the gravitational theory must have seemed to suggest attraction.

3 Huygens (1690), 'Discours sur la cause de la pesanteur'; Huygens (1888–1950), vol. 21, pp. 472, 474. This and the following extracts are translated in Koyré (1965), pp. 121sq.

4 Huygens to Leibniz, 18 Nov. 1690: Huygens (1888–1950), vol. 22, p. 538; Koyré (1965), pp. 117–118. Newton's demonstration that the Cartesian vortices are inconsistent with Kepler's laws may be found at the end of bk. two of the *Principia*.

5 For his planned revision of the vortex theory, see Huygens (1888–1950), vol. 21, p. 361; and for the result, see his additions to the 'Discours', ibid., p. 471; Koyré (1965), p. 118.

6 'Varia astronomica', Huygens (1888–1950), vol. 21, pp. 437–439. In a letter to Leibniz in October 1693, Newton said: 'But some exceedingly subtle matter seems to fill the heavens' ('At caelos materia aliqua subtili[s] nimis implere videtur'). Then he repeated the argument against Cartesian vortices from the end of bk. two of the *Principia*: 'For since the celestial motions are more regular than if they arose from vortices and observe other laws, so much so that vortices contribute not to the regulation but to the disturbance of the motions of planets and comets; and since all phenomena of the heavens and of our sea follow precisely, so far as I am aware, from nothing but gravity acting in accordance with the laws described by me; and since nature is very simple, I have myself concluded that all other causes are to be rejected and that the heavens are to be stripped as far as may be of all matter, lest the motions of planets and comets be hindered or rendered irregular.' And then he returned to the possibility of a "subtle" matter: "But if, meanwhile, someone will explain gravity along with all its laws by the action of some subtle matter [siquis gravitatem una cum omnibus ejus legibus per actionem materiae alicujus subtilis explicuerit] and will show that the motion of planets and comets will not be disturbed by this matter, I shall be far from objecting.' See Newton (1959–1977), vol. 3, pp. 285–287.

7 In his 'Discours sur la cause de la pesanteur'; see n. 3.
8 Accordingly, Huygens never could have seen the full implications of the concept of mass.
9 This exists at present in the form of a computer printout prepared by I. B. Cohen, Owen Gingerich, Anne Whitman, and Barbara Welther. It is hoped to make it available to other scholars in print or on microfiche. Eventually we would like to produce an index of major words in context in both Latin and English.
10 Newton also uses the verb *trahere* (to draw, or to pull), which does not necessarily have the same overtones as *attrahere*; thus a horse may be said to "draw" (*trahere*) a wagon by means of the traces, but not to "attract" (*attrahere*) the wagon. There are, however, very few occurrences of *trahere* in book 3.
11 This corol. 3 to prop. 5, bk. three, was not part of the first edition; it was printed for the first time in the second edition (1713).

§3.5

1 These 'Rules' and 'Phenomena' were lumped together as "Hypotheses" in the first edition; see §3.6, n. 5.
2 In the first edition there is no statement of the area law and harmonic law for Saturn's satellites. These had been discovered by Cassini just before the *Principia* was published, but the English astronomers were not at first willing to admit the existence of these new satellites. In the first edition of the *Principia*, Newton refers only to the first satellite of Saturn, which had been discovered by Huygens. In the second and third editions, the area law and harmonic law are introduced for the Saturnian satellite system.
3 The "proof" or evidence in support of the area law for the planets and our moon is not very satisfying by itself. Of the planets, Newton says only that their 'motion is a little swifter in their perihelia and slower in their aphelia, in such a way that the description of areas is uniform'. He adds that this 'proposition is very well known to astronomers'. As for the area law for the motion of our moon, it 'is evident from a comparison of the apparent motion of the moon with its apparent diameter'. He notes that the moon's motion is perturbed by the sun's force but says that 'in these Phenomena I neglect imperceptible minutiae of errors'. Earlier, in the first version of the system of the world, Newton says explicitly (sect. 27) that the planets describe areas proportional to the times 'so far as our senses can tell', and in sect. 31 he takes cognizance that the moon does *not* constantly describe areas proportional to the time.

  Even the third or harmonic law is not exact, neither theoretically (as Newton will prove later on in bk. three; see Ch. 5) nor observationally. Newton tabulates the periodic times from observation, then computes the distances from these periodic times by Kepler's third law, and finally compares these results with direct determination of the planetary distances. For Mercury, the difference is about 0.3 per cent; for Saturn, it is 0.3 percent or 0.02 percent; for Mars, it is 0.1 percent. These computed and theoretical values, says Newton, 'do not differ sensibly' from each other.
4 In bk. one Newton uses the area law (props. 1–3) to establish that there is a centripetal force and then applies Kepler's third law to circular

orbits (prop. 4) to show that this centripetal force follows the inverse-square law; he does the same in bk. three for the satellite systems of Jupiter and Saturn. Then, in bk. one, Newton shows that elliptical orbits imply an inverse-square law (prop. 11), which he does not do in bk. three. Since Jupiter's satellites move in nearly circular orbits, their motion can be analyzed by the methods of prop. 4, bk. one.

5 Newton is aware that the circular orbits for planets comprise at best too crude an approximation; so he adds that the inverse-square law may be 'demonstrated with the greatest accuracy from the aphelia being at rest'.

6 That the planetary apsides should rotate like the moon's is only a loosely approximate "proof" of the inverse-square law.

7 Newton argues also (in prop. 3, bk. three) that 'this motion of the apogee arises from the action of the sun (as will be displayed below) and accordingly is to be ignored here'. The action of the sun, furthermore, 'is pretty nearly as the distance of the moon from the earth' and so (according, now, to corol. 2 of prop. 45, bk. one) 'is to the centripetal force of the moon as roughly . . . 1 to 178 29/40'. Ignoring so small an extraneous (or foreign) force, Newton says that the remaining force by which the moon is maintained in its orbit' will be found to be as the inverse square – as is 'even more fully established' by the moon test in the immediately following prop. 4, bk. three. For an analysis of Newton's theory of the advance of the lunar perigee, and the significance of this problem, see Whiteside's commentary in Newton (1967–), vol. 6, pp. 508–537; also Whiteside (1976); Waff (1975), (1976); Chandler (1975). In brief, the Newtonian analysis gives an incorrect value for the advance of the lunar perigee, by a factor of 2, as he was forced to admit in the third edition of the *Principia* (end of par. 1 of cor. 2 to prop. 45, bk. one): 'The apsis of the moon is about twice as swift [in its advance]'. Here we may see a result of bk. three (prop. 25) being applied to bk. one; for details see Whiteside's n. 260 in Newton (1967–), vol. 6, p. 380. This discrepancy was one of the reasons why Clairaut and Euler at one time suggested the modification of the inverse-square law by the addition of one or more terms of higher order. But, eventually, Clairaut (in Whiteside's words) 'broke through to the true explanation that this [discrepancy] arises as the third-order effect of both the central and transverse components of the solar perturbation of the earth-moon system'.

8 This is the famous "moon test" of the theory of gravity.

9 Prop. 4, bk. three, concluding paragraph. An alternative proof is given in a scholium to prop. 4. Here Newton considers a purely hypothetical system in which the earth is encircled by a series of moons, as Jupiter and Saturn are; the lowest of these is small and just grazes the mountain tops.

10 It is here in prop. 6 that Newton introduces his celebrated pendulum experiments to show that, at a given place, all bodies have weights proportional to their masses; see §5.7.

11 Prop. 7 reads: 'Gravity tends toward all bodies universally, and is proportional to the quantity of matter which each body contains.'

12 He proves that they act gravitationally as if their whole mass were concentrated at their geometric centers.

13 As we have seen in the use of corols. 1 and 2 to prop. 45, bk. one, and as we shall see below in other examples, bk. three is never wholly free

from considerations of the mathematical systems and imagined constructs that characterize phase one of the Newtonian style.

14 On the use of the word "attraction" in this new corollary, see the end of §3.4.

15 In the first edition Newton had actually used the word "gravity" in this way, as in the statement of prop. 5 ('Planetas circumjoviales gravitare in Jovem . . .') and corol. 1 ('Igitur gravitas datur in Planetas universos') and corol. 2 ('Gravitatem, quae Planetam unumquemque respicit . . .'), but he had not made an explicit declaration about it.

16 Prop. 11 on immovability is proved on the basis of hypoth. 1 (in the third edition); hypoth. 4 (in the first edition).

17 These rules had first been published in 1702 (see Cohen, 1975a).

18 The major sets of imagined constructs or systems and assumptions that Newton made (in the original *Principia* and its revisions) in presenting his theory of the moon in bk. three are admirably summarized in Whiteside (1976).

19 On the conflict between the exactness of mathematical theory and the approximateness of nature, see the conclusion of §3.12.

Supplement to §3.5

1 When Newton first wrote out the *Principia*, he conceived of it as being composed of two books, *De motu corporum liber primus* and *De motu corporum liber secundus*. The first of these was rewritten to become the first book of the *Principia*, where it has the same title. The imperfect and incomplete MS of this early draft was deposited in the University Library in accordance with the terms of his professorship, although it may be doubted that he had actually read this very text to students. The *liber secundus* was recast by Newton so as to become the eventual *Liber tertius* of the *Principia*, Newton having decided to write a wholly new *Liber secundus* on the motion of bodies in resisting mediums, a topic that had presumably been treated briefly at the end of the original *liber primus*.

Thus the final *Principia* contains three books: *De motu corporum liber primus*, a revised version of a text deposited by Newton in the University Library; a *De motu corporum liber secundus*; and a *Liber tertius*: *De mundi systemate*, a recast version of that first *De motu corporum liber secundus*, which dealt with the same subject and of which the text was deposited by Newton in the University Library.

After Newton's death, English and Latin versions of the earlier and discarded *liber secundus* were published under the titles, *A Treatise of the System of the World* (1728) and *De mundi systemate liber* (1728), these being taken from the subtitle of the final *Liber tertius* of the *Principia*. To avoid confusion, I shall refer to Newton's discarded and posthumously published work as *System of the World* so as to distinguish it from bk. three of the *Principia*. For details see I. B. Cohen (1969a, 1971); Dundon (1969).

2 From a new translation by I. B. Cohen and Anne Whitman.

3 The Latin reads: '. . . quantitatem et proprietates ipsius eruere atque effectus in corporibus movendis investigare mathematice'; the full text is given in the introduction to Newton (1969).

4 The *System of the World* does, however, conclude with five numbered lemmas and two numbered problems.

5 This is the title of a widely quoted and cited article in the *Journal of the History of Ideas* (1961), vol. 12, pp. 90–110.

6 These paragraphs appeared also in subsequent editions and were continued in the widely used Cajori-Motte edition of the *Principia* and *System of the World*.

7 A recent example occurs in Wiener (1973), vol. 3, where a whole section of the article on 'Newton and the method of analysis' is called 'Newton's mathematical way', a phrase which occurs on p. 389*b*. Of course, there may turn up yet another MS copy of the *System of the World*, in which these words may appear. The probability of this may not be absolutely zero, since this phrase would occur in a part of the known MS in which Newton made serious alterations; possibly he might have essayed yet other alterations on another copy (if there should be one). The continued use of this phrase shows a lack of the caution that should be exercised in quoting English translations, which – as in this case – may prove to have no correspondence with known Latin originals.

8 But he does refer to the 'center of gravity', e.g., in sect. 28 (where he proves that 'the common center of gravity of all the planets is at rest') and in sect. 48 (where he mentions that 'the earth and moon revolve about a common center of gravity'); another example occurs in sect. 56. In the style of the *System of the World*, this is a misnomer.

9 For example, in sect. 55, he mentions the 'absolute centripetal force of the moon'.

10 On the nonneutrality of the term "attraction", see Koyré (1965), pp. 57–58.

§3.6

1 The review was published anonymously; Paul Mouy (1934), p. 256, suggested that Régis may have been the author.

2 The major parts of this review are translated in Koyré (1965), p. 115.

3 That is, Newton says that he considers 'these forces [*vires*] not physically but mathematically', but the reviewer seems to assume that this applies to the principles ('principes'), presumably the Newtonian 'principles' of natural philosophy, which, it must be confessed, are said in the title to be 'mathematical principles'.

4 End of discussion of def. 8. As we have seen above, the early part of bk. one presents essentially a one-body system, a mass-point moving about a center of force. In bk. three this is approximated by the sun and any one of the four inner planets, each of whose mass is very small compared to that of the sun, which accordingly may be considered motionless to within ordinary limits of observation.

5 In the first edition of the *Principia*, bk. three began with nine 'Hypotheses'. The first three were philosophical or methodological. By the time of the second edition, 'Hypoth. I' and 'Hypoth. II' had become the first of the 'Regulae Philosophandi'. 'Hypoth. III' had been cast out; its place was taken by a wholly new 'Regula III'. In the third edition Newton introduced an additional (fourth) 'Regula' to make up the set as they are commonly known today. 'Hypoth. IV' of the first edition, dealing with the 'center of the system of the world' being at rest, remained a hypothesis in all editions; in the second and third editions, it occurs as 'Hypothesis 1', and is placed after prop. 10, bk. three. The remaining 'Hypotheses' of the first edition (5–9) became the 'Phaenomena' of

the second and third editions, printed right after the 'Regulae Philosophandi'. 'Hypoth. V' ('Phaenomenon I') states the area law and harmonic law for the satellites of Jupiter. 'Phaenomenon II' does the same for the satellites of Saturn; this was new in the second edition, there being no corresponding 'Hypothesis' in the first edition. Numerical data are given to support the applicability of the harmonic law to these two satellite systems. 'Hypoth. VI' ('Phaenomenon III') states that the five 'primary planets' (Mercury, Venus, Mars, Jupiter, and Saturn) move in orbits around the sun; 'Hypoth. VII' ('Phaenomenon IV') states the harmonic law for the motion of these planets with respect to the sun, and of the sun with respect to the earth or of the earth with respect to the sun. 'Hypoth. VIII' ('Phaenomenon V') states that, with respect to the earth as center, the five planets do not sweep out areas proportional to the time; but they do so with respect to the sun. 'Hypoth. IX' ('Phaenomenon VI') states the area law for the motion of the moon with respect to the center of the earth.

It is to be noted that the elliptical orbits are not mentioned in these 'Phaenomena' ('Hypotheses'). On the changes from 'Hypotheses' to 'Regulae' and 'Phaenomena', see Koyré (1965), pp. 261–272 ('Newton's "Regulae Philosophandi" '); also Cohen (1966).

6 On Kepler's "laws" as "planetary hypotheses" or statements that may be true only in a phenomenological sense, see §5.8.

7 It is difficult to decide, without knowing the identity of the author, whether or not he was writing these last two sentences with tongue in cheek.

8 See Cohen (1971), ch. 6, sect. 6. While there is no direct evidence on this matter, we may observe that Newton was always ultrasensitive to any form of criticism.

§3.7

1 Prop. 4, cor. 7, bk. one; this corollary was first printed in the second edition and seems to have been suggested by Fatio de Duillier (see Cohen, 1971, pp. 182sq).

2 This model occurs at the beginning of the second discourse, 'Of refraction'; trans. in Descartes (1965), pp. 75sqq.

3 Ibid., p. 67. Descartes has also a model in which he considers a ball or a rock being deflected by bodies it encounters. Descartes made a significant but not always very clear distinction between motion itself and a tendency (*conatus* or *tendance*) to motion. Thus he could simultaneously conceive of light as having an infinite speed, or – more properly – being transmitted without loss of time, and as having different speeds in such diverse media as air, glass, and water.

In *Le monde*, ch. 12, Descartes introduces yet other models (including a crooked or doubly-curved stick and a bulging vase filled with hard balls; also a set of five cords attached to a single pulley) which in one way or another contravene his fundamental principles.

4 Descartes (1974), vol. 2, p. 206 (*Oeuvres*). Cf. the reference to "comparaisons" in *La dioptrique* ('discours premier'), ibid., vol. 6, p. 86, ('discours second'), ibid., p. 104.

5 Ibid., vol. 11, p. 102, §8. Cf. Buchdahl (1969), pp. 97–99, 118sqq.

6 *Webster's New International Dictionary of the English Language* (2d ed., 1934), s.v. comparison. In a discussion of this topic in my

graduate seminar, Peter Galison reported that in the *Dictionnaire de l'Académie française* (Paris: Jean Baptiste Coignard, 1694), metaphor occurs as the second meaning of *comparaison*. It is possible that metaphor may be more appropriate than simile in explaining Descartes's *comparaison* (see Galison, 1978).

7 The expression "atom-model" occurs several times in the opening section of Bohr (1913). Einstein did not propound a theory of photons to explain the photoelectric effect, but named his paper 'Über einen die Erzeugung und Verwandlung des Lichtes betreffenden heuristischen Gesichtspunkt' [On a heuristic point of view regarding the production and transformation of light] (1905).

8 On the literature concerning models, see n. 13 to §3.3.

9 Cf. Buchdahl (1969), pp. 96–97, 118sqq; Hesse (1967), pp. 356–357. Descartes expounded this view in rules 9, 14; see Descartes (1974), vol. 10, pp. 400–403, 438, 452.

10 In the preface to the French version of *Les principes de la philosophie*, he says: 'J'aurois aussi adjousté un mot d'advis touchant la façon de lire ce Livre, qui est que je voudrois qu'on le parcourust d'abord tout entier ainsi qu'un Roman . . .' (Descartes, 1974, vol. 9, p. 11). In *Le monde*, end of ch. 5, he says: 'Mais afin que la longueur de ce discours vous soit moins ennuyeuse, j'en veux envelopper une partie dans l'invention d'une Fable . . .' (ibid., vol. 11, p. 31).

11 *La dioptrique* ('discours premier'), Descartes (1974), vol. 6, p. 83. Descartes was imitating the astronomers, 'qui, bien que leurs suppositions soyent presque toutes fausses ou incertaines', could nevertheless 'en tirer plusieurs conséquences très vrayes & très assurées'.

12 In order to conceive the possibility of inertial motion, Gassendi, too, had to think of an imaginary world, far from the action of forces, out in a void beyond our world (see Koyré, 1965, pp. 178, 186).

13 Newton, however, does not so approximate the conditions for inertial motion. Possibly, he may have believed that we could have no direct (i.e., experiential) knowledge of such a motion, or that the only place it might occur would be out in the infinity of space.

14 Newton also referred to continuing inertial motions of not so long a duration: the inertial component of projectile motions and of the component particles of spinning hoops or trochees.

15 In the *Principia* (e.g., corol. 2 to the laws of motion), such inertial motion is usually presented as if limited. It is not amiss to observe that it was specifically in relation to inertial motion that both Descartes and Gassendi resorted to the concept of a fiction (see n. 12).

§3.8

1 I am concerned here only with the problem of the origin or cause of the "circumsolar" and "circumplanetary" forces, which later became the universal gravitating force. Newton had been exploring aspects of the aether in other contexts long before the beginning of the correspondence with Hooke in 1679.

2 He still believed in the possibilities of some kind of vortex on the eve of the *Principia*, as we shall see in Ch. 5.

3 That is, vortices could account for a tendency toward a center, and variations in aether density could account for an urging toward a body;

but neither could produce a mutual (equal and opposite) gravitating force, nor an inverse-square force.

4 That is, a force admissible to the mechanical philosophy (according to which all phenomena originate in matter and motion) and that could act over such vast distances, be mutual, and have a magnitude proportional to the quantity of matter and not the surface area of bodies.

5 In the discussion of rule 3, which was added in the second edition of the *Principia*, he said, 'I am by no means affirming that gravity is essential to bodies'.

6 First published in Hall & Hall (1962), pp. 320–347.

7 He says here that 'I do not define the manner of attraction, but speaking in ordinary terms call all forces attractive by which bodies are impelled towards each other . . . whatever the causes be'. Also: 'The force of whatever kind by which distant particles rush towards one another is usually, in popular speech, called an attraction. For I speak loosely when I call every force by which distant particles are impelled mutually towards one another, or come together by any means and cohere, an attraction'. In qu. 31 of the *Opticks*, he again said that what 'I call Attraction may be performed by impulse, or by some other means unknown to me. I use that Word here to signify only in general any Force by which Bodies tend towards one another, whatsoever be the Cause'.

8 Also first published in Hall & Hall (1962), pp. 302–308.

9 The phrase "cannot be explained" implies certain standards of understanding, such as the canons of the mechanical philosophy; G. Holton has referred to these as "themata". But it is an error to say that in an early draft of the concluding general scholium, Newton wrote: 'I have not yet disclosed the cause of gravity, nor have I undertaken to explain it, since I could not understand it from the phenomena'. The original Latin sentence of Newton's does not say 'I could not understand it from the phenomena', but rather 'ex phaenomenis colligere nondum potui', that is, 'I have not yet been able to gather [infer] it [i.e., the cause] from phenomena' (see Hall & Hall, 1962, pp. 350, 352; Holton, 1973, pp. 51–52).

10 But scientists are not in general any longer concerned with real "existence".

11 Newton to Boyle, 28 Feb. 1678/9, Newton (1959–1977), vol. 2, pp. 288–296; Newton (1958), pp. 250sqq. See also Newton's early essay, 'De aere et aethere', in Hall & Hall (1962), pp. 214–228.

12 This experiment was described by Newton in the essay, 'De aere et aethere' (Hall & Hall, 1962, pp. 227sq); it was referred to in his hypothesis of 1675, 'explaining the Properties of Light' (Newton, 1958, pp. 179–80; Birch, 1756–1757, vol. 3, pp. 249sq). For details, see Westfall (1971), pp. 336, 374, and Westfall (1970).

13 *Principia*, final paragraph of general scholium at the end of sect. 6 (ed. 1) or sect. 7 (ed. 2), bk. two. In this experiment Newton used a freely oscillating pendulum, eleven feet long, with a hollow wooden bob which could be filled with different substances. He studied the rate at which the successive oscillations diminished when the bob was empty (i.e., contained nothing but air) and when filled with metal. In this way he sought to determine whether the resistance to motion (disclosing itself in the slowing down of the pendulum) depends only on the outer surface of the bob or on the interior parts or contents as well (as would be

the case if any resistance arose from a subtle matter, such as aether, which might permeate through the substance of the hollow bob and act on its solid contents). He concluded that the resistance to motion arising from any 'aethereal and exceedingly subtle medium that quite freely permeates the pores and passages of bodies is either nil or completely insensible'.

14 Newton discusses vortex motion in (1959–1977), vol. 2, pp. 310, 322, 331, 337, 338, 341, 360.

15 On the additions to *De motu*, see Hall & Hall (1962), pp. 256sq, 261sq, 280sq, 285sq; Herivel (1965a), pp. 297–299, 301–303.

16 The argument about the small or null resistance of the aether in the revised *De motu* is based on the progression of resistance of mediums according to density (or 'the quantity of their solid matter'): quicksilver, water, air (and so down to aether, which will have the same resistance as air that is rarefied 'until it reaches the tenuousness of aether'). Newton then compares the way that horsemen 'feel the resistance of the air strongly' with the experience of 'sailors on the open seas' who, 'when protected from the winds, feel nothing at all of the continuous flow of aether'. He then argues: 'If air flowed freely between the particles of bodies and thus acted not only on the external surface of the whole body but also on the surfaces of its individual parts, its resistance would be much greater. Aether flows between [the parts] very freely; and yet does not sensibly resist'.

This idea was also expressed earlier (but not in reference to sailors) in the essay beginning 'De Gravitatione et aequipondio fluidorum . . .', sect. 8, par. 3; published in both Hall & Hall (1962) and in Herivel (1965a). This same basic idea led to the pendulum experiment in the *Principia* (see n. 13); surely, if Newton had already made such an experiment, he would have referred to it in this addition to *De motu*.

17 In the later revisions to *De motu*, Newton actually says: 'Aetheris enim puri resistentia quantum sentio vel nulla est vel perquam exigua' ('For the resistance of pure aether, in my judgment, is either nil or extremely small'). This text is printed in Hall & Hall (1962), pp. 261, 286, and in Herivel (1965a), pp. 297, 301(C).

The verb *sentio, sentire*, used by Newton, literally means "to perceive by the senses", which might seem to imply that Newton had already performed an experiment, such as the pendulum experiment. But this verb has also the general sense of think, deem, judge, propose. Hence, by the phrase "quantum sentio", Newton would have meant no more than "in my judgment" or "in my opinion" or even "I think". I believe that if Newton had already made an experiment of the sort he described in the *Principia*, he would not have used a phrase such as "quantum sentio" but rather something like "as experiment shows" or "as I have found by experiment". This interpretation would place the date of the experiment late in 1685, during the actual writing of the *Principia*. On the other hand, should the phrase "quantum sentio" have been intended to refer to an experiment, with the meaning of "so far as I am aware by sense perception", then the experiment would presumably have been made in December 1684 or early in 1685, that is, after the first version of *De motu* (November 1684) and before the revision. It should be noted, however, that had Newton intended to indicate "so far as the senses can perceive", he would probably have used the adverb *sensibiliter*, as he did in prop. 48, bk. two, of the *Principia*.

There is one aspect of the experiment, furthermore, which supports the later date. The principle on which the experiment rests is not at all simple, but depends on a level of analysis of the dynamics of pendulum motion of which we have no trace whatever in Newton's papers earlier than the composition of bk. two of the *Principia*. This fact would seem to rule out the possibility of any date earlier than late in 1685.

18 Apart from references to a medium, such as the 'medium, if there should be any, freely pervading the interstices between the parts of bodies' (in def. 1), there are a number of direct references to aether in the *Principia*. Toward the end of the scholium to the Laws of Motion, Newton writes (without raising any question of existence) about 'the whole earth, floating in the free aether'. In corol. 2, prop. 6, bk. three, he writes: '. . . if the aether or any other body whatever either were entirely devoid of gravity or gravitated less in proportion to the quantity of its matter . . .' Toward the conclusion of the scholium at the end of sect. 8, bk. two, he decides that sound consists in 'the agitation of the whole air', and not 'in the motion of the aether or of a certain more subtle air'. At the end of the scholium with which sect. 11, bk. one, concludes, he says: 'I use the word *attraction* here in a general sense for any endeavour whatever of bodies to approach one another, whether that endeavour . . . arises from the action of the aether or of air or of any medium whatsoever – whether corporeal or incorporeal – in any way impelling toward one another the bodies floating therein'. In discussing the tails of comets in lemma 4, bk. three, he states his opinion that they 'arise either from the reflection of smoke [arising from the head and] being scattered through the aether, or from the light of the head'.

Finally, in prop. 41, bk. three, he explains the fact that the tails of comets ascend from the heads and extend away from the sun by an analogy to ascent of smoke in a chimney 'by the impulse of the air in which it floats', the air being 'rarefied by heat' so as to gain a 'diminished specific gravity'. He assumes that 'the tail of a comet' may 'ascend from the sun in the same way'. The sun's rays, he argues, will 'warm the reflecting particles' of the medium through which they pass, and these 'reflecting particles, warmed by this action [of the sun's rays], will warm the aethereal aura in which they are entangled'. It is not immediately clear whether these reflecting particles compose the comet's tail or are a part of the celestial matter (see D. Gregory's comments in Newton (1959–1977), vol. 3, pp. 311, 316, n. 9). The aethereal aura will accordingly become rarefied and have a decrease in the 'specific gravity with which it was formerly tending toward the sun', and so 'that aura will ascend and will carry with it the reflecting particles of which the comet's tail is composed'. Motte (Newton, 1729*b*) translated Newton's "aethereal aura" by "aethereal air". But Newton has just been using "aer" for "air" in this paragraph, as was his wont elsewhere in the *Principia*. The word "aura" at that time rather meant a subtle or tenuous exhalation, and had been used by Kepler (whose name is mentioned by Newton at the head of this paragraph) as a subtle medium or very fine matter in space. This is, with regard to the sun, very much like the 'aethereal medium' introduced in Newton's hypothesis for 'explaining the Properties of Light' in 1675; 'an aethereal medium much of the same constitution with air, but far rarer, subtler, and more strongly elastic'.

19 This part of the definition does not occur in the first redaction of Newton's MS, which he later deposited in the University Library as if it had been the text of his professorial lectures (see Cohen, 1971, ch. 4, sect. 2, and suppl. 4).

20 In the case of the spaces between the sun and planets, or the earth and the moon, the air could certainly be eliminated, along with 'spirits emitted', etc., leaving only the possibility of the 'aether or any medium'.

21 And also the long-lasting orbital motions of planets and of comets, which would be slowed down if space were filled with an aether offering a sensible resistance to motion (see n. 23).

22 See Bopp (1929), Fatio (1949), and Newton (1959–1977), vol. 3, p. 69. Fatio's hypothesis was presented to the Royal Society on 27 June 1688. On Fatio, there is an informative thesis by Charles Domson, completed at Yale in 1972.

23 Newton (1959–1977), vol. 4, pp. 1, 3. What Newton actually says is that as a result of the theory of gravity, the older concept of 'solid spheres' carrying planets round must be rejected; 'not only are the solid spheres to be resolved into a fluid medium, but even this medium is to be rejected [sed etiam hanc materiam rejiciendam]'. The reason is that any such medium (or fluid matter) would 'hinder or disturb the celestial motions that depend on gravity'.

24 See the articles on this subject by Henry Guerlac, conveniently summarized in Guerlac's article on Hauksbee in the *Dictionary of Scientific Biography*.

25 See Hall & Hall (1959a), supplemented by Koyré & Cohen (1960). On Newton and electricity see the studies by Guerlac and Hawes.

26 In the later queries of the *Opticks* published first in English in 1717/1718, Newton introduced the concept of an 'aethereal medium', presumably made up of mutually repelling particles, which, inter alia, was supposed to give a clue as to the cause and mode of transmission of universal gravity. The vicissitudes of Newton's beliefs in an aether have been studied by Koyré, Guerlac, and Westfall.

27 This occurs in Huygens's *Discours sur la cause de la pesanteur* (Huygens, 1888–1950, vol. 21, p. 471; see Koyré, 1965, p. 118).

28 With respect to Newton's views on the forces associated with the particles of matter as opposed to the gravitational forces of gross bodies, and the possible relations between the two kinds of forces, see the conclusion of §5.5.

§3.9

1 D'Alembert uses the word "révolte" rather than "révolution" to describe how Descartes had shown 'intelligent minds how to throw off the yoke of scholasticism, of opinion, of authority' (see §2.2).

2 In an essay of the 1750s, 'On universal history', trans. in Turgot (1973), p. 94; see Turgot (1808–1811), vol. 2, p. 277.

3 For Bailly's concept of a "Copernican revolution", see Cohen (1977a).

4 Bailly (1785), vol. 2, bk. 12, sect. 9, p. 486. Bailly is using the confusing concept of a planet (or satellite) having a "force" of inertia (which is an internal, and hence nonaccelerating force) and being accelerated by an external centripetal force (see §4.).

5 Maupertuis (1736), p. 474; this paper was read at the Académie des Sciences in 1732.

6 *Histoire*, p. 158. In another essay of about the same time, 'Discours sur les différentes figures des astres' (first published in 1732), Maupertuis (1756) compared the investigations made by Huygens and by Newton on the form of the earth. Later on, Maupertuis set out on an expedition of his own to Lapland, in order to determine the true shape of the earth. In a "Discussion métaphysique sur l'attraction' (sect. 2 of the *Discours*), Maupertuis contrasted the two points of view. Huygens, he says, declares that weight is 'the effect of a centrifugal force of some kind of matter that, circulating around bodies toward which others have weight, pushes them toward the center of circulation'. Newton, 'without searching for the cause of weight, regards it as if it were an inherent property of bodies'. The 'word attraction', according to Maupertuis, 'has scared away certain minds' who have feared 'to see reborn in natural philosophy the doctrine of occult qualities'. Then, with clearness and elegance, he expounded the point of view that characterizes the Newtonian style: 'But in fairness to Newton it must be admitted that he has never regarded the attraction as an explanation of the gravity of bodies, the ones towards the others: he has often asserted that he only employed this term to designate a fact and not a cause; that he only employed it to avoid systems and explanations; that it was even possible that this endeavour was caused by some subtle matter which issued from the bodies, and was the effect of a true impulsion: but however this might be, it was always a first fact, from which one could proceed to explain the facts dependent on it. Every regular effect, although its cause is unknown, can be the object of the mathematicians, because all that is more or less susceptible [to analysis] is in their province, whatever is its nature; and the application they will make will be just as certain as that which they could make of objects whose nature was known absolutely.'

For an account of this work of Maupertuis, see Aiton (1972), pp. 201–205, where the extract quoted above is given on p. 202. On Mauptertuis's life and work, see Brunet (1929).

§3.10

1 Tracing the effects of the *Principia* must be done on several different levels: Newton's concept of mass (and the distinction between gravitational and inertial mass), his particular formulation of the laws of motion, his treatment of Kepler's laws in sect. 2 and 3 of bk. one, his alternative approach (as in prop. 41, bk. one, which, as E. J. Aiton has shown, was far more influential on the Continent), his development of the properties of the inverse-square law, his development of the theory of perturbation (as in prop. 66, bk. one, and its twenty-two corollaries), his presentation of the resistance to motion in bk. two and his general theory of wave motion (including his erroneous law of the speed of sound), his system of the world (as expounded in bk. three), his mode of computing the orbits of comets, his rules for computing the motion (and positions) of the moon, and so on.

2 In this respect, the only predecessor that was similar in its net effect (although on a different level altogether) would have been Ptolemy's

Almagest; and a successor would be Laplace's *Mécanique céleste.*

3 Although it is common to talk today about "the Ptolemaic system", there were in fact a number of different systems in the *Almagest*: for the moon, for Mercury, for Venus, for the sun, for the superior planets, and for the fixed stars.

4 Much more research is needed in order to ascertain how extreme it is. But A. I. Sabra has shown that among Islamic writers on astronomy there was a duality of conception and explanation, one being astronomical and mathematical (geometrical) and the other philosophical (or physical). The same MS text may even contain two sets of drawings that illustrate these two different levels of prediction or explanation. Hence this might be the original locus of the Newtonian style (see Sabra, 1976).

5 This essay was printed in the same style and format as Copernicus's dedication, so that there could be no possible reason to suppose that Copernicus had not been the author. In the early nineteenth century, Delambre still thought this was an essay by Copernicus himself; see the introduction to the 1969 reprint of Delambre (1821), p. xviii, and pp. 139sq of vol. 1 of Delambre's text.

6 Kepler, in commenting on his discovery, said: 'It is a most absurd fiction . . . that the phenomena of nature can be demonstrated by false causes. But this fiction is not in Copernicus. He thought that his hypotheses were true, no less than did those ancient astronomers. . . . And he did not merely think so, but he proves that they are true.' See Rosen (1971), p. 24, n. 68; this affair is discussed by M. Caspar in Kepler (1929), p. 399.

7 This point of view may be exemplified in Kepler's rejection of the law of motion of planets, according to which – instead of using the law of areas – the speed (and the position) is determined by the uniform rotation of a radius vector centered in the empty focus of the ellipse. This law would attribute the regulation of a planet's motion to an empty point in space, to a geometric position rather than a physical body. Since Kepler held that planetary motions are caused by, and regulated by, forces and that forces must originate in physical bodies, he could not but reject the law of speed he had found, depending on the empty focus. Kepler, as I am reminded by D. T. Whiteside, liked to have "causes" for planetary motions, but he required that they yield observationally true paths. In 1601 he developed his *hypothesis vicaria*, which was in fact a very accurate computing scheme for relating true motion to mean motion, and used it in the elimination of various attempts to construct the orbit of Mars until only the sun-focal elliptical orbit survived.

8 At least, he did not do so in his *Dialogue on the Two Great World Systems*, presumably since he was writing for the generally educated public and not for either the trained or the would-be astronomer.

9 The performance of such an experiment in a tube evacuated of air, after Boyle had improved the vacuum pump, shifted the ontological state of Galileo's imagined system to the reality of experience.

10 Galileo also had to introduce an imagined system or construct in introducing a version of inertial motion in the *Two New Sciences*. He imagined an infinitely extended plane on which a body can move without friction. Cf. Galileo (1974), beginning of fourth day, p. 217; (1890–1909),

vol. 8, p. 268. Gassendi similarly had to imagine a special world for pure inertial motion; cf. Koyré (1965), ch. 3, app. G.

11 A possible influence on Newton in this regard may have been Hooke. Newton might well have heard discussions of Galileo's methods without having actually read Galileo's *Two New Sciences*, for example, by hearing these topics discussed by Barrow. For my reasons why Newton most probably did not invent his method, see Ch. 4.

12 See McGuire & Rattansi (1966). Newton believed that the ancient Pythagoreans had known the inverse-square law, having obtained it from the "Chaldeans". In the seventeenth century the law would have thus been discovered once again. But Newton apparently did not believe that any predecessor (in any age) had demonstrated by the inverse-square law that the planetary orbits are elliptical and that cometary orbits are elliptical or parabolic.

13 Although Newton does not make much use of a specific algorithm for the calculus in the *Principia* (except in sect. 2, bk. two), a number of propositions in bk. one state as a condition that it is possible to find the area under certain curves (or to perform integration). Newton's constant recourse to limits could easily reveal to an alert reader (such as the Marquis de l'Hôpital) how much of the *Principia* was really an exercise in the calculus.

14 See the end of §3.8, where some of Newton's attempts to find such a cause are listed. After publishing the *Principia* in 1687, the three chief explanations of gravity which Newton successively explored with some real degree of commitment were: the motion of aethereal particles (proposed by Fatio de Duillier), the action of an electrical "spirit", and the varying density of an "aethereal medium".

§3.11

1 Wolfson (1934), vol. 1, pp. 41, 42, 48sq, 52; ch. 2 ('The geometrical method') discusses this topic at length.

2 Of course Newton had used mathematical techniques in obtaining some of his results. Hence, in the posthumous 'fourth edition, corrected' (London, 1730), the editor added supporting references for many of Newton's mathematical statements by citing pertinent propositions of the (then recently published) posthumous *Lectiones opticae* (London, 1728). For instance, in computing the result of an experiment in prop. 7, bk. one, pt. 1 (p. 95 of Newton, 1952, *Opticks*), he gave a result in which a certain diameter had the value $(R^2/I^2) \times (S^3/D^2)$ 'very nearly', 'as I gather by computing the Errors of the Rays by the Method of infinite Series, and rejecting the Terms, whose Quantities are inconsiderable'. The editor adds: '*How to do this, is shewn in our* Author's Lect. Optic. Part I. Sect. IV. Prop. 31.'

3 Editor's preface to Newton (1728c), p. vi. The editor did not reveal his name, nor did he give any clue as to his identity. Whiteside, in Newton (1967–), vol. 3, p. 440, proposes Pemberton as his candidate for the anonymous editor-translator. A new edition and translation of these lectures has been undertaken by Alan Shapiro.

4 There are two MS versions in the University Library, Cambridge. The earlier one has been reproduced in facsimile in Newton (1973). The later one was printed in its Latin original in 1729. Portions of both sets of

lectures have been given in vol. 3 of Newton (1967–), pp. 435sqq, together with the preface to the English translation, a comparison of the two MS versions, and a critical discussion of Newton's optical discoveries and methods.

5 See, e.g., Whiteside's n. 42 in Newton (1967–), vol. 3, p. 471.

6 In a scholium following prop. 96, Newton observed that these 'attractions are not much dissimilar to the reflections and refractions of light'. And he concluded: 'Therefore because of the analogy that there is between the propagation of rays of light and the motion of bodies, I have decided to subjoin the following propositions for optical uses, meanwhile disputing not at all about the nature of the rays (that is, whether they are bodies or not), but only determining the trajectories of bodies, which are clearly similar to the trajectories of rays.' Accordingly, this may be an example of a model.

7 See Whiteside's analysis in Newton (1967–), vol. 6, p. 429, n. 21.

8 That is, in the *Opticks* as we know it in the English and Latin editions of 1704, 1706, 1717/18. If Newton had stuck to his original program, the *Opticks* would not have been a wholly separate entity, but its contents would have been a part of the larger work which was to contain also the mathematical proofs presented in his Lucasian lectures. This particular model, however, does not appear in the lectures, and was first elaborated during the composition of the *Principia*. It was then incorporated into the *Opticks* in 1692/93.

9 Newton (1952), p. 79. Newton resolves the motion of each incident ray into a component perpendicular to the refracting surface and another parallel to the surface. Then, using a variant of the mathematical construct or "model" used in sect. 14 of bk. one of the *Principia*, he considers 'any Motion or moving thing whatsoever' (so phrased so as to avoid the charge that he is dealing exclusively with any particular kind of particle), incident on a broad but thin space bounded by two extended parallel planes, and urged perpendicularly toward the farther plane with a force depending in some manner on the distance from that farther plane. He supposes that an incident ray only undergoes a change in velocity in the perpendicular direction, and not in the horizontal direction, or direction parallel to the plane. Newton assumes that his demonstration is 'general, without determining what Light is, or by what kind of Force it is refracted, or assuming any thing farther than that the refracting Body acts upon the Rays in Lines perpendicular to its Surface'; he takes this 'to be a very convincing Argument of the full truth of this Proposition'.

10 A different point of view has been expressed in Guerlac (1973), p. 389*b*, where it is said that although the '*Opticks*, unlike the *Principia*, consists largely of a meticulous account of experiments', it 'can hardly be called non-mathematical, although little more than some simple geometry and arithmetic is needed to understand it. In spirit it is as good an example of Newton's "mathematical way" as the *Principia*: light is treated as a mathematical entity, as *rays* that can be represented by lines; the axioms with which he begins are the accepted laws of optics; and numbers – the different refrangibilities – serve as precise tags to distinguish the rays of different colors and to compare their be-

havior in reflection, refraction, and diffraction. Wherever appropriate, and this is most of the time, his language of experimental description is the language of number and measure. It is this which gives Newton's experiments their particular cogency'. It need only be said that the representation of rays by lines is only to be expected in geometrical optics, and that the issue is not whether Newton uses 'the language of number and measure'. What seems to me to determine the character of this book is whether a theory of physical optics is developed in a mathematical style, as in the above-mentioned prop. 6, bk. 1, pt. 1, of the *Opticks* or as in sect. 14, bk. one, of the *Principia*. On the distinction between quantitation and mathematics as exemplified in the *Principia*, see Ch. 1 and the earlier part of Ch. 3. On 'Newton's "mathematical way" ', see suppl. to §3.5.

11 This assumption had also been made in Descartes's *Dioptrique*, in the presentation of the law of refraction, with which Newton was familiar (see Cohen, 1970, pp. 150sq). On Descartes's proof of the law, see Sabra (1967), ch. 4.

12 Newton, however, was able to do much more with respect to the mathematics of refraction than he let the readers of the *Opticks* and the *Principia* know; see the discussion of this topic by D. T. Whiteside in Newton (1967–), vol. 3, pp. 514–528, and vol. 6, pp. 422–444.

13 Newton does not himself set forth these conditions clearly and coherently. Lohne remarks: 'Newton did not openly admit that he followed such principles, but they are very apparent in his manuscripts, and, when we look for them, also in his printed works'.

14 This comes from Lohne's list, given in Lohne (1961), p. 393.

15 See Bechler (1973). The title of Bechler's article is 'Newton's search for a mechanistic model of colour dispersion'. The imagined systems for optical phenomena may be similar to Newton's proposed "models" to explain Boyle's law (see end of §3.3) and to account for gravity in terms of an aether shower or an aether of varying degrees of density and dissimilar to the imagined systems or mathematical constructs invoked in bks. one and two of the *Principia*. The latter (see §3.2) tended to be mathematicizations of a simplified and idealized situation occurring in nature, and insofar are not "models" in the sense understood today by scientists and philosophers of science. It is clear that in terms of applying mathematics to deduce consequences of the initial conditions, it does not matter whether the system to which the mathematics is applied (phase one) is a "model" or a construct based ultimately on nature simplified and idealized. In this sense the Newtonian style is manifested in the treatment of "models". In the optical examples, Bechler uses the term "models", which seems very appropriate.

16 See Bechler (1974*b*), esp. p. 117: 'The actual argument was far more complicated than the one which Newton chose to make public and which looked so innocent, but it was no less rigorous. Every single point was taken care of and separately proved: that the stationary position which the refracted beam reaches indicated minimum deviation; that in this position the mean refraction is equal to the mean incidence; that in this position the alternate incidence and refraction angles of the two extreme rays defining the pencil are equal; that, therefore, the angles of

divergence of this pencil at incidence and refraction are equal, and, as a consequence, that the refracted image must be geometrically similar to the form of the light-source.'

17 Probably Oldenburg was responsible for the omission. The sentiment may have been justified by the mathematical underpinning, but in the absence of such underpinning, this sentence was grossly out of place and overly dogmatic (see Bechler, 1974b, sects. 1, 2).

18 This appears in the beginning of bk. two, pt. 2; Newton (1952), pp. 227, 240.

19 Newton (1959–1977), vol. 1, pp. 187sq. It is to be noted that Newton was saying in the 1670s that 'the science of colours' would become a part of mathematical science rather than pure experimental science, and that in 1704 he had given up this hope and merely said (in the *Opticks*) that color science would become as 'mathematical as any other part of Opticks'.

§3.12

1 For a sample of how fruitful such an investigation may be, see Hankins (1967); (1970), pp. 175–190. Hankins shows how d'Alembert's highly original point of view led him to a transformation of Newton's three laws of motion.

2 Prop. 41: 'Supposing a centripetal force of any kind and granting the quadrature of curvilinear figures, it is required to find both the trajectories in which bodies will move and the times of their motions in the trajectories so found.' On the influence of this proposition and its neighbors in sect. 8, bk. one, see Aiton (1964b).

3 Hankins (1970), p. 177, points out that 'd'Alembert had taken care to excuse this "obscure and metaphysical" terminology beforehand'. In the preface to the *Traité de dynamique* he warned the reader that he had 'often used the obscure term "force" in order to avoid circumlocutions'. Newton, too (see §4.5), treated the "force" of inertia as if it were only a convenient name and not a true force at all (in the sense of its not having the ability to change the state of motion or of rest).

4 See §4.5. D'Alembert, who was considering dynamics more as a branch of geometry than of physics, did not introduce mass until the third law, when the concept could no longer be avoided (in the second law for a mass point, the law could be simply $\Phi = dv/dt$). But the concept of mass that he uses is Newtonian.

5 On the title page of the second edition (Cambridge, 1713; Amsterdam, 1714), the *Principia* is said to have been written by 'Isaaco Newtono, equite aurato', a knight, Sir Isaac Newton.

6 In this poem, the only other scientist mentioned (other than Hermann, in the opening and closing verses) is Ptolemy.

7 In def. 14, Hermann avoided what has often seemed to be a circularity in Newton's definition of mass by defining it as the 'complex (or aggregation) of all the particles of which the body is composed', which is what Newton himself most likely believed. In this case, the totality of all such particles in a body represents the mass, while the number per unit volume represents the density. Although Hermann does not specifically refer to Newton, the definition follows the Newtonian def. 1 closely, even to Hermann's saying that he will call "quantitas materiae" by the name "mass".

8 On pp. 6–7, Hermann writes: 'Among philosopher-geometers [i.e., mathematical physicists], an elegant property of bodies is celebrated, consisting in the fact that the weights of bodies [*pondera*] increase in exactly the same ratio as their masses or quantities of matter, or, putting the same thing in a geometrical phrasing, *that the gravities [gravitates] or weights [pondera] of bodies are proportional to their masses.* The most illustrious Newton, by making most accurate experiments with pendulums, has always found this a property of heavy bodies, as is seen on page 305, Princ. Phil. Nat. Math., first edit., which he has elegantly drawn forth from corol. 1, prop. 24, book 2.' Hermann again refers to this contribution of Newton's in a "scholion" on p. 9.

9 It must be said that in a private letter (4 Nov. 1715) to Bernoulli, Leibniz said that there were many good things in Hermann's book but too much of the English way of thinking. And in a review in the *Acta Eruditorum* he felt he had to restore the 'true' balance by emphasizing the 'analytical' discoveries in dynamics made by himself, the Bernoullis, and Varignon.

10 Even Leibniz adopted this aspect of the style. In a MS preface to his *Tentamen de motuum coelestium causis* (or, *Essay on the Causes of Celestial Motions*) written some time after the *Principia* had been published, it is said that the *Tentamen* 'is not based on hypotheses, but is derived from phenomena' ('. . . non constant Hypothesibus, sed ex phaenomenis . . . concluduntur'), which sounds much as if he had been reading the general scholium (which was published by Newton in 1713), where Newton said that 'whatever is not deduced from phenomena is to be called a "hypothesis" ' ('Quicquid enim ex phaenomenis non deducitur, *hypothesis* vocanda est') (see Leibniz, 1849–1863, vol. 6, p. 166, trans. in Koyré, 1965, p. 136). Leibniz then says: '. . . whether there is an attraction of the planets by the sun, or not, it is sufficient that we [*sufficit a nobis*] are able to determine their access and recess, that is, the increase or decrease of their distance [from the sun] which they would have if they were attracted according to the prescribed law. And whether in truth they do circulate around the sun, or do not circulate, it suffices [*sufficit*] that they change their positions with respect to the sun as if they moved in a harmonic circulation . . .' Koyré remarked in 1965), p. 136, that this is 'a hyperpositivistic pronouncement that leaves those of Newton far behind'.

The general tenor of Leibniz's MS preface and of Newton's general scholium was far from unique in the late seventeenth and early eighteenth centuries. And, according to the doctrine of transformation of ideas, we should expect to find other somewhat similar statements of point of view. Here is one from John Wallis's *Mechanica* (1670), def. 12, which was certainly known to both Newton and Leibniz: 'Gravity is motive force downward, or toward the center of the earth. We do not here inquire what the principle of gravity is from the viewpoint of physics nor even whether it ought to be called a quality or an attribute of a body or what other term may properly be used to classify it. For . . . it is enough that by the term 'gravity' we understand that force which the senses perceive to be moving downward both the heavy body itself and the less effectual obstacles which stand in its way.'

11 See Cohen (1964*b*). The French summary was based on the 1704 edition (in English) of the *Opticks*; the Latin *Optice* was published in 1706, and the

first French translation (by Pierre Coste) in 1720, followed in 1722 by a second (and more correct) edition.

12 Guerlac (1977) also discusses the stages by which Malebranche altered his theoretical position on physical optics so as to accommodate the Newtonian discoveries on the dispersion and composition of white light. See, further, Duhem (1916); Mouy (1938); Brunet (1931). As Lohne has shown conclusively, there are many difficulties that arise in trying to repeat Newton's optical experiments of the 1672 communication in accurate detail. They actually work only "in the broad", so to speak, and such remedies as placing a lens between the two prisms are apt to produce additional difficulties (as from chromatic aberration).

13 It is to be observed that in ch. 1 of bk. 2 of the *Mécanique céleste* ('On the law of universal gravitation, deduced from observation'), Laplace concludes: 'observations of the heavenly bodies, compared with the laws of motion, lead therefore to this great principle of nature, namely, that all the particles of matter attract each other in the direct ratio of their masses, and the inverse ratio of their distances.' And he does not at all discuss whether the concept of such a force acting at a distance may give rise to philosophical difficulties. Nor is there even a hint as to this problem in the *Méchanique analytique*, even in the 'Septième section' of the 'Seconde partie' (devoted to 'La dynamique'), where Lagrange introduces orbital motions and perturbations resulting from forces of attraction.

14 See n. 10. D'Alembert followed Newton in insisting, as Hankins (1970), p. 166, has pointed out, 'that gravitation was merely an observed phenomenon that could be described mathematically without any need for causes'. Furthermore, in his article on attraction in the *Encyclopédie*, d'Alembert wrote that if Newton 'claimed that he had considered impulsion to explain gravity, one can believe that this was a kind of tribute that he wished to pay to prejudice or to the general opinion of his century'. Were d'Alembert forced to choose between gravity as a product of impulsive forces or as an innate force, he would choose the latter, even though (as Hankins notes, ibid.) it would go against his own inclination. The reason, said d'Alembert, is that 'it is not yet possible to explain the celestial phenomena by the principle of impulsion; and because even the impossibility of explaining it by this principle is based on very strong proofs [*preuves*], if not on demonstrations'.

15 This aphorism comes from an essay on 'Biogenesis and abiogenesis' in Huxley (1894), p. 244.

16 Of course, it must be kept in mind that Newton may have supposed that there were other forces (or factors relating to the computations and actions of forces) that had not been taken into consideration and which would eventually be uncovered so that the "fudge factor" would be an anticipation of some additional "causes" remaining to be found.

17 Quoted from the Herbert Spencer Lecture at Oxford, *On the Method of Theoretical Physics*, reprinted in Einstein (1954), pp. 270–276; cf. Holton (1973), pp. 233sq.

18 Quoted in Holton (1973), pp. 236sq. The final sentence ('Da könnt' mir halt der liebe Gott leid tun, die Theorie stimmt doch') is all the more significant in that it is, as Holton says, a 'semi-serious remark of a person who was anything but sacrilegious'. Apparently it was Lorentz

who sent the telegram; Eddington was the scientist responsible for the eclipse expedition.

Holton calls attention to the expression of a somewhat similar point of view by Dirac, in relation to 'Schrödinger's wave equation, for describing atomic processes'; Schrödinger, 'who worked from a more mathematical point of view', 'got this equation by pure thought, looking for some beautiful generalization of de Broglie's ideas and not by keeping close to the experimental development of the subject in the way Heisenberg did'. But, according to Dirac, when Schrödinger applied the equation 'to the behavior of the electron in the hydrogen atom, . . . he got results that did not agree with experiment . . . because at that time it was not known that the electron has a spin'. But 'he noticed that if he applied the theory in a more approximate way, not taking into account the refinements required by relativity . . . , his work was in agreement with observation. He published his first paper with only this rough approximation, and in this way Schrödinger's wave equation was presented to the world'. Later on, when the spin of the electron was found, 'the discrepancy between the results of applying Schrödinger's relativistic equation and the experiment was completely cleared up'. Whence Dirac concludes: 'I think there is a moral to this story, namely, that it is more important to have beauty in one's equations than to have them fit experiment' (Dirac, 1963, pp. 46sq). It may not be amiss to remark that Dirac was a successor of Newton's as Lucasian Professor.

19 Lest I be misunderstood, I remind the reader that by "style" I do not mean the geometric form of Newton's argument, nor his special use of limit methods. Nor do I have in mind the degree to which he was limited by the state of the mathematical art, including his own innovations. I am referring here only to what I have called the three phases of investigation which comprise a general "style" of doing mathematical physics. Basic to this "style" is the systematic reduction of physical problems to mathematical analogues, so that they may be solved as mathematical problems. In the twentieth century this part of the Newtonian style may be seen used in a dramatic and effective fashion by Henri Poincaré and G. D. Birkhoff.

4. *The transformation of scientific ideas*

§4.1

1 In logic, synthesis is 'the action of proceeding in thought from causes to effects, or from laws or principles to their consequences' (*OED*). In qu. 31 of the *Opticks*, Newton says: '. . . the Synthesis consists in assuming the Causes discover'd, and establish'd as Principles, and by them explaining the Phaenomena proceeding from them, and proving the Explanations'. He also (qu. 31) refers to 'the Method of Composition', rather than "synthesis", as being opposed to 'the Method of Analysis'.

2 On "inertia" and "force of inertia", see §4.5; on Kepler's physics, see Koyré (1973), Krafft (1973), and Aiton (1972).

3 A convenient introduction to Kepler's ideas may be found in Koyré (1973).

4 On this point see Duhem (1954), pp. 140–145; Popper (1957); Cohen (1974d). I am not concerned here with the question as to whether Newton's

modification of these laws is merely "theoretical" or is of "practical" (i.e., observational) significance.

5 When the MS of bk. one of the *Principia* was officially presented to the Royal Society, it was described as 'a mathematical demonstration of the Copernican hypothesis as proposed by Kepler' (Birch, 1756–1757, vol. 4, pp. 479sq; Newton, 1958, pp. 489sq). Newton himself wrote of the 'Hypothesis, quam Flamstedius sequitur, nempe Keplero-Copernicanaea' (the 'Hypothesis which Flamsteed follows, namely the Keplero-Copernican') (see Cohen, 1971, p. 241, n. 8).

6 The simple elliptical orbits occur as exact laws for the imagined system or construct of a single body moving in a central force field; on the qualifications or modifications for both the elliptical orbits and the law of areas in the solar system, see prop. 13, bk. three.

7 For Newton these relations were not exact because of the nonsphericity of the earth and the effects of the earth's rotation.

8 It may be argued that many of the scholars who have written about "synthesis" in science have implicitly assumed a doctrine of the "transformation" of scientific ideas in the sense which I have adopted here. This is certainly true of the writings of Alexandre Koyré. As I see it, however, "synthesis" (like "revolution") implies a loosely defined large-scale event which is often very difficult to analyze in a precise fashion. The doctrine of "transformation" directs our attention to the smaller-scale individual events which in their totality constitute the "synthesis" (or "revolution") and which, it seems to me, make possible a better understanding of the creative act.

§4.2

1 I am using the term "idea" here in a general sense, to include a concept, a principle, a law, a method, a mode of explanation, and so on.

2 My discussions of transformations of ideas in science are not generically linked with any aspects of transformational grammar (or linguistic transformations), and especially not with generative transformational grammar. I would not deny, however, the potential fruitfulness of a linguistic analysis of science and its development. On the origins of the concept of transformation of scientific ideas, see the supplement to Ch. 5, §§4.5, 4.6.

3 Stillman Drake has argued that Galileo's debt to these predecessors is minimal. In particular, he would insist that the medieval mean-speed reasoning does not appear in his notes on motion and in his books: 'The open-ended motions that concerned him did not lend themselves to the concept of a mean, as did the necessarily bounded motions that concerned Aristotelians.' Drake has concentrated his argument (against a strong medieval influence on Galileo) on the alleged contributions of impetus theory to Galileo's physics (see, notably, Drake, 1975*a*, 1975*b*, 1976).

4 Galileo is alone in this achievement insofar as the public record and the historical succession leading up to Newton are concerned. But even before Galileo there was another scientist of transcendent genius, Thomas Harriot, who around 1590 investigated the laws of motion and constructed mathematically the paths of projectiles according

to his premise of resistance. Galileo would then be perhaps the first to present such notions in a systematic "deductive" way, which apparently was not to Harriot's taste at all.

5 Historians and philosophers often write about conflicts between ideas, as if ideas were persons. For example, in his *Popular Scientific Lectures* (1898), Ernst Mach (p. 63) compared ideas to animal species: 'Slowly, gradually, and laboriously one thought is transformed into a different thought, as in all likelihood one animal species is gradually transformed into new species. Many ideas arise simultaneously. They fight the battle for existence not otherwise than do the Ichthyosaurus, the Brahman, and the horse.' A chapter of this book, 'On Transformation and Adaptation in Scientific Thought', invokes the metaphor and analogy of Darwinian evolution to explain the development of scientific thought; but it is not always clear as to whether Mach conceived of ideas as species or as individuals in competition with one another (this essay is discussed further in the supplement to Ch. 5). A rebuttal of the notion that ideas may be in conflict – and even be conceived to have a life of their own, apart from the men who conceived them – is given by Miguel de Unamuno's pseudo-Hippocratic precept for historians: 'No hay opiniones sino opinantes' ('There are no ideas, but only holders of ideas'), which he said he had adapted from 'some doctors' who said 'No hay enfermedades sino enfermos' ('There are no illnesses, but only people who are ill') (Unamuno, 1951, 'Mi religión', vol. 2, p. 371; cf. Marichal, 1970, p. 104).

6 In this way, the concept of transformation of ideas mediates between the role of the creative individual and the pressure of the intellectual environment. Contrasting Newton's theory and the theories of Galileo and Kepler, Popper has concluded that 'logic, whether deductive or inductive, cannot possibly make the step from these theories to Newton's dynamics. It is only ingenuity which can make this step' (see Popper, 1957, p. 33). The 'step' in question cannot be logical derivation, and must be a transformation. Popper's statement is an admirable guide for all practicing historians of science; I would improve it only by suggesting that "insight" or "exercise of the creative imagination" might serve better than his reference to mere "ingenuity" (see Cohen, 1974d, esp. pp. 321sq).

7 See Poincaré (1912); Birkhoff (1913), (1926) and (1931). The conditions for invariance are that the transformation (one-one and continuous) admit of an area-preserving integral and that points on the inner boundary of the ring be advanced and points on the outer boundary be regressed.

8 An example is given in §4.5, with respect to Newton's use of a "force" of inertia.

§4.3

1 Herein may lie a fundamental difference between scientific and artistic or literary creativity. The scientist is apt to get his ideas while seeking to solve a specific problem or to explain a given effect or to correlate apparently disparate phenomena, and so on. The scientist generally does not indulge in the free exercise of his creative imagination.

2 C. Darwin (1960, 1967), pt. 6, pp. 134sq; cf. Herbert (1971), pp. 209sq. This example was brought to my attention and interpreted for me by Ernst Mayr.

3 Herbert (1971), p. 217. This article contains references to some of the literature and opinions concerning Darwin and Malthus, both before and after the publication of the 'Notebook'.

4 Ibid., pp. 216sq: 'Lyell's conflation of species and individuals misled Darwin in his search for a mechanism for species change . . .'

5 Ibid., pp. 214, 217. For a particularly enlightening account of the influences that led to Darwin's chief concepts, see Schweber (1977). By means of an analysis of the role of intellectual sources not hitherto fully exploited by scholars, Schweber has cast much new light on Darwin's intellectual development and has in fact provided a fully documented example of the stages of transformation that produced Darwin's majestic view of nature and the process of the evolution of species. Schweber does not, however, use the term "transformation". On Malthus and Darwin, see Bowler (1976).

6 See n. 1. We shall see below (Ch. 5) that Newton's most significant contribution to science, the concept of universal gravity, arose in the course of his wrestling with a knotty scientific problem: the application of his third law to a model of planetary motion.

7 The word "fluid" in those days signified not only a material liquid (such as water, oil, or alcohol), but also an "elastic fluid" or compressible gas (such as air).

8 Franklin's studies of Newtonian experimental science are delineated and analyzed in Cohen (1956), ch. 7.

9 This paper was printed in all editions of Franklin's book on electricity; it appears in Franklin (1941), pp. 213sqq.

10 For an analysis of these ideas in their historical context, see Cohen (1956), ch. 8.

11 A number of scientists, among them Watson and Nollet, were then coming to the point of view that electrical effects were not due to the creation of "something" during the act of rubbing; this would lead to the idea of something being collected or transferred or concentrated. Franklin made conservation an active principle, however, showing in numerous experiments that positive and negative charges always appear and disappear in equal quantities. The Leyden jar, an early form of capacitor, thus had equal quantities of positive and negative charge on its two conductors (separated by glass) and thus had no more net "electrical fluid" when charged than when uncharged.

§4.4

1 For a convenient summary, see Dugas (1950), pt. 2, ch. 5; Dugas (1954), ch. 10, sects. 1–3; also Moscovici (1967), pp. 165sqq.

2 In the first (and in the second) edition of the *Principia*, Newton referred to Wren, Wallis, and Huygens as 'the foremost geometers of this age'; but by 1726 this was no longer appropriate and so the text was changed from 'hujus aetatis' to 'aetatis superioris' (see Newton, 1972, vol. 1, p. 66).

3 On Newton's second law, see Cohen (1970). The Latin text reads: 'Mutationem motus proportionalem esse vi motrici impressae . . .'

4 That is, if a single impulse Φ is the vector sum of impulses $\Phi_1 + \Phi_2$ ...

$+ \Phi_n$, and if each impulse produces a change of momentum $\Delta m V_1$, $\Delta m V_2$, ..., $\Delta m V_n$, it follows that the change of momentum $\Delta m V$ produced by the impulse Φ is $\Delta m V_1 + \Delta m V_2 + \cdots \Delta m V_n$.

5 Some authors have misread Newton's statement of the second law so as to make it appear to be the continuous form of the law. Thus Ball (1893), p. 77, adds 'per unit of time' in square brackets so as to make the law read 'The change of momentum [per unit of time] is always proportional to the moving force impressed . . .'

6 Since Newton conceived that gravity might possibly be caused by "impulses", a distinction must be made between "acts" and "appears to act".

7 This example shows how, in the *Principia*, the 'axioms or laws of motion' have been anticipated in the prior definitions. The quoted paragraph comes from the discussion following def. 8. In def. 7 Newton says: 'The accelerative quantity of centripetal force is the measure of this force that is proportional to the velocity which it generates in a given time.' In def. 8 the wording is the same for 'The motive quantity . . . proportional to the motion [i.e., quantity of motion, momentum] which it generates in a given time.'

8 A very different point of view is expounded in Westfall (1971).

9 This proposition has often been discussed but was not completely analyzed until the work of D. T. Whiteside. Among other things, he has shown that the intervals are not only non-finite, but must be second-order infinitesimals (say dt/n, $n \to \infty$) for the proposition to be valid: that is, after n instants of time ($n \to \infty$), it is necessary that the total time ($n \times (dt/n) = dt$) be still infinitesimal. There is a question here as to 'whether Newton himself realised this'.

10 Although the form of the second law (as it appears in the axioms of the *Principia*) and the first proposition of bk. one start out with impulses, we have no way of telling whether in his actual thinking processes Newton began with impulses and got to continuous forces or vice versa. Certainly, he was greatly concerned with continuous forces (particularly centripetal forces), and he doubtless saw early on that these could be studied by being broken down into parts, infinitesimal parts. This would be, for a mathematician, analysis, resolving an entity into components; the transition from the component impulses back to the original force would be synthesis. It is the latter that appears in the beginning of the *Principia* and elsewhere, but Newton's real genius lies in his vision of analyzing continuous motion (and continuous forces) into fragments which he can then treat mathematically. I am indebted for this point of view to D. T. Whiteside.

11 See the extracts from the *Waste Book* in Herivel (1960*b*), (1965*a*), p. 130.

12 See *De motu*, theor. 1; this tract has been published in Rigaud (1838), Ball (1893), Hall & Hall (1962), Herivel (1965*a*), and by Whiteside in Newton (1967–) vol. 6.

13 Sect. 2, bk. one; sect. 1 deals with the mathematical theory of limits.

14 The long sequence of such transformations is the subject of a separate study in progress; an aspect of this topic is discussed in Cohen (1964*b*).

15 How Newton encountered the actual expression "inertia" and transformed it is discussed in §4.5.

16 In this sense, Newton is the true or sole discoverer or inventor of the second law in its quantitative completeness. Before Newton, many writers had been groping toward the concept of mass but had confused it with weight.

17 Thus Ernst Mach (1960), ch. 2, pt. 3, sect. 11: 'Perhaps the most important achievement of Newton with respect to the principles [of dynamics] is the distinct and general formulation of the law of the *equality of action and reaction*, of pressure and counter-pressure'.

18 See also Gabbey's review of our edition of the *Principia*, Gabbey (1974), esp. p. 242.

19 This may be taken as an indication that Newton had not read Galileo's *Two New Sciences*, or that he had either read that book carelessly or had forgotten what he had read (see Cohen, 1967c).

20 In the statement of the laws, it is only in the paragraph following the first law that Newton gives supporting experiential evidence. For the third law, such evidence is given in the scholium following the laws, but only for a restricted aspect of the law (collisions). The evidence for the first law consists of the inertial component of projectiles, the inertial (or tangential) component of the motion of the parts of spinning hoops or "trochees", and the inertial (or tangential) components of the rotational and the orbital motions of planets. In this Newton somewhat anticipates the parallelogram law for motions and forces in the following corollaries to the laws.

21 U.L.C. MS Add. 4004, fols. 10^r–15^r, 38^v, 38^r; printed in Herivel (1965a), pp. 133–182. See Whiteside's comments in Newton (1967–), vol. 5, pp. 148sq.

22 This essay was published by Hall & Hall (1962), pp. 157–164, and again by Herivel (1965a), pp. 208–218.

23 On the background of interest in this problem, see A. R. Hall (1966), and Dugas (1954), ch. 7, sect. 20,d, and ch. 8, sect. 2. An early paper by Wren ('Lex naturae de collisione corporum') was published in the *Philosophical Transactions* for 11 Jan. 1668/9, vol. 4, pp. 867–888. The conservation of momentum was announced in Wallis's paper (1699) in the *Philosophical Transactions*, followed by Wren's; but Huygens's paper was published elsewhere. For details, including Huygens's priority, see Dugas (1954), ch. 10, sect. 2.

24 Here he said that when two bodies moving toward one another in the same straight line encounter each other and are reflected, 'none of theire motion [*read* momentum or quantity of motion] shall bee lost . . . For at their occursion [in the hypothesis of perfect elasticity] they presse equally uppon one another and [upon impact] they shall bee reflected, soe as to move as swiftly frome one another after the reflection as they did to one another before it'; quoted from Newton (1967–), vol. 5, p. 149, n. 153, corresponding to U.L.C. MS Add. 4004, fols. 10^v–11^r and Herivel (1965a), pp. 142sq.

25 Newton (1967–), vol. 5, pp. 148sq. These lectures were rewritten before they were deposited in the University Library.

26 In Newton's experiment, the lodestone and the piece of iron float on separate vessels, apparently in such a way that the lodestone and iron cannot make actual contact. The two vessels eventually come to rest. While this is not exactly an illustration of Newton's three-body system (in

which there are two attracting bodies with a third body or obstacle
between them), the principle is the same.

27 Classically, induction could be a process of showing that a property or
law applied to each of a given set of items (as dealing with each possible
variety or species of triangle seriatim and hence saying that something
is true of all triangles).

28 He notably omits from the list the gravitation of planets toward the sun.

29 These, in turn, were corrections and extensions of Descartes's laws of
collision in his *Principia*, pt. 2.

§4.5

1 Leibniz published a 'Specimen dynamicum' in the *Acta Eruditorum*
of 1695, pp. 145–157. It is translated in Leibniz (1956), vol. 2, pp. 711–738.
Leibniz appears first to have used the word "dynamics" in 1691 or 1692.

2 The development of a physics of motion based on inertia may be
traced in the writings of Clagett, Dugas, Herivel, Koyré, Westfall,
and others listed in the Bibliography.

3 In fact, Aristotle's concept of "motion" was a general "process", in which
a potentiality achieved actuality. On Aristotelian views of motion, see
the writings of Clagett and Grant listed in the Bibliography.

4 In Huygens's *Horologium oscillatorium* (1673), pt. 2 opens with three
hypotheses, of which the first states a version of the law of inertia for
the condition of there being no gravity and no impediment to the motion
from air. In this case the body will continue to move 'with uniform
velocity along a straight line' ('velocitate aequabili, secundum lineam
rectam'). The limited conditions of applicability probably arose
from the restricted subject matter ('On the descent of heavy bodies and
their motion in a cycloid'). It may be noted that Huygens does not
refer to a "state" of motion, as Newton (following Descartes) does (see
Huygens, 1888–1950, vol. 18, p. 125).

5 It must be kept in mind, however, that in 1713, when the second edition
was published, there was a new vortex theory that had been introduced
by Leibniz in his *Tentamen* (1689c) (see Aiton, 1960, 1962, 1964,
1972, ch. 7).

6 In the commentary ('Exposition abrégée du système du monde, et
explication des principaux phénomènes astronomiques tirée des Principes
de M. Newton'; Newton 1759, vol. 2, p. 9 of the second numeration), it is
said: 'Ce second Livre . . . paroît avoir été destiné à détruire le
système des tourbillons'. An accompanying postil reads: 'M. *Newton*
a composé ce Livre pour détruire les tourbillons de *Descartes*'.

7 This is the essay beginning 'De gravitatione et aequipondio fluidorum'
(Hall & Hall, 1962, pp. 89sqq).

8 Newton (1967–), vol. 1; other influences included Wallis and Oughtred.
See also Whiteside (1964a).

9 Cohen (1964b). It implies the sense of "in so far as it can, of and by itself".

10 In fact, the designation "axiomata" is used along with "leges motus"
in the heading, but in stating the laws and in references throughout the
text of the *Principia*, Newton uses "lex" rather than "axioma".

11 Only Wallis's paper was entitled 'The laws of motion', the title being
in English, though the text was in Latin. Wren's paper was called 'The

law of nature concerning the collision of bodies'. Huygens's paper was not published in the *Philosophical Transactions* but was mentioned in an editorial comment (see §4.4, n. 23).

12 This text has been published in Hall & Hall (1962), pp. 157–164; in Newton (1959–1977) vol. 3, pp. 60–64; and in Herivel (1965a), pp. 208–215.

13 U.L.C. MS Add. 3958, fols. 81–83. I am indebted to D. T. Whiteside for an opinion concerning the date. From the form, size, and shape of the words, and the fact that this essay is written in English, Whiteside concludes that it may be firmly dated in the middle 1660s, say 1666–1667, a couple of years before he could have read Wallis's article. Whiteside finds a virtually absolute identity between this hand and that in the "October 1666 tract" (MS Add. 3958, fols. 48–63), and calls attention to the fact that the second and third paragraphs on fol. 81ʳ are a straight borrowing out of the 1666 tract. As far as I know, this is the first time that Newton used the term "law" in a scientific context. Shortly afterwards, in his *Lectiones opticae* (beginning in 1669), he would write '. . . non vobis displiceat si de legibus refractionum nonnulla praesternam' (U.L.C. MS Add. 4002, fol. [70], Lect. 9, dated July 1670) (see Newton, 1973).

14 Newton's third law of "motion" is perhaps a more general principle of nature than any of Descartes's laws of "nature".

15 On this feature of Descartes's law, see Koyré (1939), pp. 319sqq.

16 Newton to Cotes, 28 March 1713; Newton (1959–1977), vol. 3, pp. 396sq. Here, as in the concluding general scholium of the *Principia*, the word "deduce" (or the Latin word *deducere*) means nothing more than "derive". In a prior draft of the letter to Cotes, Newton in fact referred to 'deducing things mathematically from principles'; he also referred to 'Laws . . . being deduced from Phaenomena' (ibid., p. 398). Newton certainly was aware of the difference between logical deduction and induction, although he did write to Cotes (31 Mar. 1713) that 'Experimental philosophy . . . deduces general Propositions . . .by Induction'.

17 This evidence is contained, as was seen in the concluding part of §4.4, in the scholium to the laws of motion, in the beginning of the *Principia*.

18 Galileo, in the *Two New Sciences*, wrote of a quasi-inertial motion that was supported on a vast plane; Kepler's "inertia" caused bodies to come to rest rather than to continue moving.

19 But this may have been in itself a transformation of an older tradition of doing just that in optics. For example, Kepler's *Dioptrice* has "axiomata", "definitiones", and "problemata". These "axiomata" are both geometrical and experiential. For example, axiom 6: 'The refractions of crystal and glass are approximately the same'; axiom 9: 'The greatest refraction of crystal is around 48 degrees'.

20 On Kepler's concept of inertia, see Koyré (1973), pt. 2, sect. 2, ch. 4, 6; Krafft (1977). The pre-Newtonian history of inertia is is discussed in Cohen (1978b).

21 Newton refers to 'Part 2 Epist 96 ad Mersennum' (Hall & Hall, 1962, p. 113): obviously the second part (or volume) of the Latin edition in three parts of the letters of Descartes (1668). Newton must have read the first edition in Latin, of which vol. 2 was issued in both Amsterdam and London in 1668, based on Clerselier's French edition of vol. 2 (1659),

for details of which see the beginning of vol. 1 of Descartes's *Oeuvres*
(1974). A second Latin edition was published in Frankfurt am Main
(sumptibus Friderici Knochii) in 1692. In letter 94 of this pt. 2 or
vol. 2, Descartes says, 'Inertiam nullam aut tarditatem naturalem in
corporibus agnosco'; this is a Latin translation of Descartes's letter to
Mersenne printed by Adam & Tannery in (1974), vol. 2, as Letter
152, where the foregoing sentence reads, 'Je ne reconnois aucune Inertie ou
tardiveté naturelle dans les cors'. Later on in the same paragraph of
this letter, Descartes refers to 'Inertia ista naturali' ('cette Inertie
naturelle'). Since the terms "inertia" and "natural inertia" occur in a
letter (no. 94) only two removed from the letter (no. 96) to which Newton
refers in his essay, and since both letters are written to Mersenne and
discuss the same topics, it may be assumed that Newton would have read
both letters and that this is the source of his acquaintance with Keplerian
inertia in concept and by name – but without any association with
Kepler as original progenitor.

Other references to "inertia" are to be found in the Latin pt. 2
or vol. 2 of the *Epistolae* in letter 25 and letter 34. These are, respectively,
from Descartes to Debeaune (30 Apr. 1639; Descartes, 1974, vol. 2, letter
161, pp. 543sq) and to Mersenne (25 Dec. 1639; ibid., letter 179, p. 627).

22 See Newton (1972), p. 40. In Cohen (1971), pp. 27–29, there is printed a
note of Newton's (also in Latin) saying much the same thing.

23 Gregory's *Elements of Astronomy* was, as he said in the preface, intended
to make available 'the Celestial Physicks, which the most Sagacious
Kepler had gotten scent of, but the Prince of Geometers Sir *Isaac Newton*,
brought to such a Pitch as surprized all the World'. There were Latin
editions (1715, 1726) and two English versions (1715, 1726), of which
the 1726 revised English translation is available in a reprint. Gregory
describes Kepler's physics (basing his account on bk. 4 of Kepler's
Epitome astronomicae Copernicanae) in prop. 66 of bk. 1, where he describes the 'conflict between the Vectory Power of the Sun, and the
sluggishness of the Planet'. The English translator (1726, p. 185) rendered
Gregory's 'inertia materiae' as 'the sluggishness of matter'.

24 See Leibniz & Clarke (1956), pp. 111sq. In Leibniz's fifth paper he
says that 'the inertia of matter . . . , mentioned by Kepler, repeated by
Cartesius, mentioned in my Theodicy . . . has no other effect, than
to make the velocities diminish, when the quantities of matter are increased: but this is without any diminution in the force' (p. 88). In an
appendix Clarke collected four examples in which Leibniz spoke of a
Keplerian inertia. They are: 'Nay rather, matter resists motion, by a certain
natural inertia, very properly so styled by Kepler; so that matter is
not indifferent to motion and rest as is vulgarly supposed . . .'; 'A natural
inertia, repugnant to motion'; 'A certain sluggishness, if I may so
speak, that is a repugnancy to motion'; 'A sluggishness, or resistance
to motion, in matter'. These were taken by Clarke from '*Acta Erudit. ad
Ann.* 1698, Pag. 434.' and '*Acta Erudit. ad Ann.* 1695. Pag. 147.' and were
quoted in the 1717 edition of the *Correspondence* edited by Clarke after
Leibniz's death.

25 Presumably Newton wrote out his two notes contrasting his view of
inertia and Kepler's either while Clarke was writing out the replies to

Leibniz or after Clarke published the edition of the *Correspondence* in 1717.

26 Notably in the *Astronomia nova* and the *Epitome astronomiae Coperni-canae*. We have no evidence (either direct or indirect) that Newton may have read either of these two works before writing the *Principia*.

27 Goclenius (1613), p. 321, s.v. "vis", says: 'Vis Insita est, vel Violenta. Insita, ut naturalis potestas.' That is, 'Force is inherent or violent. Inherent, as a natural power.' Cf. also pp. 322 and 722.

28. In a college notebook, U.L.C. MS Add. 3996, marked on the title page 'Isaac Newton/Trin: Coll Cant/1661', p. 3 (fol. 89ʳ), Newton refers to a statement made by the 'excellent Dʳ Moore [i.e., More] in his booke of the soules immortality', and he particularly notes More's opinion that 'the first matter must be atoms' and 'that matter may be so small as to be indiscirpible'. The word "indiscirpible" was introduced by More. See Koyré & Cohen (1962), pp. 123–126.

In H. More (1679), p. 192, it is mentioned that there are "adversarii" who posit an 'innate force or quality (which is called heaviness) implanted in earthly bodies' ('innatam quandam vim vel qualitatem (quae Gravitas dicitur) corporibus terrestribus insitam'). In a supplement to H. More (1659), written and printed after Newton had made notes on his youthful reading of the book, the word "implanted" occurs in this same sense, and is translated in Latin (not by More, however) as "insita".

This literal sense of "implanted" was not the only traditional one given to "insitus"; it was also used in a general way to mean a quality that is "inherent" or "natural". It occurs in this sense in Cicero's *De natura deorum* and other works; see the lengthy discussion of this term by Arthur Stanley Pease in his edition of *De natura deorum* (1955), pp. 298sq. The phrase "vis insita" also occurs in an ode of Horace.

29 From Magirus (1642), *Physiologia*, bk. 1, ch. 4, sect. 28, Newton wrote down concerning motion, that it is 'either *per se* or *per accidens*'. In the following sect. 29, Magirus says: 'Motion is *per se* or proper, when a movable body moves by its own power [sua virtute]: thus man is said to move *per se* because he moves wholly and by his inherent force [insita vi sua]'. Although Newton's summary includes portions of sect. 29, it contains only the part of the above extract up to the colon, thus omitting the actual words 'insita vi sua'. There are, however, yet further occurrences of "insitus" or "insita" in Magirus's writings, sometimes in association with "vis" and at other times with "virtus". The word "insitus" also occurs in certain Latin versions and epitomes of Aristotle's *Ethics* (bk. 2, ch. 1: 1103a), including the Latin translation printed by Magirus.

Newton's notes on Magirus are to be found in MS Add. 3996; the portions dealing with sects. 28 and 29 occur on fol. 17ᵛ. If these are notes made by Newton while actually reading Magirus's *Physiologia*, then he would necessarily have encountered the above-mentioned occurrence of 'insita vi sua'; but it is possible that Newton may merely have copied an outline from a MS then available from a tutor or from another student.

30 He does point out that this force is exerted as a resistance only during the course of a change of state, when an external force is being impressed on a body. And in this case, it is a force of resistance opposing the

force producing the change of state, but it may also be an impulse (as changing the state of a body colliding with the given body, since the given body then tends to try to change the state of that colliding body by exerting a force on it).

31 By the Newtonian principles of dynamics, only an external force can produce a change in a body's state (of resting or of moving uniformly straight forward).

32 This is illustrated in the beginning of the scholium to the laws, where the motion arising from the projection of a body 'is compounded with the motion arising from gravity' to produce a parabolic path. The first corollary to the laws compounds motions, not forces. For Newton was fully aware that the "force of inertia" can neither generate nor alter motions, and so cannot be combined vectorially with forces that do generate or alter motions.

33 Newton does not always refer to inertia as a "vis inertiae" of matter, but sometimes merely as "inertia materiae".

34 The separation of mass from weight is one of Newton's supremely original contributions to science; of course it was the last in a series of transformations.

35 I have not gone into the question here of the way in which mass (and momentum, based upon mass) enters into the third law.

36 In considering some of the major steps of transformation whereby Newton reached the first of his 'axioms or laws of motion', I have not gone into one part of the question which has been discussed by R. S. Westfall (1971), *Force in Newton's Physics*. This concerns Newton's possible lack of continuity in his allegiance to a true inertial physics. This question is complicated somewhat by the complexities of his belief in absolute space, but Westfall has indicated that Newton may very well have had a period prior to the *Principia* in which he did not fully believe in inertial physics and the law of inertia as we would understand it. In part this question hinges on things which Newton did not say, just as much as those which he did say. For example, when he wrote down certain statements in the *Waste Book*, he did not necessarily say that there was any "cause" for such inertial motion. But we do not know what was in his mind, and must restrict ourselves to what he wrote down. Of course, even when Newton conceived there was some kind of "cause" for inertial motion, which he called a "vis inertiae" or "vis insita", it may not necessarily mean that he was thinking of a "true cause" in the sense that would have been true of his medieval predecessors. As pointed out in §3.12, n. 3, others in the Newtonian age wrote about a "force of inertia" even when they were aware that in terms of ordinary understanding the word "force" had no meaning in this context, if it had any at all. This question is discussed further in Cohen (1978*b*).

§4.6

1 Presumably, what came into Newton's mind was the idea of comparing the falling of the apple with the "falling" of the moon, which is 60 times as far from the earth's center as the apple is. If the force and consequent acceleration diminish as the square of the distance, the moon should fall in one second $1/3600$ as far as the apple does.

There is reliable evidence that Newton himself told the story of the apple; it was made available in 1936, when J. Hastings White published a biography of Newton, based in part on a personal interview, that had lain unknown and unused by scholars since the eighteenth century (see Stukeley, 1936, pp. 19sq). For further information on this topic, see McKie & De Beer (1952) and Cohen (1946).

We have no idea as to when this alleged incident would have occurred. It is exceedingly improbable, however, that Newton could have made such a moon test in the 1660s.

2 On the fruitfulness of transformation by misinterpretation, see Robinson (1962), p. 4, in relation to Plotinus. Robinson concludes: 'Many of the great advances and novelties of human thought have arrived in the form of misinterpretations of the past.'

3 That is, Galileo may be said to have known this law in certain limited circumstances (as in the motion at terminal velocity of a body falling freely in a resisting medium) or in a restricted form (such as a body moving on an indefinitely extended frictionless plane).

4 This is the form in which Newton states the second law in the *Principia*. Galileo certainly did not know the continuous form of the law either, which Newton uses in the *Principia*, since in this form of the law the mass also provides the proportionality between a continuous force and the acceleration it produces.

5 This paragraph was added to the one printed above only in the third edition of the *Principia*; versions may be found in Newton's personal interleaved copy of the second edition (see Newton, 1972, vol. 1, pp. 64sq).

6 See Millikan (1947), pp. 15, 24. The opening historical chapters are unchanged from the earlier editions (1935, 1937).

7 Such transformations, based on an imperfect understanding of a prior author's intentions, often have served as a major step in the intellectual leaps forward by scientists; but in history the same process seems to lead to error by obscuring the intentions, aims, and limitations of those scientists of the past whose thought we are trying to reconstruct. A similar problem exists in the history and criticism of literature.

8 Possibly this same association in his mind of himself with Plato could explain why Galileo attributed to Plato what appears to have been his own highly original conceit about the creation of the world by God's letting planets fall toward the sun; see §1.5, esp. n. 8.

9 On the potentialities of Plato's writings for misinterpretation, see Robinson (1962), pp. 1–4.

10 Quoted from Marx & Engels (1947), pp. 41–46; Marx & Engels (1949), pp. 101–104; Marx & Engels (1967), vol. 1, pp. 155–159; Marx & Engels (1954), pp. 313–319. The latter work (pp. 320sq) quotes several statements by Marx concerning Balzac. See further, Lukacs (1948) and (1955), pp. 63sqq. Marx himself wrote a critical article on Balzac, in which he called him a 'prophetic creator'. See, further, Marx (1947), p. 41; Marx and Engels (1937), vol. 3, p. 449.

11 Steiner (1970), p. 321; cf. p. 306. Hence we may have a "Balaam effect", so named after Balaam, who was induced to curse Israel and was rebuked by the ass he was riding. By God's inspiration, his utterance became a blessing rather than a curse (Numbers, xxii:8–xxiv).

For a discussion of the contrary idea, that there is a true meaning to all literary work, see Hirsch (1967) and (1976).

12 A striking example of literary transformation occurs in Dante's *Inferno* (canto 25: 49–78), in the description of how a serpent with six feet entwines itself around the body of Agnello or Agnolo de' Brunelleschi and merges with it. This horrible image appears to have been derived from the beautiful account in Ovid's *Metamorphoses* (4:373–379) of the gratification of the wish of the nymph Salamacis that she never be separated from her lover Hermaphroditus (son of Hermes and Aphrodite); their bodies were joined together, 'were merged in one, with one face and form for both'. Their bodies were so 'knit in close embrace' that 'they were no longer two, nor such as to be called, one, woman, and one, man'. For Dante's transformations of Ovid's images in the *Metamorphoses*, see Dobelli (1897), and Charles Singleton's commentary to the *Inferno*, in Dante Alighieri (1970), pp. 437sq. The *Metamorphoses* itself, as its title reveals, is devoted entirely to transformations, though of a different sort than those under consideration in the present book.

13 In addition to setting forth the sources of Coleridge's imagery, and showing how Coleridge transformed what he encountered in his reading into extraordinary poetry, Lowes also advocated a theory of unconscious creation, referring (Lowes, 1927, pp. 59–60) to 'the "deep well of unconscious cerebration" [that] underlies your consciousness and mine'. In particular, Lowes claimed that it was this unconscious element in the creative process which was the key to Coleridge's power of imagery. This latter aspect of Lowes's work has not fully stood the test of time. Werner W. Beyer argues that the stress given by Lowes to Coleridge's unconscious creation 'has given such wide currency to the concept of unconscious metamorphosis [in Coleridge] that its conscious counterpart has threatened to be ignored' (1963, p. 43). According to Beyer, R. C. Bald (another scholar specializing in Coleridge's creative process) has come to the same conclusion after studying Coleridge's later notebooks; he too 'stressed what Lowes had seemed to minimize: the conscious element in the creative process, the deliberateness of Coleridge's reading, for purposes of poetry, and the recency of some of it which could not have been long submerged in the subconscious' (Beyer, 1963, p. 66, referring to Bald, 1940; see also Beer, 1959).

The example of Coleridge, as interpreted by Lowes, was used by Theodosius Dobzhansky (1959, pp. 204, 205) to account for Darwin's lack of explicit references to some of his predecessors, e.g., Edward Blyth. In the present context it is particularly pertinent to find Dobzhansky stating explicitly that 'it is not illegitimate to compare the creative processes of a poet, Coleridge, with those of a scientist, Darwin', since the two are not fundamentally different. Loren Eiseley (1965) replied to Dobzhansky. A protagonist of Blyth, Eiseley could not accept the theory of unconscious creation in the case of either Darwin or Coleridge.

Whether the creative activity of Darwin or of Coleridge was primarily unconscious or conscious, there can be no doubt that in both cases a primary aspect of the creative process was transformation, in the sense in which I have developed it in this chapter and applied in this book. Indeed, it is this aspect of the nature of creativity in poetry and in science that gives an essential unity to such divergent points of view as those expressed by Dobzhansky and Eiseley and by Lowes and Beyer.

14 The image of a star 'Within the nether tip' of the 'hornèd Moon', apparently based on the report in the *Philosophical Transactions*, was a late version occurring in print in 1817 (see Bald, 1940, p. 9). In 1798 Coleridge had made a voyage on the Mediterranean and had observed the crescent moon with 'the evening Star almost crowning its upper Tip' (ibid., p. 8). The 1800 printed text (ibid., p. 11) read 'The hornèd moon, with one bright star / Almost atween the tips.'

15 'Dans le champs de l'observation', Pasteur said, 'le hasard ne favorise que les esprits préparés' (see Vallery-Radot, 1911, ch. 4, p. 88). The statement was made on 7 December 1854, in the inaugural lecture Pasteur gave when he was made professor and dean of the newly established Faculté des Sciences at Lille.

16 For a general account, see Movius (1948), pp. 390–393.

17 For a number of years of excavation, no 'actual human implements were recorded'; later, when such tools began to be found, 'the enormous quantity of artificially broken stones, chips, waste, and trimming flakes, etc.' emphasized by comparison that 'the total number of worked implements is very small' (ibid., p. 393*a*). According to Clark (1946), p. 33, 'the crudely fractured pieces of stone from Choukoutien would never, in the vast majority of instances, have been recognized as showing traces of artificial work had they been recovered isolated in a geological deposit'.

18 The example of *Homo pekinensis* is further instructive. It not only calls attention to the preparation of the mind that grasps the occasion and recognizes the potentiality of a transformation; it also indicates that such recognition occurs in relation to other factors, to a background in which the idea or artifact may be encountered. This suggests an important role of the *Gestalt* or field of conscious or unconscious associations.

19 For a discussion of the real differences between Darwin's and Wallace's concepts of evolution, see Romanes (1895).

§4.7

1 An example has already been given in §4.6, nn. 16–18, in relation to the artifacts found in association with *Homo pekinensis*.

2 Even though there was close attention paid to some of the realistic aspects of the aborigines and their customs, Columbus and other explorers also tended, as Harry Levin has pointed out, to 'draw upon a rich backlog of fabulous lore about aborigines, namely the myth of the golden age'. Whenever they encountered aborigines, the literary tradition of a golden age was apt 'to come into play, almost as if it had been touched off by a reflex action'. As a result, according to Levin, 'Life at its barest and least sophisticated was somehow decked out with a set of trappings inherited from the learned conventions of literature' (see Levin, 1969, p. 60).

3 The same transformation occurred with respect to auditory experience, since both Columbus and his sailors recorded how they heard nightingales, which at that time were not indigenous in the West Indies. Harry Levin has called attention to the question of 'the psychological process

whereby, on that occasion . . . presuppositions colored the first encounters of the Europeans with the New World' (see Levin, 1969, p. 60).

4 In the *Opticks*, prop. 6 (prob. 2), bk. 1, pt. 2, Newton divides the circle into 'seven Parts . . . proportional to the seven Musical Tones or Intervals' in an octave, and has these parts represent 'all the Colours of uncompounded Light gradually passing into one another, as they do when made by Prisms', so that the circumference represents 'the whole Series of Colours from one end of the Sun's coloured Image to the other'. Then by assigning masses to the colored rays, the result of color mixtures is found by determining the center of gravity of the parts; thus the center of gravity of all the colors being at the center of the circle produces white. There are references to the analogy between color and musical theory in Newton's 1675 letter to Oldenburg (on light and colors) and in his Lucasian *Lectiones opticae.*

In prop. 2 (theor. 2), bk. 1, pt. 2, he refers to the prismatic spectrum as formed by rays that 'appear tinged with this Series of Colours, violet, indigo, blue, green, yellow, orange, red, together with all their intermediate Degrees in a continual Succession perpetually varying'. In prop. 3 (prob. 1), bk. 1, pt. 2, Newton reported how 'an Assistant, whose Eyes for distinguishing Colours were more critical than mine', drew lines across the spectrum in order to 'note the Confines of the Colours'; the ratios of these lengths were found to be related to the seven tones of the diatonic scale. In his original paper on the composition and dispersion of white light (1672), Newton had merely said 'The Original or primary colours are, *Red, Yellow, Green, Blew,* and a *Violet-purple,* together with Orange, Indico, and an indefinite variety of Intermediate graduations'.

5 The color circle was designed in order to determine the hue and saturation of any mixture of colors; see Mach (1926), pp. 96sq.

6 These are easily found to be "blind" pits, as may be seen by the attempt to pass a fine straw through them or to force water through them.

7 The quotations from Vesalius are taken from Singer (1922), pp. 27–28.

8 Even if a Western record should be found, the fact will remain that this extraordinary celestial spectacle was not "seen", or was not considered to have been worthy of notice, by a great number of writers. (In June 1978 a report in *Nature* called attention to the observation of this supernova by an Islamic scientist; see Brecher, Lieber, & Lieber, 1978).

9 Tycho recounts the story of his difficulty in believing that there was a new star in Cassiopeia, and the later difficulty he had in convincing Pratensis and Dancey about the new star, in his report *De nova stella* (1573) and the later *Astronomiae instauratae progymnasmata* (1602, 1610); reprinted in Brahe (1913–1929), vol. 1, pp. 16–19; vol. 2, pp. 307–329; vol. 3, pp. 93–96. These autobiographical texts are summarized and paraphrased in Dreyer (1890), ch. 3, pp. 38 sq. A partial English translation of *De nova stella*, by John H. Walden, appears in Shapley & Howarth (1929), pp. 13–19.

10 Galileo presented his discoveries about the moon in his *Sidereus nuncius* (1610), reprinted in Galileo (1890–1909), vol. 3, translated by Stillman Drake in Galileo (1957).

11 On Galileo's actual observations and new 'light on Galileo's actual process

of reaching the conclusion that he was seeing bodies which literally circled around Jupiter', see pp. 146–153, esp. pp. 148–149, of Drake (1978).

12 On the sunspot debate, see Galileo's *History and Demonstrations concerning Sunspots and their Phenomena*, pp. 59–144 of Galileo (1957), esp. pp. 91–92, 95–99.

13 On the question of Gestalt in relation to science, see (in addition to the works of Hanson and Nash), Kuhn (1970), pp. 64, 85, 111, 122, 150; also Kuhn (1977), p. xiii.

§4.8

1 See, further, the supplement to Ch. 5 on the history of the concept of transformation.

2 Freud later wrote another paper on Popper-Lynkeus (Freud, 1932). On Popper, see Blüh (1952); Löwy (1932).

3 Freud first drew attention to the coincidence between his views and those of Popper-Lynkeus in a postscript added to ch. 1 of his *Traumdeutung* in the 1909 edition, and in a new footnote to ch. 6 (see Freud, 1923, p. 263n, and 1966–1974, vol. 4, pp. 94sq, 308sq).

§4.9

1 In a Postscript–1969 to the 'second edition, enlarged', Kuhn says (1962, p. 174, n. 2) that he has introduced only two alterations to the text, other than correcting of typographical errors; of these, one 'is the description of the role of Newton's *Principia* in the development of eighteenth-century mechanics'.

2 Canguilhem (1955), p. 172: 'En matière d'histoire des sciences aussi, il y a une échelle macroscopique et une échelle microscopique des sujets étudiés'.

3 'What has been learned in physics stays learned. People talk about scientific revolutions. The social and political connotations of revolution evoke a picture of a body of doctrine being rejected, to be replaced by another equally vulnerable to refutation. It is not like that at all. . . . There is no good analogy to the advent of quantum mechanics, but if a political-social analogy is to be made, it is not a revolution but the discovery of the New World' (Physics Survey Committee, 1973, pp. 61sq).

4 Weinberg (1977), pp. 17sq. Weinberg, however, admits that there have been revolutions in the history of science, and he refers to 'the development of special relativity and of quantum mechanics' as 'great revolutions'.

5 The author of the most ambitious psychological study of Newton to date declares that he will not presume 'to unlock the secret of Newton's genius or its mysterious energy' (cf. Manuel, 1968, p. 2). But Manuel (p. 84) does nevertheless suggest a possible origin of universal gravitation in 'the fact . . . that Newton . . . was in a critical period of his childhood powerfully drawn to distant persons, that he was hungry for communion with the departed ones in an elementary, even primitive sense. Since this yearning never found an object in sexuality, it could have achieved sublime expression in an intellectual construct whose configuration was akin to the original emotion.' Yet, although Manuel suggests that skepticism should be our proper response to a 'coupling of the emotions of a child being drawn to distant and absent ones with the idea of a

natural force that as an adult he could never define', he does nevertheless
propose in conclusion that what Newton achieved was the 'giant step'
of 'translation of longing, [of] one's passion for persons, into a systematic
inquiry of a mathematical-astronomical character'. He argues the
reasonableness of this statement by noting that 'Newton lived in an
animistic world in which feelings of love and attraction would be
assimilated to other forces'. Each reader will have to decide for himself
whether such a line of thought is helpful in understanding Newton's
scientific creation.

5. Newton and Kepler's laws: stages of transformation leading toward universal gravitation

§5.1

1 The alleged cause of the delay was Newton's inability to prove the later
 theorem that a homogeneous sphere (or a sphere composed of homogeneous shells) gravitates as if all its mass were concentrated at its
 geometric center. See Cajori (1928); Glaisher (1888) says that the whole
 "explanation was pointed out to me" by Adams.

2 For examples, see Cohen (1974*d*), p. 300.

3 In the first edition of the *Principia*, at the beginning of bk. three, Newton
 referred to Kepler's laws under the rubric of "Hypotheses"; he later
 made a change from "Hypotheses" to "Phenomena" (see §3.5 and §5.8).
 This change is discussed in Koyré (1955*b*), (1956), (1960*c*); Cohen (1956),
 pp. 131sq, (1966), (1970), ch. 6, sect. 6.

4 The first two laws were announced in 1609 in Kepler's *Astronomia nova*;
 the third law in 1619 in his *Harmonice mundi*.

5 In the immediate pre-*Principia* stage of Newton's thinking (as we shall see
 below in §5.6), this result was not recognized at first. The main outlines
 of this development have been sketched in Ch. 3.

6 The one-body system (a body circulating about a center of force) is
 identical to a two-body system in which the circulating body does not
 attract the central body. The latter would be very closely approximated in
 the real world by the system of the sun and of the planets Venus,
 earth, or Mars. For these planets have so small a mass compared to that
 of the sun that their gravitational action on the sun is all but nonexistent
 in terms of causing the sun to move. Or, if we consider that both the
 sun and the earth circulate about their common center of gravity, then
 (since the earth–sun distance is about one hundred million miles
 and the earth's mass is about one-three hundred thousandth of the sun's
 mass) the center is but a negligible three hundred miles from the sun's
 center. For Jupiter, however, this simple model would not serve so
 well, since its mass is about one-thousandth of the sun's mass. Actual
 planetary orbits suffer perturbations resulting from the gravitational
 attractions of the other planets.

7 *Principia*, bk. three, prop. 12; from the point of view of exact mathematical
 science, what matters is not so much the magnitude of such a difference
 as the fact that there is a difference at all. In this sense, Kepler's

third law is a hypothesis, or it is a statement that is phenomenologically true – true within certain limits of observational accuracy.

§5.2

1 *Principia*, bk. three, phen. 4; in the first edition, this was hypoth. 5. On the shift from 'hypotheses' to 'regulae philosophandi' and 'phaenomena', see Cohen (1974*d*).

2 See Cohen (1962), esp. pp. 76–79, containing the correspondence between Leibniz and Fontenelle on the new ovals of Cassini (whose history and properties are described on pp. 79–81).

3 See Wing (1651), bk. 3, ch. 5, p. 44, where this construction is attributed to Bullialdus, who '(to make the operation more easie) shews how to performe the same by an Epicycle, whose motion is double to the motion of a Planet in his Orbite, and so by the solution of right lined Triangles, it may be found with more ease'. On Bullialdus's own preferred way to represent planetary motions kinematically, see Wilson (1970), pp. 111–113.

4 On this topic, however, see Russell (1964); but additionally, see the works cited in §5.5, n. 5.

5 Streete (1661) contains both the law of elliptical areas and the third law; on Newton's recording of Kepler's third law from this book (in U.L.C. MS Add. 3996, fol. 29), see Whiteside (1964*b*), p. 124.

6 For details, see Aiton (1969).

7 The center of the equating circle is not the ellipse's center, but is a point on the major axis of the ellipse (or line of aspides) that is 'at a distance from the sun roughly 5/8 times the doubled eccentricity'; he 'sets this ratio exactly equal to the "divine section" $(\sqrt{5}-1)/2)$'; see Whiteside (1974*b*), p. 310*b*.

8 On this topic see Maeyama (1971); Wilson (1970), esp. pp. 106–123.

9 The best discussion of Newton's theorem is given by D. T. Whiteside in Newton (1967–), vol. 6, pp. 302–309, in an extensive commentary (in nn. 119–129).

10 Wren's essay on this subject ('De recta tangente cycloidem primariam') was published as an appendix to Wallis (1659), pp. 70–74 [= 82]; the concluding section (pp. 80sqq) is entitled 'De problemate Kepleriano per cycloidem solvendo'. See, further, n. 11.

11 Wren writes that from Kepler's rule that the orbital speed of a planet is inversely proportional to its distance from the sun, 'he [Kepler] ingeniously contrived the following hypothesis. That is, by lines drawn [from the ellipse] to the sun, he cuts the area of a planetary ellipse into equal mixtilineal triangles, whence it results that the curve of the ellipse is divided into unequal portions [or arcs], smaller ones in the neighborhood of aphelion and larger ones in the neighborhood of perihelion; and he supposes that the planet is carried through these [unequal arcs or] portions in equal times. Wherefore, in order to obtain the co-equated anomaly from the mean motion, a semi-ellipse has to be cut through a focus in a given ratio; or (which he demonstrates to be the same thing) a semicircle is to be cut through any point of its diameter in a given ratio. It is remarkable how much Kepler has sweated over this problem, "revolving" his orbits "with great exertion and he gets

nowhere at all". At length, puffing and panting, he tearfully implores
the aid of geometers, fearing meanwhile lest the problem be found
unresolvable [i.e., undevelopable, *in-explicabile*] because of the ἑτερογένεια
[heterogeneity] of arcs and sines. Which problem we ourselves, nevertheless, demonstrated long ago with the aid of a protracted cycloid, as
follows'. The quotation given by Wren is from a misquotation of a piece
of poetry in Cicero's *Tusculan Disputations*. This paragraph is translated
from Wren (1669), p. 80. Wren's solution of Kepler's problem was
essentially reproduced by Newton in the scholium following prop. 31,
bk. one, of the *Principia*.

12 Newton was apparently familiar with Wren's essay in the late 1660s; see
Newton (1967–), vol. 2, pp. 191sqq.

13 But Horrox does make an implicit appeal to Kepler's area law in one of his
rules for lunar motion.

14 For this reason, Newton appears to have believed that he deserved credit
for the first two laws of planetary motion, or the first two Keplerian
"planetary hypotheses" (see nn. 9–11 to §5.5). Conduitt was undoubtedly
echoing a Newtonian opinion.

15 Lest my conclusion be read out of context, I remind the reader that the
"Systema Copernicana" laid out by Kepler himself in his 1627 *Tabulae
Rudolphinae* (in the prefatory material and tables) is just such an
astronomical system based on his three laws. And in at least one astronomical treatise, Mercator's 1676 *Institutiones astronomicae*, all three
of Kepler's laws are accepted as true. Furthermore, in a strict sense, while
Kepler's three laws appear in the beginning of the *Principia* (in sects.
2 and 3 of bk. one, the first two "sections" dealing with mathematico-
physical topics), Newton is not there erecting an astronomical system
on those laws so much as showing how the laws were consequences
that follow from conditions in his mathematico-dynamical theory, give
or take a few small "interactions". And in bk. three, 'On the system of the
world', the foundations expressed at the start ('hypotheses' in the
first edition, 'phaenomena' in the second and third editions) include the
area law and harmonic law but not the law of elliptical orbits; the
latter do not actually appear until prop. 13. Nevertheless, while Mercator
does display Kepler's three laws, he also gives (and uses) an equant
plus auxiliary circle to replace the area law. And it was only after the
Principia, as far as I know, that Kepler's three laws were taken together as
a unified group as the primary basis for physical astronomy, but subject
to departures from the strict law owing to "interactions" between
the members of the solar system. I believe, furthermore, that this
became the case after the *Principia* for the same reason that made
Newton use these three laws in so fundamental a way, namely, that
he had found their dynamical significance and had made of them more
than mere phenomenological generalizations or rules of convenience.

§5.3

1 This derivation, of course, presupposes a belief in the possibility that solar
forces (or "solipetal" forces) can act on planets (see §3.4).

2 This is essentially what Newton does in prop. 4, bk. one, of the *Principia*
and in the antecedent tract *De motu*.

3 In bk. three of the *Principia*, scholium to prop. 4, Newton introduces
the hypothesis of several moons encircling the earth. In the beginning of
the first version of bk. three (published posthumously as *A Treatise
of the System of the World*; see Newton, 1728c) there is also a discussion
of an artificial terrestrial satellite, with a diagram showing how it might be
put into orbit.

4 U.L.C. MS Add. 3968, sect. 41, f. 85. On this and similar autobiographical
statements, see Cohen (1971), suppl. 1, pp. 290sqq. Whether the year
in question was intended to have been 1665 or 1666 is not of much
consequence, since it is only the relative chronology of discovery or invention that is at issue. In 1718 these were recollections based on "memory"
of events that would have occurred about fifty years earlier.

5 It is to be observed that even in this statement, Newton does not say
that he had at that time identified the inverse-square force acting on
the planets with gravity; the latter had been conceived to extend only as far
out as the moon's orbit.

6 Herivel (1960b), (1965b), pp. 7–13, 130; cf. Whiteside's succinct account
of the first stages of development of Newton's principles of dynamics
in Newton (1967–), vol. 6, pp. 3sqq.

7 As late as the 1670s Newton tended to assume (as in his letter to Hooke in
November 1679) a constant solar "gravity".

8 In a document of the 1660s, first published in A. R. Hall (1957) and
published again in Newton (1959–1977), vol. 1, pp. 297–300, and in
Herivel (1965a), pp. 192sqq, Newton did apply Kepler's third law to the
planets, to find that 'their endeavours of recess from the Sun will be
reciprocally as the squares of their distances from the Sun'. Newton
then gave numerical examples for each of the six primary planets (see,
further, n. 13). It is important to note that this document does not contain
an actual calculation of the moon test, nor does it accordingly assert
a balance between a solar centripetal "gravity" and planetary centrifugal
forces; Newton is concerned with a "conatus" or endeavor and not
with balanced forces (cf. Whiteside, 1964b, esp. p. 120, n. 13).

Newton's paper (U.L.C. MS Add. 3958, sect. 5, fol. 87) contains a
calculation that the force of terrestrial gravity is 4275 times that of the
moon's "endeavour to recede" (at a distance of 60 earth radii). Had he then
made the connection between the two, and so had in fact linked the
"cause" of the moon's 'endeavour to recede' with terrestrial gravity
acting according to an inverse-square law, which involves a great many
suppositions, the alleged 'pretty nearly' would have been out by a little
more than 18 percent $(4275 - 60^2 / 60^2)$.

9 But we may be sure that the 1718 memorandum attempts to set out an
unhistorical myth. There is no factual evidence that even as late as
the early 1680s, Newton conceived of a simple balancing of solar "gravity"
and the "vires centrifugae" of the planets. On the contrary, there is
evidence of a number of kinds that shows that Newton then believed
that there must be a different law of variation for these "vires centrifugae" and solar "gravity". In Whiteside (1964a) it is suggested that even
in the 1679–1680 Hooke letters, Newton still wanted an inverse-square
"gravity" and an inverse-cube "outwards pull", so as to yield $\ddot{r} =
c^2 / r^3 - c'^2 / r^3$.

10 Glaisher based his suggestion on an idea given to him by John Couch Adams; see Glaisher (1888), p. 7; cf. Whiteside's discussion of this affair in Newton (1967–), vol. 6, pp. 19–20, n. 59.

11 Whiteside (1964a), p. 119. Newton's correspondence shows that he was still making use of the theory of vortices in relation to planetary motion on the very eve of writing the *Principia*.

12 By 1669, as we know from the MS end-notes in his copy of Wing's *Astronomia*, Newton no longer even accepted Kepler's third law as an exact observational law.

13 See n. 8. Referring to this paper in a letter to Halley (22 May 1686), Newton said that the calculations he had made at that time show 'that I had then my eye upon comparing the forces of the Planets arising from their circular motion & understood it' (see Newton, 1959–1977, vol. 2, p. 431).

14 This document may be found in U.L.C. MS Add 3958, sect. 5, fol. 45$^{v/r}$.

15 In bk. 2 of Descartes's *Principia* (see, further, Herivel, 1965a, ch. 2.2, ch. 3). A development of this concept, one with which Newton was familiar, was made by Borelli (see Koyré, 1973, pt. 3).

16 U.L.C. MS Add. 3968, sect. 29, fol. 415v; this text was first published in Koyré & Cohen (1962).

17 This question is discussed in Herivel (1965a) and in Westfall (1971).

18 The expression "vis centrifuga" appears in all three editions of the *Principia*.

Supplement to §5.3

1 This MS may be dated by means of a statement made by Newton (1959–1977, vol. 3, p. 331) to David Gregory, to whom he showed the MS in May 1694. It was written, so Gregory reported, 'ante annum 1669 (quo tempore Auctor D. Newtonus factus est professor Matheseos Lucasianus)', that is, 'before the year 1669 (at which time the author, Mr Newton, was made Lucasian professor of Mathematics)'. Some years earlier, on 20 June 1686, Newton referred to this MS in a letter to Halley (Newton, 1959–1977, vol. 2, p. 436) as 'one of my papers writ (I cannot say in what year but I am sure some time before I had any correspondence with Mr Oldenburg & that's above fifteen years ago)'. Oldenburg's first letter to Newton is dated 2 Jan. 1671/2, so that this reference to 'some time before' would agree with Gregory's 'before the year 1669'. Newton dated this MS once again in another letter to Halley (Newton, 1959–1977, vol. 2, p. 445), written on 14 July 1686, in which he referred to 'that very paper which I told you was writ, some time above 15 yeares ago & to the best of my memory was writ 18 or 19 years ago'. This would set the date of composition at 1667 or 1668.

The existence of the MS under discussion and its significance were announced in Turnbull (1953); the MS was published and analyzed, together with other documents, by A. R. Hall (1957), by H. W. Turnbull in Newton (1959–1977), vol. 1, pp. 297–303, and by J. Herivel (1965a), ch. 4 and pp. 192–198. See also Newton (1959–1977), vol. 3, pp. 46–54, for a related document with similar computations, also published and analyzed in Herivel (1965a), pp. 183–191.

2 Since the distance, by Galileo's rule, varies as the square of the time, the conversion of distance per day to distance per second requires successive division by 24^2, 60^2, and 60^2.

3 In both this MS and in a related sheet of computations (see n. 1), Newton computes the ratio of the centrifugal force of the earth from the sun (due to its orbital motion) to the centrifugal force at the earth's surface on the equator (due to its daily rotation) and to the earth's gravity.

4 In arguing for the plausibility of Newton's allegation of having made a moon test in the 1660s, Herivel (1965a, p. 70) maintains that 'Otherwise one seems forced to conclude that Newton not only misinformed both Whiston and Pemberton, but also fabricated the very circumstantial account in the *Portsmouth Draft Memorandum* [i.e., the memorandum written for Des Maizeaux in 1718]'. Herivel repeats this argument on p. 74, concluding that 'the assumption that there was no true test of the law of gravitation during the Plague Years argues for a degree of duplicity on Newton's part, both in casual conversations with Whiston and Pemberton . . . and in his *unpublished* account in the *Portsmouth Draft Memorandum*, [which] is difficult to credit'. Herivel does insist, however, that (p. 72) Halley would have gained the misleading impression from Newton's letter of 20 June 1686 'that Newton thought this paper proved his early familiarity with the 'duplicate proportion' of *universal gravitation*'.

As mentioned in §5.3, Newton may very well have made a mental comparison of the theoretical value of 3600 and the computed value of more than 4000, and so the account of the matter that Newton gave to Whiston and to Pemberton may not have been an example of misinforming or of duplicity so much as the normal transformation of events that occurs in the memory when a later point of view is imposed on earlier events. But there can be no doubt that the Des Maizeaux memorandum is indeed misleading and directly contrary to the facts in many respects. And this is true of many of Newton's statements in the second decade of the eighteenth century, when he was so deeply embroiled in the question of priority with Leibniz. In any event, this particular memorandum was never published by Newton, it has been cancelled and – although it is widely published and discussed today – it may never have been seen by anyone but Newton himself during his lifetime.

5 Newton was not alone in reinterpreting this document in the light of his later discoveries. David Gregory recorded in 1694 that he saw a MS which, from its description, is the MS being discussed here. Gregory says that in the MS 'all the foundations of his philosophy are laid: namely the gravity of the moon to the earth, and of the planets to the sun'. He adds that 'in fact all these are . . . subjected to calculation' (see Newton, 1959–1977, vol. 3, pp. 331–333). Of course, Newton had not yet conceived of gravity as a force that exists between sun and planets; he had not written of a 'gravitas planetarum versus solem' (as Gregory put it), but rather of a 'conatus recedendi [planetarum] a sole'. Nor had he expressly declared that there was a 'gravitas lunae versus terram'; he not only had restricted himself to a 'conatus recedendi lunae a centro terrae' but did not even hint in the MS that the earth's gravity might extend to the moon.

§5.4

1 This copy is in the Trinity College Library, where it is press-marked NQ.18.36 (see Whiteside, 1964a, pp. 124sqq). On the methods of approximation used in place of Kepler's second law, and based – in the first instance – on the uniform rotation of a radius vector about the empty focus of a planetary elliptical orbit, see §§5.2 and 3.3.

2 The communications dealing with dispersion and the composition of white light and with the new reflecting telescope were published in the *Philosophical Transactions* in 1672; they are reprinted in facsimile together with the comments they aroused in Newton (1958), sect. 2, §§2, 6–16.

3 Hooke to Newton, 24 Nov. 1679, Newton (1959–1977), vol. 1, p. 297. On this topic, see Koyré (1952b) and Lohne (1960). The problem in question is an interesting one. Newton is proposing a test to prove that the earth has a diurnal motion or rotates from west to east. He supposes that initially a heavy body is suspended in the air so as to move around with the earth and to be continually located above the same point on the earth. Then it is let fall, Newton says, and 'its gravity will give it a new motion towards the centre of the earth without diminishing the old one from west to east'. The more distant the body is from the earth at the moment it is let fall, the greater will be its motion from east to west. Accordingly, it will not descend along a straight line from its original position toward the earth's center, 'but outrunning the parts of the earth will shoot forward to the east side'. This, said Newton, is 'quite contrary to the opinion of the vulgar who think that if the earth moved, heavy bodies in falling would be outrun by its parts and fall on the west side of the perpendicular'. Newton was stating a result in proof of the earth's rotation that we today would obtain by applying the principle of the conservation of angular momentum; the proof of the earth's rotation is given by a falling body's landing ahead (in the sense of the rotation) of the place from which it is dropped rather than being left behind, since if $mr^2 \omega$ is constant, ω must increase as the body falls (and r decreases). Although the effect would be necessarily small in a descent of twenty or thirty yards, Newton conceived that the experiment could be made and would prove this point. Unfortunately, he hastily and not quite correctly also assumed that the path of the falling body (onto a moving earth) would be 'a spiral line' below the earth's surface, which – as Hooke said in reply – was 'a kind of spirall which after sume few revolutions [would] Leave it [i.e., the falling body] in the Centre of the Earth'.

4 Hooke to Newton, 9 Dec. 1679, Newton (1959–1977), vol. 1, p. 305. On this diagram see Whiteside (1964), p. 132, n. 52, and (1966b), p. 117, n. 10; for a comprehensive study of Newton's diagrams and their subsequent corruption see Lohne (1967).

5 Hooke could well have been misled by Kepler's procedure in developing the law of areas, as, for example, described by Wren (see his statement in §5.2, n. 11). But Hooke was evidently unaware that Kepler had eventually rejected the law of speed in inverse proportion to the distance (see Aiton, 1969).

6 In prop. 16 it is stated that a body P moves in orbit about a center of

force *S*; this force varies inversely as the square of the distance *SP* (note

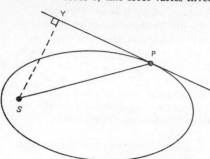

that *S* stands for *Sol* and *P* for *Planeta*). A perpendicular *SY* is let fall on the tangent *PY* drawn to the ellipse at *P*. Newton proves that if a series of bodies move in elliptical orbits and there is a common focus or center of force *S*, then their velocities will be in a ratio compounded of such perpendicular distances (as *SY*) inversely and the principal *latera recta* directly. Hence in any given elliptical orbit, the velocity at any point *P* is inversely proportional to the perpendicular distance *SY*: that is, if the ellipse is given, the principal *latus rectum* is fixed.

7 It was reported by Halley to Newton in a letter of 22 May 1686 (see Newton, 1959–1977, vol. 2, pp. 431sq).

8 In his correspondence with Halley, Newton also cited his "hypothesis" of 1675, in which it is assumed that anything spreading out from a central body will vary inversely as the square of the distance.

9 These are discussed below. Possibly, also, if Newton had replied to Hooke by sending his own original solution of the problem of motion according to Kepler's laws, it might have seemed to Hooke as if Newton had merely been working out the mathematical consequences of Hooke's suggestion. That this may have been the case may be seen in a letter Newton wrote when he had heard that Hooke had raised the question of his priority in the invention of the inverse-square force, 'pretending I had all from him'. Newton said that Hooke 'has done nothing & yet [has] written in such a way as if he knew & had sufficiently hinted all but what remained to be determined by the drudgery of calculations & observations . . . . Now is not this very fine? Mathematicians that find out, settle & do all the business must content themselves with being nothing but dry calculators & drudges & another that does nothing but pretend and grasp at all things must carry away all the invention as well of those that were to follow him as of those that went before' (Newton to Halley, 20 June 1686; Newton, 1959–1977, vol. 2, p. 438). On Hooke and Newton, and their correspondence in 1679–1680, see (in addition to the works cited in n. 3) Patterson (1949, 1950) and Whiteside (1964*a*), pp. 131–137.

10 In a number of autobiographical documents Newton dated his 'demonstration of Kepler's Astronomical Proposition, viz. the Planets move in Ellipses' in '1679', at the close of the year 1679 ('anno 1679 ad finem vergente'); but he also (and mistakenly) put this discovery 'in the winter between the years 1676 & 1677'. Most probably, he found this demonstration in the winter between 1679 and 1680, some years before Halley's famous visit of 1684, which is discussed below. A number of such statements of Newton's are printed (largely from Newton's MSS) in Cohen (1971), suppl. 1.

11 These letters are printed in Newton (1959–1977), vol. 1; see, further, Baily (1835).

12 Kepler believed that comets move in straight lines (at least over most

of their orbits) but with nonuniform speeds (see Ruffner, 1966, 1971).
Newton used Wren's simplified hypothesis of uniform rectilinear cometary
motion (see Newton, 1967–, vol. 5, p. 299, n. 400).

13 See D. T. Whiteside's edition of the lectures on algebra in Newton
(1967–), vol. 5, pp. 210sqq; and ibid., pp. 298sqq.

14 See Newton (1959–1977), vol. 2, pp. 387–393. Brattle's information was sent
in letters to Flamsteed. His data are recorded in the *Principia* under
"Boston" rather than under his name; he also appears as 'observator in
Nova-Anglia'. In the *Principia* the major cometary example discussed
is the comet of 1680.

§5.5

1 U.L.C. MS Add. 3968, fol. 101; see Cohen (1971), suppl. 1, p. 293. It is
to be observed that Newton states that before his exchange with Hooke in
1679–1680 he had inferred from Kepler's third law that the solar
force keeping the planets in their orbits must diminish as the square of
the distance (see §5.3); and that he now found that a central force would
be a sufficient condition for the law of areas (as in prop. 1, bk one,
of the *Principia*).

2 Newton, of course, had no way of knowing for sure when Huygens had
found the law of centrifugal force; his statement that Huygens might have
had the priority has a "grudging" quality of reluctantly admitting
that Huygens probably was ahead of him in finding the measure of "vis
centrifuga", but he is not certain and he still insists that he himself
had done it independently (as indeed he had done). On Newton's
early discovery of the law of centrifugal force, see §5.3.

3 Centrifugal force and its law were announced in 'De vi centrifuga ex motu
circulari theoremata' in Huygens (1673), pp. 159–161; also Huygens
(1888–1950), vol. 6, pp. 315–318; vol. 18, pp. 366–368. In th. 3 Huygens
refers to the 'vis centrifuga'.

4 For details, see Westfall (1971).

5 On this topic see Whiteside (1964*b*). See also a series of letters in
Nature: by D. T. Whiteside (vol. 248, 19 April 1974, p. 635); by J. W.
Herivel (ibid.); and a rebuttal by I. B. Cohen (vol. 250, 19 July 1974, p 180).
It has been mentioned that with respect to the third law, however, in
the notes on Wing's treatise on astronomy, Newton (*c.* 1670) rejects
Kepler's third law as an observationally exact description of planetary
phenomena.

6 Here, as earlier, I am assuming that the significance of the area law
would have dawned upon Newton during the course of wrestling with
the problem of planetary motion in elliptical orbits according to a
centripetal force acting on a body with inertial motion, where that force
varies as the inverse square of the distance: in other words, the problem
posed to him by Hooke. There is no evidence of any kind as to the
actual thought processes whereby Newton achieved the revelation that
changed the whole course of thinking about celestial dynamics, but
Newton himself said that the realization of the generality of the Keplerian
area law did occur as a response to Hooke's intellectual stimulus 'in
the end of the year 1679'. As far as we know, almost all Newtonian
scholars (R. S. Westfall being an outstanding exception) now agree that
the way in which Newton made his great discovery with regard to

elliptical orbits and the inverse-square law was more or less by the series
of steps that are illustrated in both the tract *De motu* and sects. 2 and 3
of bk. one of the *Principia*.

7 In prop. 3 of the *Principia* the results are extended to a moving center
of force.

8 See n. 1. The other law of Kepler's to which Newton refers is the harmonic
law; this law, in combination with the law of centrifugal force, had
led Newton to the concept of an inverse-square solar force – assuming
circular planetary orbits.

9 In the autobiographical statement just quoted, however, Newton does
mention Kepler as the originator of the elliptical orbits (see also n. 11).

10 From John Conduitt's 'Memorandum relating to Sir Isaac Newton given
me by Mr Abraham Demoivre in Nov[embe]r 1727'; this MS, formerly
in the possession of Mr. Joseph Halle Schaffner, is now in the University of
Chicago Library. A nineteenth-century transcript by H. R. Luard is in
the University Library of Cambridge (MS Add. 4007, fols. 706sq). The
portion relating to the *Principia* is published in full in Cohen (1971),
pp. 297sq.

11 In the tract *De motu* (see §4.4, n. 12), Newton attributed both the law
of elliptical orbits and the area law to Kepler (scholium following prop. 3),
but saying that this was 'as Kepler supposed' ('ut supposuit Keplerus')
(see §5.6). In a letter to Halley, 20 June 1686, Newton stated that he
himself had concluded that the 'cause of gravity towards the earth, Sun,
and Planets' must follow the law of the inverse square. Then he said:
'But grant I received it afterwards from Mr Hook, yet have I as great
a right to it as to the Ellipsis. For as Kepler knew the Orb to be not
circular but oval & guest it to be Elliptical, so Mr Hook without knowing
what I have found out since his letters to me, can know no more but
that the proportion was duplicate *quam proxime* [very nearly] at great
distances from the center, & only guest it to be so accurately & guest amiss
in extending that proportion down to the very center, whereas Kepler
guest right at the Ellipsis. And so Mr Hook found less of the Proportion
than Kepler of the Ellipsis.' What Newton 'found out' was apparently
that a uniform sphere (or a sphere composed of uniform concentric
shells) gravitates as if all its mass were concentrated at the center (see
Newton, 1959–1977, vol. 2, pp. 436sq).

12 A wholly different interpretation of Newton's approach to the physics
of forces is given in Westfall (1971), esp. p. 377. Here it is said that 'What
he [Newton] proposed was an addition to the ontology of nature'. Westfall
argues that 'the ultimate ontological status of forces in Newton's
conception of nature is a complex and involved question. In his published
writings, he chose to refer to his true opinion, as I understand it, only
obliquely, and we must consult unpublished manuscripts to seize
his meaning. As far as the works he published are concerned, and above
all, as far as the conceptual tools he employed in scientific discussion
are concerned, he treated forces as entities that really exist.'

13 Westfall's argument (see n. 12 and Westfall, 1971, p. 377) would seem
to apply unquestionably to the short-range forces of particles, if not
necessarily to the long-range forces of gross bodies (such as universal
gravity). On the relations between the two kinds of forces, see n. 15.

14 This point has emerged from recent studies by Westfall, Dobbs, and Figala.

15 In this regard, it is of some importance to note that in the *Principia* Newton goes beyond mere considerations of analogy and existence in treating the relations between particulate forces and the forces of bodies such particles compose. In his treatment of this problem (sect. 12, bk. one) he shows us his maximally creative spirit. Here he is able to prove by rigorous mathematics (and not by analogy) that in the 'two principal cases of attraction', gross bodies will exert 'centripetal forces [that] observe the same law of increase or decrease in receding from the center as the forces of the particles themselves'. These two cases are the square of the distance inversely and the distance directly, the only two that occur in nature. The first refers to the attraction exerted on an external body or particle by a homogeneous sphere or a homogeneous spherical shell (or a sphere composed of concentric homogeneous shells), and the second refers to the attraction exerted by a homogeneous solid sphere on an internal particle. That this should be true of the 'two principal cases of attraction' is, as Newton says, 'very remarkable'.

In prop. 8, bk. three, he considers two mutually gravitating spheres (in which 'the matter in places on every side equidistant from the centers is similar') and proves that their 'weight' toward each other will be inversely as the square of the distance between their centers. This is what he says:

'After I had found that the force of gravity towards a whole planet did arise from and was compounded of the forces of gravity towards all its parts, and towards each individual part was as the inverse square of the distance from that part, I was yet in doubt as to whether the inverse square proportion did accurately hold, or but nearly so, in the total force compounded of so many partial ones. For it might possibly be that the proportion which is exact enough at greater distances might be far from the truth near the surface of the planet, where the distances from the particles are unequal and their relative situation dissimilar. By the help of props. 75 and 76, bk. 1, and their corollaries, I was at last satisfied of the truth of the proposition as it lies before us.'

In this example we may, I believe, see how Newton would ideally proceed by strict mathematical proof to associate mathematically specified particulate forces with the more apparent ones of gross bodies. Accordingly, I would see the *Principia* and the dynamics of gross bodies leading Newton in the direction of an exact or mathematical theory of matter, an aspect of Newton's exact science that is independent of, or at least supplementary to, conjectures as to whether prior considerations of short-range particulate forces may or may not have had a creative effect on the concept of universal gravity and the science of the *Principia*.

16 The most positive statements of Newton's with respect to the existence of particulate forces in general are to be found in the published queries of the *Opticks* and their MS drafts or versions. These are in the form of negative rhetorical questions, such as (qu. 31): 'Have not the small Particles of Bodies certain Powers, Virtues, or Forces, by which they act at a distance . . .'?

17 That is, they are not developed in the Newtonian style of the *Principia*.

18 The association of forces with particles was a radical step, going against the
canons of acceptable philosophy, as Newton well knew. In the MS,
Newton is concerned with the possible duality: the force of gravity
('whereby great bodies attract one another at great distances') and 'certain
kinds of forces whereby minute bodies attract or dispell one another
at little distances'. He says:

'How the great bodies of the earth Sun moon & Planets gravitate towards
one another what are the laws & quantities of their gravitating
forces at all distances from them & how all the motions of those bodies
are regulated by those their gravities I shewed in my Mathematical
Principles of Philosophy to the satisfaction of my readers: And if Nature
be most simple & fully consonant to her self she observes the same
method in regulating the motions of smaller bodies which she doth in
regulating those of the greater. This principle of nature being very
remote from the conceptions of Philosophers I forbore to describe it in that
Book least I should be accounted an extravagant freak & so prejudice
my Readers against all those things which were the main designe
of the Book: but & yet I hinted at it both in the Preface & in the book
it self where I speak of the inflection of light & of the elastick power
of the Air but the design of that book being secured by the approbation
of Mathematicians, I had not scrupled to propose this Principle in plane
words. The truth of this Hypothesis I assert not, because I cannot
prove it, but I think it very probable because a great part of the
phaenomena of nature do easily flow from it which seem otherways
inexplicable: such as are chymical solutions, precipitations, philtrations,
. . . volatizations, fixations, rarefactions, condensations, unions, separations,
fermentations, the cohesion, texture, fluidity and porosity of bodies,
the rarity & elasticity of air, the reflexions & refraction of light, the rarity
of air in glass pipes & ascention of water therein, the permiscibility of
some bodies & impermiscibility of others, the conception & lastingnesse
of heat, the emission & extinction of light, the generation & destruction
of air, the nature of fire & flame, the springinesse or elasticity of hard
bodies.' (Quoted from U.L.C. MS 3970.3, fol. 338.)

19 Or of resisting the effort of an external force to change its state.

20 A quite different point of view is taken in Westfall (1971).

21 I remind the reader that the transformation was multiple and complex.
Newton began with a traditional Aristotelian concept of "vis insita", which
he used in a wholly new way. He then equated this "vis insita" with
a "vis inertiae" of matter, which he had invented. Kepler's "inertia
materiae" is a property of bodies which brings them to rest whenever the
motive force ceases to act; it was transformed by Newton so that it
would maintain bodies in their state of motion or of rest (see §4.5 and
Cohen, 1972.) Then, this new "inertia" was coupled with a special
kind of internal "force" ("vis insita") to become the "vis inertiae", which
was engrafted onto Descartes's "law" of inertia.

Since this "vis insita" or "vis inertiae" does not change a body's state
of motion or of rest, and does not affect other bodies (or particles)
as would be the case for forces of cohesion or attraction or repulsion,
it is a "non-force" kind of "force". In any case, it really produces no
distortion of Newton's thought to translate his "vis inertiae" by "inertia".

22 See §4.5; and, for details, Westfall (1971).

23 From the point of view set forth here, the essential ingredient that
enabled Newton to move ahead with such incredible speed toward the
resolution of the problem of planetary motion in accordance with
Kepler's laws, and then to move on to a celestial mechanics based upon
a mutually acting universal gravity, was the ability to reduce the complex
physical situations of nature to simple systems or constructs, to which
he could apply mathematical techniques. These were "mathematical"
in two senses. First, they consisted of a series of conditions that
could generally be expressed in terms of mathematical relations, and
to which both the traditional and the new methods of mathematics could
be applied. Second, they were "mathematical" because (to paraphrase
Newton's own words) he was not at that stage of his enquiry raising
any questions as to the nature of the forces, their mode of action, or
even their possible existence. The success of this "Newtonian style" is
evident in each book of the *Principia*, and is displayed here in Ch. 3 and
Ch. 5. By beginning with systems or constructs that were simpler
than physical reality, and to which he could apply his mathematical
techniques, Newton prevented himself from becoming prematurely mired
in meta-physical, or at least meta-scientific, questions about the nature
of the forces, reserving all questions about the nature of the forces
until the end of his enquiry, when they could no longer be avoided.

Westfall (1971), esp. pp. 377sqq, would rather stress the 'significance
of the redirection' of Newton's 'philosophy of nature undertaken . . . about
1679', as has been mentioned here on a number of occasions. Westfall,
in particular, would connect this new direction in Newton's philosophy
with an experiment that he has 'tentatively placed about 1679', in
which Newton essentially shows that a comparison of the resistance of
air to the motion of a pendulum with an empty bob and one with a
bob filled with metal is not commensurate with the hypothesis of an
all-pervading aether. Accordingly, Newton would have turned to the
reality of forces, giving up the aether with its potentiality of explaining all
the actions of nature in terms of the traditional categories of the "mechanical philosophy": matter (with its characterizations of size, shape, and
solidity) and motion. Newton would have introduced forces of
attraction and repulsion into the then-accepted physical reality of
material particles and motion. This 'addition to the ontology of nature',
as Westfall points out (p. 377), is not easily apparent in Newton's
published writings, for (to quote Westfall) 'he chose to refer to his true
opinion, as I understand it, only obliquely, and we must consult unpublished manuscripts to seize his meaning'.

I do not wish to minimize the importance of the point made by
Westfall. But I believe that the pendulum experiment in question could
hardly have been made in 1679, for a number of reasons (the details
are presented in nn. 16–17, §3.8). Furthermore, as I have mentioned
elsewhere in the present book, even in Newton's unpublished writings
he tends to introduce the subject of such particulate forces with some
diffidence, rather than certainty. Additionally (as pointed out in §5.4), there
is a fundamental question of considerable mathematical difficulty
which Newton would have had to solve in order to be able to shift easily
from the scale of particulate attractions to the attractions of close
bodies. Not until Newton was writing the *Principia* did he prove that

only if the force of attraction varies either as the square of the distance inversely or as the distance directly, will the sum of the attractions of the particles (i.e., the attraction of the body made up of those particles considered as a whole) follow the same law as that of the particles themselves. Furthermore, only if the force varies as the square of the distance inversely or as the distance directly, will a thin homogeneous spherical shell attract as if all of its mass were concentrated at its geometric center.

While I am aware of the importance for Newton of being able to consider forces as if they were real, it does seem to me that it was also absolutely necessary for him to have been able to think in terms of mathematical systems and imagined constructs, in the manner I have described in Ch. 3 and illustrated again in Ch. 5, in relation to the Newtonian style. Perhaps Newton could not have been successful in this mode of thinking about mathematics in relation to physical problems if he had not begun to give serious consideration to the possible reality of centripetal forces. Nevertheless, I believe that any serious reading of Newton's writings on dynamics, and especially of his *Principia*, would show that whatever he had thought about forces, and their possible or ultimate reality, he would never have achieved the greatness of his celestial dynamics if he had not been able to apply mathematics to the physics of nature in the Newtonian style.

§5.6

1 This version was published in Rigaud (1838).

2 The tract *De motu* (see §4.5, n. 22) is very short, occupying some twenty pages; the contrast makes sense especially if one confines attention to sects. 2 and 3 of bk. one of the *Principia*, which are of comparable length and deal with similar subjects.

3 In Newton's personal interleaved copy of the first edition, and in a personal annotated copy, he has added the letters L and T to designate the body (L) that moves around another body (T); this occurs in both the proposition itself and its four corollaries. The change was then introduced into the second and third editions. Newton evidently wanted to alert his readers as early as possible that, despite the generality and abstraction, his aim was to explain natural philosophy and in particular the phenomena of our solar system. These letters would remind the reader (in Latin) of the astronomical problem of the moon (Luna) moving about the earth (Terra), itself in motion (see Newton, 1972).

4 In the *Principia*, but not *De motu*, Newton adds to prop. 1 that the areas which are proportional to the times of description all lie in a single plane.

5 In the *Principia* the corollaries are more numerous; in the second edition (1713) and in Newton's personal copy of the first edition (in a note of about 1691), the ratio of velocity (V) to radius (R) is generalized to any power n, so that if the periodic time be as R^n, and the velocity accordingly as $1/R^{n-1}$, the centripetal force will be as $1/R^{(2n-1)}$; and contrariwise. This generalization was apparently the result of a suggestion made by N. Fatio de Duillier (see Cohen, 1971, ch. 7, sect. 9).

6 In the *Principia*, however, the applications to astronomy appear separately in bk. three, 'The System of the World'.

7 See Whiteside (1970*b*), esp. pp. 122sq. Newton's measure of a centripetal
force depends on the basic area law. Consider the motion of a body
(a particle or mass point) in an orbit of which the indefinitely small arc
PQ is traversed in time *dt*; let *S* be the fixed center of centripetal force.
Then, during this indefinitely small time *dt*, the force will have produced
a total deviation *RQ* from what would have been its inertial rectilinear

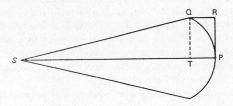

path along the tangent *PR* to the curve at *P*. This deviation is computed by the second law of motion and Galileo's rule, $RQ = \frac{1}{2} f \cdot dt^2$.
But, since the force is directed to a center, the area law must hold, so
that *dt* is proportional to the focal or central sector *SPQ*, which is equal to
one-half of the altitude (*QT*, drawn perpendicular to *SP*) times the
base (*SP*), or $\frac{1}{2} SP \cdot QT$. It thus follows that $RQ \propto f \cdot SP^2 \cdot QT^2$, or that
$f \propto [RQ/(SP^2 \cdot QT^2)]$, a measure of the force that is exact if $RQ/(SP^2 \cdot QT^2)$
is the value that this magnitude (prop. 6, corol. 1, bk. one) 'ultimately
acquires when the points *P* and *Q* coincide', that is, the limit of this
ratio as *Q* comes to coincide with *P*.

    This measure enables Newton to solve a number of problems (props.
7–13), including the computation of the force by which (1) a body in a
circular orbit has a force directed to a point on the circle, (2) a body moving
in an equi-angular or logarithmic spiral has a force directed to its pole,
and (3) a body moving in a conic section has a force directed to either
its center or to a focus.

    In formulating this measure, Newton assumes that the deviation *RQ*
is a straight line parallel to *SP*, rather than a curve, so that the force
is considered to be constant in magnitude and direction, an assumption
justified by the fact that the arc *PQ* is said to be indefinitely small
(it is, to use Newton's expression, "just now nascent"); which is another
way of saying that the result is valid only in the limit. Furthermore, the
area of the focal sector *SPQ* is reckoned as if it were a linear triangle, or as
if the chord *QP* could be substituted for the arc *QP*, which again
(as Newton has already proved) is true in the limit. D. T. Whiteside
has shown, furthermore, that the deviation *QR* is of 'second-order
infinitesimal magnitude', so that no real error is introduced by considering
it to be a straight line parallel to *SP*, rather than a curve. The second
law of motion, $f \propto mA$, may be used because the mass may be taken as unity
(for a mass point) or absorbed into the constant of proportionality,
whence *f* may be substituted for *A* (as in Newton's "accelerative measure"
of a force, according to def. 7).

8 These are given in Herivel (1965*a*), pp. 294–303; also in Hall & Hall
(1962) and in Newton (1967–), vol. 6.

9 Herivel (1965*a*), p. 294. Among the 'conceptual advances in Newton's
dynamical thought' that are found in the emendations to the original *De*

motu, Herivel notes particularly 'the change in status of the parallelogram law from an hypothesis . . . to a derived lemma'.

10 A "line" is a twelfth part of an inch.

11 This astonishing agreement between theory and observation has been a source of a belief that Newton may have "cooked" or "fudged" his data (see Westfall, 1973).

12 Herivel (1965a), p. 302. It is to be noted that in this statement Newton does not explicitly identify (in so many words) the centripetal forces acting on the moon with gravity. He also is aware 'that gravity is diminished by an increase in our distance from the centre of the earth' (as shown by 'the slower motion of pendulum clocks in the tops of high mountains than in valleys'), but he says that the proportion of this change in gravity 'has not yet been observed'.

13 See Whiteside's documentary discussion at this point in Newton (1967–), vol. 6, pp. 58–59, n. 79; pp. 81–85; pp. 481–507. Newton was apparently using a 'modified rectilinear technique' to approximate 'the curved path of the comet of 1680–1, as late as October 1685.

14 Using the "parabolic" approximation, one cannot of course find all the elements of a cometary orbit; for instance, not the main axis.

15 He was the first astronomer to do so.

16 We have no way of dating this stage of Newton's thought. By the time he wrote the two addenda to *De motu,* he was aware that the sun exerts an inverse-square force on planets and comets, as do planets on their satellites, that the planets may exert forces on one another, that the force exerted by the earth on the moon agrees with a diminution of terrestrial gravity according to the law of the inverse square, and that the above forces of attraction are mutual. When was this?

   Newton thanked Aston on 23 February 1684/5 for having entered a transcript of *De motu* in the Register of the Royal Society; Newton had sent it after Halley's second visit (December 1684). Since this version does not contain the additions, it seems likely that they were made no earlier than 1685. Bk. one of the *Principia* was written out in a fair copy (in a revised version of the text he deposited in the University Library as his professorial lectures) and sent to Halley for printing by April 1686; presumably, therefore, he would have begun writing the *Principia* no later than some time in 1685. Hence it would seem that Newton's new thoughts, added to *De motu,* were written before he started to work on the *Principia,* which would appear to be some time in 1685. On the other hand, Newton would on occasion return to early manuscripts to make revisions long after they had been copied in a later version, and also had other methods of work that confuse attempts to set up definitive chronologies of his work. Hence it is possible that these additions may have been made to *De motu* before a version had been sent to the Royal Society but had not been incorporated in that version. Whiteside, in his edition (Newton, 1967–, vol. 6, p. 74), dates these addenda 'December 1684?' and gives the date 'Autumn 1684' to the original tract *De motu.*

17 The first two of these rules were hypotheses in the first edition; they state that 'no more causes' should be admitted 'than are both true and sufficient' to explain phenomena, and that 'causes assigned to natural effects of the same kind must be, so far as possible, the same'. Rule 3, new

in the third edition, justifies the extension of qualities found in our experience by the senses to bodies beyond the range of our sensory experience. Hence, 'if it is universally established by experiments and astronomical observations that all bodies on or near the earth are heavy [or gravitate] toward the earth, in proportion to the quantity of matter in each body, and that the moon is heavy toward the earth in proportion to the quantity of its matter, and that our sea in turn is heavy toward the moon, and that all planets are heavy toward one another, and that there is a similar heaviness of comets toward the sun, we shall have to say by this third rule that all bodies gravitate mutually toward one another'.

§5.7

1 In celestial mechanics, only gravitational forces are of any major importance; but of course the measure of resistance to a change of state applies to any force whatever.

2 "Mass" is introduced by Newton as a synonym for "body" or the traditional "quantity of matter"; see def. 1 of the *Principia*. In def. 3, and again in rule 3 (in the second and third editions of the *Principia*), the resistance to being accelerated or undergoing a change of state (the *vis inertiae*) is made the equivalent of the traditional *vis insita* (for which see §4.5). See, further, Cohen (1970), (1978).

3 Hence it was written in 1685 (see Cohen, 1971, pp. 62sqq). The text of this MS was published in Herivel (1965a), pp. 304–315. A new and more correct Latin text (but without a translation) is given by Whiteside, together with a running commentary, in Newton (1967–), vol. 6, pp. 189–194; for an emendation of the order in which Herivel has printed the parts of this text, see Cohen (1971), pp. 93–95.

The quoted pair of sentences (Herivel, 1965a, pp. 306sq) were originally part of the discussion of "Quantitas motus" (in an early version in which this definition was number 11). In the revised version, "Quantitas motus" should be assigned the number 12, and these two sentences suppressed. They are replaced by a discussion which goes into what now becomes number 7, the definition of "pondus" (see Whiteside's analysis; Newton, 1967–, vol. 6, pp. 189sq, esp. nn. 2 and 13).

4 Herivel (1965a), pp. 306, 311, 'Pendulis aequalibus numerentur oscillationes . . . et copia materiae in utroque erit reciproce ut . . .'

§5.8

1 According to prop. 12, bk. three, the sun's mass is to Jupiter's mass as 1067:1, and the distance from Jupiter to the sun is to the sun's radius in almost the same proportion, so that the center of gravity of the sun and Jupiter will be just outside the surface of the sun. But the center of gravity of the sun and Saturn, Newton finds, will be just within the surface of the sun. Even if the earth and all the planets were on the same side of the sun, according to Newton, the distance from the center of gravity of the whole system to the center of the sun would never be more than a sun's diameter; in all other cases, this distance would be less. Since the center of gravity must be at rest, the sun must move according to the various positions of the planets; but the sun will never move far away from that center.

2 It was here that Newton introduced the concept of attraction for the convenience of mathematical readers.

3 Here S and P are the masses of the central body S and of the body P in orbit. It is obvious that he had a system of Sol and Planeta in mind. Newton actually says that 'the principal axis of the ellipse that one of the bodies P describes . . . about the other body S will be to the principal axis of the ellipse that the same body P would be able to describe in the same periodic time about the other body S at rest as the sum of the masses of the two bodies S & P to the first of two mean proportionals between this sum and the mass of that other body S'. The algebraic transformation leads to the following law, $\bar{a}_i{}^3/T_i{}^3 = K(S+P_i)$, where T_i and P_i are respectively the period and mass of the ith planet and $\bar{a}_i$ is the semimajor axis of each of the new ellipses for the sun and a planet moving about the common center of gravity. Whence

$$\frac{\bar{a}_1{}^3/T_1{}^2}{\bar{a}_2{}^3/T_2{}^2} = \frac{S+P_1}{S+P_2} = \frac{1+(P_1S)}{1+(P_2/S)}.$$

This law for a two-body system (ignoring interplanetary perturbations) reduces to Kepler's third law when (as for most planets in the solar system) the planetary masses are so small that P_i/S can be neglected (see Cohen, 1974c, pp. 316–319).

4 Newton has not yet shown that there is a general or universal force of *gravity*, that the force of terrestrial gravity extends to the moon and is the same as the solar and planetary forces and is thus to be taken as a force of and acting on all bodies. These stages of generalization do not occur until bk. three.

5 Once again, we cannot (here in bk. one) speak of "gravitating" planets. Newton has been considering the "accelerative quantities" or "measures" of forces in relation to the accelerations they produce (according to def. 7); now he shifts to the actual forces, or the "absolute quantity" or "measure" of such forces (according to def. 6). The conditions are that in a system of bodies A, B, C, D,..., any one of the bodies (A) attracts all the others (B, C, D, ...) with inverse-square "accelerative forces", and so does another body (B); then the 'absolute forces of the attracting bodies A and B' will be to each other as the masses of A and B.

6 The exchange of letters between Newton and Flamsteed on the possible perturbation of Saturn's orbit by Jupiter may be found in Newton (1959–1977), vol. 2.

7 Newton's late introduction of the elliptical orbits has been discussed in §3.5.

8 See Westfall (1973). The OED traces the lineage of this sense of the word "fudge" back to the seventeenth century. The word "fudge" was first applied to Newton (as far as I know) in Truesdell (1970).

9 That is, in most of Newton's lunar theory, he too proceeded by trial and error and like his predecessors made extensive use of geometric computing schemes. In this sense, Newton's lunar theory was a failure and hardly revolutionary. But what was revolutionary was the vision of a lunar theory based wholly on applications of physical principles or causes, namely gravitational forces; and this was not wholly visionary and programmatic since Newton had used gravitation theory with some success in certain aspects of the problem.

Supplement

1 This 131-page volume contained the texts of two lectures: Lecture 1, 'Laws of Motion and Principles of Philosophy', and Lecture 2, 'From Hypotheses to Rules'. The latter has been published twice (see Cohen, 1966). An expanded version is scheduled for publication in a volume tentatively entitled *Newton's Natural Philosophy*.

2 Aiton's and Pogrebysski's comments are printed in the journal, *Acta Historiae Rerum Naturalium necnon Technicarum*, special issue 4 (Prague, 1968), pp. 67–69, 44–50. A summary of my paper and amplification of the doctrine of transformation and revolution appears, with some additions and corrections to the main article, on pp. 35–41 of that same issue (but under the incorrect title, 'Optics in the 17th century').

3 My communication, entitled 'The concept and definition of mass and inertia as a key to the science of motion: Galileo-Newton-Einstein' (XIII International Congress of the History of Science, USSR, Moscow, August 18–24, 1971) was distributed at the Congress in a preprint edition (Moscow: Nauka Publishing House, Central Department of Oriental Literature, 1971), but it was not published in the proceedings or acts of the Congress, where it was summarized in Russian. A revised version is scheduled for publication in Cohen (1978a).

4 This is manifest in many of his studies on the filiation and development of scientific ideas, e.g., the chapter on Newton and Descartes in Koyré (1965). Cf. his statement of 'Orientation et projets de recherches', in Koyré (1966), pp. 1–5.

5 Duhem (1954), ch. 6, sect. 4, p. 191. Duhem also referred (ibid.) to 'the experimental laws established by Kepler and transformed by geometric reasoning'.

6 Mach (1898), p. 61. The German words are 'in die physikalische Technik überall umgestaltend eingreift'.

7 Ibid., p. 63; in German, 'Langsam, allmählich und mühsam bildet sich ein Gedanke in den andern um, wie es wahrscheinlich ist, dass ein Tierart allmählich in neuen Arten übergeht'.

8 Ibid., pp. 214sqq; in German, 'Über Umbildung und Anpassung im naturwissenschaftlichen Denken'. Writing in 1883, Mach (ibid., p. 216) said: 'Scarcely thirty years have elapsed since Darwin first propounded the principles of his theory of evolution. Yet already we see his ideas firmly rooted in every branch of human thought, however remote. Everywhere, in history, in philosophy, even in the physical sciences, we hear the watchwords: heredity, adaptation, selection'. In the German edition the original of 'processes of transformation' (p. 218) is 'Umbildungsprozesse' (p. 242); 'evolution and transformation' (p. 218) is a translation of 'Entwickelung und Umbildung' (p. 241).

9 The feelings of Freud and Ellis toward each other are portrayed graphically in Wortis (1954).

10 I quote from the extracts given by Ellis and reprinted by Freud (1920), pp. 263sq.

11 As noted by Freud (1920), p. 264.

12 This example had been called to Freud's attention, many years earlier, by Otto Rank; it had been quoted by Freud in his *Traumdeutung* of 1900, as he was careful to note in his 1920 reply to Ellis.

13 Quoted in Freud (1920), p. 265.

14 Freud relates that Ferenczi's attention had been drawn to this essay by Dr. Hugo Dubowitz.

15 This definition is taken from *Webster's New International Dictionary of the English Language* (2d ed., 1939), under the new words.

16 For another aspect of Freudian cryptomnesia, not apparently recognized by Freud himself, see §4.7.

17 Foucault (1972), p. 21; see also pp. 13–15, 38, 44, and an example on pp. 33sq.

18 Ibid., p. 4. It is of more than passing interest that the two authors whom Foucault cites with evident high approval are relatively uninfluential on Anglo-American readers. None of Canguilhem's books has been translated into English, and only popular books by Bachelard are available to readers in English.

19 For example, ibid., pp. 62, 69, 71, 74, 117, 120–122, 124, 200. Grammatical transformations are mentioned on pp. 81, 99. Transformations of history as a mode of thought also occur on pp. 136, 140, 141; this may be likened to his reference to 'the transformation from an incinerating to an inhuming culture' in Foucault (1973), p. 166.

20 Foucault (1972), p. 188. Cf. the discussion of 'the formation and transformation of a body of knowledge' on p. 194.

21 In (1973), Foucault writes about the possibility of a 'concept . . . capable of transforming itself', and observes that the resulting 'conceptual transformation was decisive' (p. 97).

22 An additional example of cryptomnesia: after the typescript of the present work had been completed and sent to the publisher, I came by chance upon the essay (1856) by William Whewell, 'On the transformation of hypotheses in the history of science'. What is remarkable about this incident in the present context is that I had selected this very essay for an anthology of nineteenth-century British scientific prose which I edited with Howard Mumford Jones and Everett Mendelsohn a decade and a half ago; I had "forgotten all about it" until a reprint of our anthology brought it to my mind's attention once again (see Jones, Cohen, & Mendelsohn, 1963).

BIBLIOGRAPHY

The following Bibliography comprises the works which are cited in
the notes, and major primary and secondary sources that are related to the
topics of the several chapters. In addition to the obvious works bearing
on the scientific thought of Newton and on Newtonian science, the
Bibliography contains entries on philosophical questions bearing on the
main themes of this book, and in particular the historical relations
between mathematics and the sciences, notably by Salomon Bochner,
Léon Brunschvicg, Ernst Cassirer, Pierre Duhem, Albert Einstein, Paul
Feyerabend, Jacques Hadamard, Mary Hesse, E. W. Hobson, Gerald Holton,
T. S. Kuhn, Imre Lakatos, Ernst Mach, Peter B. Medawar, Ernest Nagel,
Henri Poincaré, Karl Popper, Stephen Toulmin, Steven Weinberg,
Hermann Weyl, and A. N. Whitehead – all of which have been of
importance in the development of my thinking about the fundamental
questions addressed in this book. This category should also include certain
eighteenth-century scientists, notably d'Alembert, Bailly, Euler, Lagrange,
Laplace, Maclaurin, and Montucla.

I am, of course, fully aware of my great indebtedness to other
historians of science and of philosophy, and especially certain Newtonian
scholars and specialists on the seventeenth century, whose names may
often appear in the Notes primarily at times when I disagree with
them on particular issues. In this category, I would particularly
single out the writings of E. J. Aiton, Carl B. Boyer, Gerd Buchdahl,
Georges Canguilhem, E. P. Dijksterhuis, Stillman Drake, René Dugas,
Michel Foucault, Henry Guerlac, A. R. Hall, Marie Boas Hall, J. W.
Herivel, A. N. Kriloff, Alexandre Koyré, Fritz Krafft, J. A. Lohne, J. E.
McGuire, Serge Moscovici, L. Olschki, Walter Pagel, P. M. Rattansi,
Ferd. Rosenberger, Leon Rosenfeld, Paoli Rossi, A. I. Sabra, Clifford
Truesdell, Charles Webster, Richard S. Westfall, Robert S. Westman,
Derek T. Whiteside, Curtis Wilson, Harry Austryn Wolfson, and
A. P. Youschkevitch.

AITON, ERIC J. (1954). 'Galileo's theory of the tides'. *Annals of Science*, vol. 10,
pp. 44–57.
  (1955). 'The contributions of Newton, Bernoulli and Euler to the theory of
  the tides'. *Annals of Science*, vol. 11, pp. 206–223.
  (1960). 'The celestial mechanics of Leibniz'. *Annals of Science*, vol. 16, pp. 65–82.

(1962). 'The celestial mechanics of Leibniz in the light of Newtonian criticism'. *Annals of Science*, vol. 18, pp. 31–41.

(1963). 'On Galileo and the earth-moon system'. *Isis*, vol. 54, pp. 265–266.

(1964a). 'The celestial mechanics of Leibniz: A new interpretation'. *Annals of Science*, vol. 20, pp. 111–123.

(1964b). 'The inverse problem of central forces'. *Annals of Science*, vol. 20, pp. 81–99.

(1969). 'Kepler's second law of planetary motion'. *Isis*, vol. 60, pp. 75–90.

(1972). *The vortex theory of planetary motions.* London: Macdonald; New York: American Elsevier.

(1975a). 'The elliptical orbit and the area law'. *Vistas in Astronomy*, vol. 18, pp. 573–583.

(1975b). 'Infinitesimals and the area law'. *Vistas in Astronomy*, vol. 18, pp. 585–586.

D'ALEMBERT, JEAN LE ROND (1743). *Traité de dynamique.* . . . Paris: chez David l'aîné. Reprinted (1967), Brussels: Culture et Civilisation.

(1963). *Preliminary discourse to the Encyclopedia of Diderot.* Translated by Richard N. Schwab with the collaboration of Walter E. Rex; with an introduction by Richard N. Schwab. Indianapolis, New York: Bobbs-Merrill Co., Library of Liberal Arts.

ANDRADE, E. N. DA C. (1947). 'Newton'. Pp. 3–23 of The Royal Society's *Newton tercentenary celebrations, 15–19 July 1946.* Cambridge (England): at the University Press.

BABSON COLLECTION (1950). *A descriptive catalogue of the Grace K. Babson Collection of the works of Sir Isaac Newton and the material relating to him in the Babson Institute Library, Babson Park, Mass.* With an introduction by Roger Babson Webber. New York: Herbert Reichner. A supplement, compiled by Henry P. Macomber, was published by the Babson Institute in 1955.

BACON, FRANCIS (1889). *Bacon's Novum organum.* Edited with introduction, notes, etc., by Thomas Fowler. 2d ed., corrected and revised. Oxford: at the Clarendon Press.

(1905). *The philosophical works of Francis Bacon.* . . . Reprinted from the texts and translations, with the notes and prefaces of Ellis and Spedding. Edited with an introduction by John M. Robertson. London: George Routledge and Sons; New York: E. P. Dutton & Co.

(1960). *The new organon and related writings.* Edited with an introduction by Fulton H. Anderson. Indianapolis, New York: The Bobbs-Merrill Company [The Library of Liberal Arts].

BAILLY, JEAN-SYLVAIN (1781). *Histoire de l'astronomie ancienne, depuis son origine jusqu'à l'établissement de l'Ecole d'Alexandrie.* 2d ed. Paris: chez De Bure fils aîné.

(1785). *Histoire de l'astronomie moderne depuis la fondation de l'Ecole d'Alexandrie, jusqu'à l'époque de M.D.CC.XXX.* New ed., 3 vols. Paris: chez de Bure.

BAILY, FRANCIS (1835). *An account of the Revd John Flamsteed, the first Astronomer-Royal; compiled from his own manuscripts, and other authentic documents, never before published. To which is added, his British catalogue of stars, corrected and enlarged.* London: printed by order of the Lords Commissioners of the Admiralty. In 1837 there was a *Supplement to the account of the Revd John Flamsteed, the first Astronomer-Royal.* London: printed for distribution amongst those persons and institutions only, to whom the original work was presented. Photo-reprint (1966), London: Dawsons of Pall Mall; with the omission of Baily's revision of the British catalogue of stars.

BALD, R. C. (1940). 'Coleridge and *The Ancient Mariner*: Addenda to *The Road to Xanadu*', pp. 1–45 of Herbert Davis, William C. DeVane, & R. C. Bald, eds., *Nineteenth-century studies*. Ithaca: Cornell University Press.

BALL, W. W. ROUSE (1893). *An essay on Newton's "Principia"*. London, New York: Macmillan and Co. Reprinted (1972), New York, London: Johnson Reprint Corporation.

BARROW, ISAAC (1669). *Lectiones XVIII. Cantabrigiae in scholis publicis habitae; in quibus opticorum phaenomenωn genuinae rationes investigantur, ac exponuntur*. Annexae sunt lectiones aliquot geometricae. London: typis Gulielmi Godbid, & prostant venales apud Johannem Dunmore, & Octavianum Pulleyn Juniorem.

(1916). *The geometrical lectures*. Translated, with notes and proofs, and a discussion on the advance made therein on the work of his predecessors in the infinitesimal calculus, by J. M. Child. Chicago and London: The Open Court Publishing Company. A much better (i.e., more literal) translation by Edmund Stone (1735) has also the virtue of completeness, which Child's does not.

BECHLER, ZEV (1973). 'Newton's search for a mechanistic model of colour dispersion: A suggested interpretation'. *Archive for History of Exact Sciences*, vol. 11, pp. 1–37.

(1974a). 'Newton's law of forces which are inversely as the mass: A suggested interpretation of his later efforts to normalise a mechanistic model of optical dispersion'. *Centaurus*, vol. 18, pp. 184–222.

(1974b). 'Newton's 1672 optical controversies: A study in the grammar of scientific dissent'. Pp. 115–142 of ELKANA, ed. (1974).

(1975). ' "A less agreeable matter": The disagreeable case of Newton and achromatic refraction'. *British Journal for the History of Science*, vol. 8, pp. 101–126.

BEER, ARTHUR, & PETER BEER, eds. (1975). *Kepler, four hundred years*. Proceedings of conferences held in honour of Johannes Kepler. Oxford, New York: Pergamon Press (*Vistas in Astronomy*, vol. 18).

BEER, JOHN BERNARD (1959). *Coleridge, the visionary*. London: Chatto and Windus. Reprint ed., New York: Collier Books, 1962.

BELTRÁN DE HEREDIA, VICENTE (1961). *Domingo de Soto: estudio biográfico documentado*. Madrid: Ediciónes Cultura Hispánica.

BERTHELOT, MARCELIN (1890). *La révolution chimique: Lavoisier*. Paris: F. Alcan.

BEYER, WERNER W. (1963). *The enchanted forest*. Oxford: Basil Blackwell; New York: Barnes & Noble.

BIERNSON, GEORGE (1972). 'Why did Newton see indigo in the spectrum?' *American Journal of Physics*, vol. 40, pp. 526–533.

BIOT, J.-B., & F. LEFORT. *See* COLLINS ET AL. (1856).

BIRCH, THOMAS (1756–1757). *The history of the Royal Society of London for Improving of Natural Knowledge*. London: printed for A. Millar. Reprinted (1968), New York, London: Johnson Reprint Corporation.

BIRKHOFF, GEORGE DAVID (1913). 'Proof of Poincaré's geometric theorem'. *Transactions of the American Mathematical Society*, vol. 14, pp. 14–22.

(1926). 'An extension of Poincaré's last geometric theorem'. *Acta Mathematica*, vol. 47, pp. 297–311.

(1931). 'Une généralisation à n dimensions du dernier théorème de géométrie de
Poincaré'. *Comptes Rendus des Séances de l'Académie des Sciences*, vol. 192,
pp. 196–198.

BLÜH, OTTO (1952). 'The value of inspiration, a study on Julius Robert Mayer
and Josef Popper-Lynkeus'. *Isis*, vol. 43, pp. 211–220.

BOAS, MARIE (1952). 'The establishment of the mechanical philosophy'. *Osiris*, vol. 10,
pp. 412–541. Section IX deals with Newton. *See also* HALL & HALL; HALL,
MARIE BOAS.

BOCHNER, SALOMON (1966). *The role of mathematics in the rise of science.*
Princeton: Princeton University Press.

(1969). *Eclosion and synthesis: perspectives on the history of knowledge.* New
York, Amsterdam: W. A. Benjamin.

BOHR, NIELS (1913). 'On the constitution of atoms and molecules', *Philosophical
Magazine*, ser. 6, vol. 26, pp. 1–25, 476–502, 857–875.

BOPP, K. (1929). 'Die wiederaufgefundene Abhandlung von Fatio de Duillier: De
la cause de la pesanteur'. *Schriften der Strassburger Wissenschaftlichen
Gesellschaft in Heidelberg*, vol. 10 (n.s.), pp. 19–66.

BOWLER, PETER J. (1976). 'Malthus, Darwin, and the concept of struggle'. *Journal of
the History of Ideas*, vol. 37, pp. 631–650.

BOYER, CARL B. (1939). *Concepts of the calculus: A critical and historical discussion of the derivative and the integral.* New York: Columbia University Press.

(1949). 'Newton as an originator of polar coordinates'. *American Mathematical
Monthly*, vol. 56, pp. 73–78.

(1954). 'Analysis: Notes on the evolution of a subject and a name'. *The Mathematics
Teacher*, vol. 47, pp. 450–462.

(1956). *History of analytic geometry.* New York: Scripta Mathematica.

(1968). *A history of mathematics.* New York, London, Sydney: John Wiley & Sons.

(1970). 'Boulliau, Ismael'. *Directory of scientific biography*, vol. 2, pp. 348–349.

BOYLE, ROBERT (1744). *The works of the Honourable Robert Boyle.* 5 vols.
London: printed for A. Millar. A 'new edition' in 6 vols. (1772), London:
printed for J. and F. Rivington, L. Davis, W. Johnston.

BRAHE, TYCHO (1913–1929). *Tychonis Brahe Dani opera omnia.* Edidit J. L. E.
Dreyer. Auxilio Joannis Raeder. 15 vols. Copenhagen: in libraria Gyldendaliana
[typis Nielsen & Lydiche (Axel Simmelkiâer)].

(1946). *Tycho Brahe's description of his instruments and scientific work as given
in "Astronomiae instauratae mechanica"* (Wandesburg: 1598). Translated
and edited by Hans Raeder, Elis Strömgren, and Bengt Strömgren. Copenhagen:
Munksgaard.

BRASCH, F. E. (1931). *Johann Kepler 1571–1630, a tercentenary commemoration
of his life and work.* A series of papers prepared under the auspices of the History
of Science Society in collaboration with the American Association for the Advancement of Science. Baltimore: The Williams & Wilkins Company.

ed. (1928). *Sir Isaac Newton, 1727–1927. A bicentenary evaluation of his work.*
A series of papers prepared under the auspices of the History of Science
Society. Baltimore: The Williams & Wilkins Company.

BRECHER, KENNETH, ELINOR LIEBER, & ALFRED E. LIEBER (1978). 'A near-eastern
sighting of the supernova explosion in 1054'. *Nature*, vol. 273, pp. 728–730.

BREWSTER, SIR DAVID (1855). *Memoirs of the life, writings, and discoveries of Sir
Isaac Newton.* 2 vols. Edinburgh: Thomas Constable and Co. Photo-reprint from
the Edinburgh ed. of 1855 with a new introduction by Richard S. Westfall, New

York, London: Johnson Reprint Corporation [The Sources of Science, no. 14], 1965.

BRONFENBRENNER, MARTHA ORNSTEIN. *See* ORNSTEIN, MARTHA.

BROUGHAM, HENRY, LORD, & E. J. ROUTH (1855). *Analytical view of Sir Isaac Newton's Principia.* London: Longman, Brown, Green, and Longmans, [&] C. Knight; Edinburgh: A. and C. Black; Glasgow: R. Griffin. Reprinted (1972), New York, London: Johnson Reprint Corporation.

BROWN, HARCOURT (1934). *Scientific organizations in seventeenth century France.* Baltimore: The Williams & Wilkins Company.

BRUNET, PIERRE (1929). *Maupertuis.* [1,] Etude biographique. [2,] L'oeuvre et sa place dans la pensée scientifique et philosophique du XVIIIe siècle. Paris: Librairie Scientifique Albert Blanchard.

(1931). *L'introductión des théories de Newton en France au XVIIIe siècle.* Vol. 1, avant 1738. Paris: Librairie Scientifique Albert Blanchard.

BUCHDAHL, GERD (1969). *Metaphysics and the philosophy of science: The classical origins, Descartes to Kant.* Oxford: Basil Blackwell.

(1970). 'History of science and criteria of choice'. Pp. 204–230 of STUEWER (1970). An important commentary by Laurens Laudan follows on pp. 230–238.

(1973). 'Explanation and gravity'. Pp. 167–203 of TEICH & YOUNG (1973).

BURSTYN, HAROLD L. (1962). 'Galileo's attempt to prove that the earth moves'. *Isis*, vol. 53, pp. 161–185.

(1963). 'Galileo and the earth-moon system: Reply to Dr. Aiton'. *Isis*, vol. 54, pp. 400–401.

BUTTERFIELD, HERBERT (1957). *The origins of modern science, 1300–1800.* Rev. ed. New York: The Macmillan Company; London: G. Bell & Sons. Revised edition, first issued in 1949.

CAJORI, FLORIAN (1928). 'Newton's twenty years' delay in announcing the law of gravitation'. Pp. 127–188 of BRASCH, ed. (1928). *See also* NEWTON (1934).

CANGUILHEM, GEORGES (1955). *La formation du concept de réflexe au XVIIe et XVIIIe siècles.* Paris: Presses Universitaires de France.

(1968). *Etudes d'histoire et de philosophie des sciences.* Paris: Libraire Philosophique J. Vrin.

(1971). *La connaissance de la vie.* 2d ed., rev. and enl. Paris: Libraire Philosophique J. Vrin.

CASPAR, MAX (1959). *Kepler.* Translated and edited by C. Doris Hellman. London, New York: Abelard-Schuman.

CASSIRER, ERNST (1906–1920). *Das Erkenntnisproblem in der Philosophie und Wissenschaft der neueren Zeit.* Vol. 1 (1906), vol. 2 (1907), vol. 3 (1920). Berlin: Verlag von Bruno Cassirer.

(1950). *The problem of knowledge, philosophy, science, and history since Hegel.* New Haven: Yale University Press (published in German as vol. 4 of *Das Erkenntnisproblem in der Philosophie und Wissenschaft der neueren Zeit*, Stuttgart: Verlag W. Kohlhammer, 1957).

CHANDLER, PHILIP P., II (1975). Newton and Clairaut on the motion of the lunar apse. San Diego: [doctoral dissertation, University of California].

CICERO, MARCUS TULLIUS (1955). *De natura deorum liber primus.* Edited by Arthur Stanley Pease. Cambridge (Mass.): Harvard University Press.

CLAGETT, MARSHALL (1959). *The science of mechanics in the Middle Ages.* Madison: University of Wisconsin Press.

ed. (1959). *Critical problems in the history of science.* Proceedings of the Institute for the History of Science at the University of Wisconsin, September 1–11, 1957. Madison: The University of Wisconsin Press.

CLAIRAUT, ALEXIS-CLAUDE (1743). *Théorie de la figure de la terre tirée des principes de l'hydrostatique.* Paris: Duraud.

—— (1749). 'Du système du monde dans les principes de la gravitation universelle'. *Histoire de l'Académie Royale des Sciences, année MDCCXLV, avec les mémoires de mathématique et de physique pour la même année.* Paris: à l'Imprimerie Royale, pp. 329–364. Clairaut's paper is said to have been read 'à l'Assemblée publique du 15 Nov. 1747'.

CLARK, DAVID H., & F. RICHARD STEPHENSON (1977). *The historical supernovae.* Oxford, New York: Pergamon Press. *See also* STEPHENSON & CLARK (1976).

CLARK, J. GRAHAME D. (1946). *From savagery to civilisation.* London: Cobbett Press.

CLARKE, JOHN (1730). *A demonstration of some of the principal sections of Sir Isaac Newton's Principles of natural philosophy. In which his peculiar method of treating that useful subject is explained, and applied to some of the chief phaenomena of the system of the world.* London: printed for James and John Knapton.

CLAVELIN, MAURICE (1974). *The natural philosophy of Galileo: Essays on the origins and formation of classical mechanics.* Translated by A. J. Pomerans. Cambridge (Mass.), London (England): The MIT Press. Translation of *La philosophie naturelle de Galilée: Essai sur les origines et la formation de la mécanique classique,* Paris: Librairie Armand Colin, 1968.

COHEN, I. BERNARD (1940). 'The first explanation of interference'. *American Journal of Physics,* vol. 8, pp. 99–106.

—— (1946). 'Authenticity of scientific anecdotes'. *Nature,* vol. 157, pp. 196–197.

—— (1952). 'Orthodoxy and scientific progress'. *Proceedings of the American Philosophical Society,* vol. 96, pp. 505–512.

—— (1956). *Franklin and Newton: An inquiry into speculative Newtonian experimental science and Franklin's work in electricity as an example thereof.* Philadelphia: The American Philosophical Society. Reprint (1966), Cambridge (Mass.): Harvard University Press. A revised printing scheduled for 1980.

—— (1960). 'Newton in the light of recent scholarship'. *Isis,* vol. 51, pp. 489–514.

—— (1962). 'Leibniz on elliptical orbits: As seen in his correspondence with the Académie Royale des Sciences in 1700'. *Journal of the History of Medicine and Allied Sciences,* vol. 17, pp. 72–82.

—— (1964a). 'Isaac Newton, Hans Sloane, and the Académie Royale des Sciences'. Pp. 61–116 of COHEN & TATON, eds. (1964).

—— (1964b). ' "Quantum in se est": Newton's concept of inertia in relation to Descartes and Lucretius'. *Notes and Records of the Royal Society of London,* vol. 19, pp. 131–155.

—— (1966). 'Hypotheses in Newton's philosophy'. *Physis,* vol. 8, pp. 163–184.

—— (1967a). 'Dynamics: The key to the "new science" of the seventeenth century'. *Acta historiae rerum naturalium necnon technicarum.* (Czechoslovak Studies in the History of Science) Prague, Special Issue 3, pp. 79–114.

—— (1967b). 'Galileo, Newton, and the divine order of the solar system'. Pp. 207–231 of MCMULLIN, ed. (1967).

—— (1967c). 'Newton's attribution of the first two laws of motion to Galileo'. *Atti del Symposium Internazionale di Storia, Metodologia, Logica e Filosofia della Scienza 'Galileo nella Storia e nella Filosofia della Scienza',* pp. xxv–xliv. Collection des Travaux de l'Académie Internationale d'Histoire des Sciences, no. 16. Vinci (Florence): Gruppo Italiano di Storia della Scienza.

(1967*d*). 'Newton's use of "force", or, Cajori versus Newton: A note on translations of the *Principia*'. *Isis*, vol. 58, pp. 226–230.

(1969*a*). Introduction, pp. vii–xxii, to *A treatise of the system of the world* by Sir Isaac Newton (facsimile reprint of 2d ed., 1731). London: Dawsons of Pall Mall.

(1969*b*). 'Isaac Newton's *Principia*, the scriptures and the divine providence'. Pp. 523–548 of Sidney Morgenbesser, Patrick Suppes, & Morton White, eds., *Essays in honor of Ernest Nagel: Philosophy, science, and method*. New York: St. Martin's Press.

(1969*c*). 'Newton's *System of the world*: Some textual and bibliographical notes'. *Physis*, vol. 11, pp. 152–166.

(1970). 'Newton's second law and the concept of force in the *Principia*'. Pp. 143–185 of PALTER, ed. (1970). A considerably revised and corrected version of a preliminary text published in *The Texas Quarterly*, vol. 10, no. 3 [1967].

(1971). *Introduction to Newton's 'Principia'*. Cambridge (Mass.): Harvard University Press; Cambridge (England): at the University Press.

(1972). 'Newton and Keplerian inertia; an echo of Newton's controversy with Leibniz'. Pp. 199–211 of DEBUS, ed. (1972), vol. 2.

(1974*a*). 'History and the philosopher of science'. Pp. 308–373 of SUPPE, ed. (1974).

(1974*b*). 'Isaac Newton, the calculus of variations, and the design of ships: An example of pure mathematics in Newton's *Principia*, allegedly developed for the sake of practical applications'. Pp. 169–187 of R. S. COHEN et al., eds. (1974).

(1974*c*). 'Newton, Isaac'. *Dictionary of Scientific Biography*, vol. 10, pp. 42–103. Rev. and enl. ed. scheduled for publication in 1980 as a book by Charles Scribner's Sons (New York).

(1974*d*). 'Newton's theory vs. Kepler's theory and Galileo's theory: An example of a difference between a philosophical and a historical analysis of science'. Pp. 299–338 of ELKANA, ed. (1974).

(1975*a*). Bibliographical and historical introduction, pp. 1–87, to NEWTON (1975).

(1975*b*). 'Kepler's century, prelude to Newton's'. *Vistas in Astronomy*, vol. 18. pp. 3–36.

(1976*a*). 'The eighteenth-century origins of the concept of scientific revolution'. *Journal of the History of Ideas*. vol. 37, pp. 257–288.

(1976*b*). 'Stephen Hales'. *Scientific American*, vol. 234 (no. 5), pp. 98–107.

(1977*a*). 'The Copernican revolution from an eighteenth-century perspective'. Pp. 43–54 of MAEYAMA & SALTZER, eds. (1977).

(1977*b*). 'Number-relations and numerology in modern astronomy and other exact sciences'. In process.

(1977*c*). 'The origins and significance of the concept of a Copernican revolution'. *Organon*, in press.

(1977*d*). 'Perfect numbers in the Copernican System: Rheticus and Huygens'. *Festschrift for Edward Rosen*. Warsaw: in press.

(1977*e*). *Scientific revolution: History of a concept and a name*. Cambridge (Mass.), London (England): Harvard University Press, in press.

(1978*a*). 'Newton's concept of mass'. Revised and updated version of a communication to the XIIIth International Congress of the History of Science (Moscow, 1971), to appear in COHEN (1978*b*).

(1978*b*). *Newton's principles of philosophy: Inquiries into Newton's scientific work and its general environment*. Cambridge (Mass.): Harvard University Press, in process. This volume will contain revised and expanded versions of COHEN (1964*a*), (1964*b*), 1967*b*), (1967*c*), (1969*b*), (1970), (1972), and 1974*d*).

COHEN, I. BERNARD, & ROBERT E. SCHOFIELD. *See* NEWTON (1958).

COHEN, I. BERNARD, & RENÉ TATON, eds. (1964). *Mélanges Alexandre Koyré.* Vol. 1, *L'aventure de la science*; vol. 2, *L'aventure de l'esprit.* Paris: Hermann, Histoire de la Pensée, nos. 12 and 13. *See also* KOYRÉ & COHEN (1960), (1961), (1962).

COHEN, R. S., J. J. STACHEL, & M. W. WARTOFSKY, eds. (1974). *For Dirk Struik.* Scientific, historical and political essays in honor of Dirk J. Struik. Dordrecht, Boston: D. Reidel Publishing Company *Boston Studies in the Philosophy of Science,* vol 15.

COLLINS, JOHN, ET AL. (1856). *Commercium epistolicum J. Collins et aliorum de analysi promota, etc., ou, Correspondance de J. Collins et d'autres savants célèbres du XVIIe siècle, relative à l'analyse supérieure, réimprimée sur l'édition originale de 1712 avec l'indication des variantes de l'édition de 1722, complétée par une collection de pièces justificatives et de documents, et publiée par J.-B. Biot et F. Lefort.* Paris: Mallet-Bachelier.

COPERNICUS, NICHOLAS (1543). *De revolutionibus orbium coelestium.* Nuremberg: apud Joh. Petreium. Facsimile reprint, Paris: M. J. Hermann, 1927.

(1873). *De revolutionibus orbium caelestium libri vi. ex auctoris autographo recudi curavit Societas Copernicana Thorunensis.* Torun: sumptibus Societatis Copernicanae.

(1952). On the revolutions of the heavenly spheres. Translated with an introduction by C. G. Wallis. Pp. 497–838 of (1952).

(1972). *The manuscript of Nicholas Copernicus' 'On the revolutions', facsimile.* London, Warsaw, Cracow: Macmillan and Polish Scientific Publishers. Nicholas Copernicus, Complete Works, vol. 1. Edited by Paweł Czartoryski, introduction by Jerzy Zathey.

COTES, ROGER. *See* EDLESTON (1850).

CROMBIE, A. C. (1953). *Robert Grosseteste and the origins of experimental science, 1180–1700.* Oxford: at the Clarendon Press.

(1957). 'Newton's conception of scientific method'. *Bulletin of the Institute of Physics,* vol. 8, pp. 350–362.

(1969). *Augustine to Galileo.* 2 vols. Harmondsworth (Middlesex, England): Penguin Books. Reprint of rev. ed. (1959) of a work first published in 1952. American ed. of the revision published as *Medieval and early modern science.* Garden City [New York]: Doubleday & Company [Doubleday Anchor Books], 1959.

ed. (1963). *Scientific change: Historical studies in the intellectual, social and technical conditions for scientific discovery and technical invention, from antiquity to the present.* Symposium on the History of Science; University of Oxford, 9–15 July 1961. New York: Basic Books.

CUVIER, GEORGES (1812). *Recherches sur les ossemens fossiles des quadrupèdes.* . . . Vol. 1. Paris: chez Deterville, Libraire.

DANTE ALIGHIERI (1970). *The divine comedy.* Translated with a commentary, by Charles S. Singleton. *Inferno,* 2: Commentary. Princeton: Princeton University Press [Bollingen Series no. 80].

DARWIN, CHARLES (1959). 'Darwin's Journal'. Edited with an introduction and notes by Sir Gavin De Beer. *Bulletin of the British Museum (Natural History),* Historical Series, vol. 2 (no. 1), pp. 1–21.

(1960, 1967). 'Darwin's notebooks on transmutation of species. Part I. First notebook (July 1837–February 1838)'. Edited with an introduction and notes by Sir Gavin De Beer. *Bulletin of the British Museum (Natural History),*

Historical Series, 1960, vol. 2 (no. 2), pp. 23–73; 'Part II. Second notebook (February to July 1838)'. Ibid., 1960, vol. 2, (no. 3), pp. 75–118; 'Part III. Third notebook (July 15th 1838–October 2nd 1838)'. Ibid., 1960, vol. 2 (no. 4), pp. 119–150; 'Part IV. Fourth notebook (October 1838–10 July 1839)'. Ibid., 1960, vol. 2 (no. 5), pp. 151–183; 'Part VI. Pages excised by Darwin'. Edited by Sir Gavin De Beer, M. J. Rowlands, & B. M. Skramovsky. Ibid., 1967, vol. 3 (no. 5), pp. 129–176.

DARWIN, FRANCIS (1917). *Rustic sounds*. London: John Murray. Contains an essay on Stephen Hales (pp. 115–139); Francis Darwin also wrote the account of Hales for the *Dictionary of National Biography*, vol. 8, pp. 916–920.

DE BEER, G. R. *See* MCKIE & DE BEER (1952).

DEBUS, ALLEN G. (1975). 'Van Helmont and Newton's third law'. Pp. 45–52 of Sepp Domandl (ed.), *Paracelsus:Werk und Wirkung* (Festgabe für Curt Goldammer zum 60. Geburtstag), Vienna: Verband der Wissenschaftlichen Gesellschaften Österreichs Verlag.

    ed. (1972). *Science, medicine and society: Essays to honor Walter Pagel*. 2 vols. New York: Science History Publications [a division of Neale Watson Academic Publications]. Vol. 1 contains an appreciation of Pagel by Debus (pp. 1–9); vol. 2 contains a bibliography of the writings of Pagel by Marianne Winder (pp. 289–326).

DELAMBRE, JEAN BAPTISTE JOSEPH (1812). 'Notice sur Lagrange'. *Mémoires de la classe des sciences mathématiques et physiques de l'Institut de France* (année 1812). Pt. 2, pp. xxxiv–lxxx.

    (1821). *Histoire de l'astronomie moderne*. 2 vols. Paris: Mme Ve Courcier, Libraire pour les Sciences (reprint ed., New York, London: Johnson Reprint Corp., 1969).

DESCARTES, RENÉ (1668). *Epistolae, partim ab auctore latino sermone conscriptae, partim ex gallico translatae*. Pt. 2. London: impensis Joh. Dunmore, & Octavian Pulleyn. Also issued with imprint, Amsterdam: apud Danielem Elzevirium, 1668.

    (1956). *Discourse on method*. Translated with an introduction by Laurence J. Lafleur. Indianapolis, New York, Kansas City: The Bobbs-Merrill Company [The Library of Liberal Arts].

    (1965). *Discourse on method, Optics, Geometry, and Meteorology*. Translated, with an introduction, by Paul J. Olscamp. Indianapolis, New York, Kansas City: The Bobbs-Merrill Company [Library of Liberal Arts].

    (1974). *Oeuvres*. Edited by Charles Adam & Paul Tannery. 11 vols. Paris: Librairie Philosophique J. Vrin. 'Nouvelle présentation, en co-édition avec le Centre National de la Recherche Scientifique'. Augmented and corrected reprint of the original ed. (1897–1913), Paris: Léopold Cerf. [Some vols. are dated 1973, others 1974; there was an earlier reprint by Vrin, also.]

DIRAC, P. A. M. (1963). 'The evolution of the physicist's picture of nature'. *Scientific American*, vol. 208 (no. 5), pp. 45–53.

DOBBS, BETTY JO TEETER (1975). *The foundations of Newton's alchemy, or "The hunting of the greene lyon"*. Cambridge, London, New York, Melbourne: Cambridge University Press.

DOBELLI, AUSONIA (1897). 'Intorno ad una fonte dantesca'. *Bulletino della Società Dantesca Italiana*, vol. 4, pp. 16–17.

DOBZHANSKY, THEODOSIUS (1959). 'Blyth, Darwin, and natural selection'. *American Naturalist*, vol. 93, pp. 200–205.

DOMSON, CHARLES ANDREW (1972). Nicolas Fatio de Duillier and the prophets of

London: An essay in the historical interaction of natural philosophy and millennial belief in the age of Newton. New Haven: [doctoral dissertation, Yale University].

DRAKE, STILLMAN (1958). 'Galileo gleanings 3. A kind word for Sizzi'. *Isis*, vol. 49, pp. 155–165.

(1975*a*). 'Free fall from Albert of Saxony to Honoré Fabri'. *Studies in the History and Philosophy of Science*, vol. 5, pp. 347–366.

(1975*b*). 'Impetus theory reappraised'. *Journal of the History of Ideas*, vol. 36, pp. 27–46.

(1976). 'A further reappraisal of impetus theory'. *Studies in the History and Philosophy of Science*, vol. 7, pp. 319–336.

(1977). 'Galileo and *The career of philosophy*'. *Journal of the History of Ideas*, vol. 38, pp. 19–32.

(1978). *Galileo at work, his scientific biography*. Chicago, London: University of Chicago Press.

See also GALLILEO (1953), (1957), (1974).

DREYER, J. L. E. (1890). *Tycho Brahe, a picture of scientific life and work in the sixteenth century*. Edinburgh: Adam and Charles Black.

(1906). *History of the planetary systems from Thales to Kepler*. Cambridge: at the University Press (reprinted (1953) and with a foreword by W. M. Stahl, as *A history of astronomy from Thales to Kepler*. New York: Dover Publications).

DUGAS, RENÉ (1950). *Histoire de la mécanique*. Neuchâtel: Editions du Griffon. Translated (1955) by J. R. Maddox as *A history of mechanics*. Neuchâtel: Éditions du Griffon; New York: Central Book Company.

(1954). *La mécanique au XVIIe siècle (des antécédents scolastiques à la pensée classique)*. Neuchâtel: Éditions du Griffon. Translated (1958) by Freda Jacquot as *Mechanics in the seventeenth century (from the scholastic antecedents to classical thought)*. Neuchâtel: Éditions du Griffon; New York: Central Book Company.

DUHEM, PIERRE (1916). 'L'optique de Malebranche'. *Revue de Métaphysique et de Morale*, vol. 23, pp. 37–91.

(1954). *The aim and structure of physical theory*. Translated by Philip Wiener. Princeton: Princeton University Press. Reprinted (1962), New York: Atheneum.

(1969). *To save the phenomena, an essay on the idea of physical theory from Plato to Galileo*. Translated by Edmund Doland and Chaninah Maschler with an introduction by Stanley L. Jaki. Chicago, London: The University of Chicago Press. Translated from 'ΣΩZEIN TA ΦAINOMENA: Essai sur la notion de théorie physique de Platon à Galilée'. *Annales de philosophie chrétienne*, vol. 6 (1908), pp. 113–138, 277–302, 352–377, 482–514, 576–592; reprinted (1908), Paris: Hermann & Fils.

DUNDON, STANISLAUS J. (1969). 'Newton's "mathematical way" in the *De mundi systemate*'. *Physis*, vol. 11, pp. 195–204.

EDLESTON, J. (1850). *Correspondence of Sir Isaac Newton and Professor Cotes, including letters of other eminent men, now first published from the originals in the Library of Trinity College, Cambridge; together with an appendix, containing other unpublished letters and papers by Newton*. London: John W. Parker; Cambridge (England): John Deighton.

EINSTEIN, ALBERT (1905). 'Über einen die Erzeugung und Verwandlung des Lichtes betreffenden heuristischen Gesichtspunkt'. *Annalen der Physik*, ser. 4, vol. 17, pp. 132–148.

(1954). *Ideas and opinions*. Based on *Mein Weltbild*, edited by Carl Seelig, and other sources. New translations and revisions by Sonja Bargmann. New York: Crown Publishers.

EISELEY, LOREN (1965). 'Darwin, Coleridge, and the theory of unconscious creation'. *Daedalus*, vol. 94, pp. 588–602.

ELKANA, YEHUDA (1970). 'The conservation of energy: A case of simultaneous discovery?' *Archives Internationales d'Histoire des Sciences*, vol. 23, pp. 31–60.

(1974). *The discovery of the conservation of energy*. London: Hutchinson International; Cambridge (Mass.): Harvard University Press.

ed. (1974). *The interaction between science and philosophy*. Atlantic Highlands, N.J.: Humanities Press.

ELLIS, HAVELOCK (1919). *The philosophy of conflict and other essays in war-time*. 2d series. London: Constable and Co.

ELLMANN, RICHARD (1977). *The consciousness of Joyce*. New York: Oxford University Press.

EULER, LEONHARD (1736). *Mechanica sive motus scientia analytice exposita*. 2 vols. St. Petersburg: ex typographia Academiae Scientiarum. Available, edited by Paul Stäckel (1911), in *Leonhardi Euleri opera omnia*. Sub auspiciis Societatis Scientiarum Naturalium Helveticae (general eds.: Ferdinand Rudio, Adolf Krazer, and Paul Stäckel); 2d series (opera mechanica et astronomica), vols. 1 and 2. Leipzig, Berlin: typis et in aedibus B. G. Teubneri.

(1770). *Lettres à une princesse d'Allemagne sur divers sujets de physique & de philosophie*. Leipzig: chez Steidel et compagnie.

(1833). *Letters of Euler on different subjects in natural philosophy, addressed to a German princess*. With notes, and a life of Euler, by David Brewster. . . . With additional notes by John Griscomb. 2 vols. New York: printed and published by J. & J. Harper. Reprinted (1975), New York: Arno Press.

FATIO DE DUILLIER, NICOLAS (1949). 'De la cause de la pesanteur. Mémoire présenté à la Royal Society le 26 février 1690. [Reconstitué et publié avec une introduction par Bernard Gagnebin]. *Notes and Records of the Royal Society of London*, vol. 6, pp. 105–160. *See also* BOPP (1929).

FEYERABEND, PAUL K. (1962). 'Explanation, reduction, and empiricism'. *Minnesota Studies in the Philosophy of Science* (Herbert Feigl and Grover Maxwell, eds.), vol. 3, pp. 28–97.

FIGALA, KARIN (1977a). *Die "Kompositionshierarchie" der Materie – Newtons quantitative Theorie und Interpretation der qualitativen Alchemie*. Munich: [unpublished Habilitationsschrift in the Technische Universität].

(1977b). 'Newton as alchemist'. *History of Science*, vol. 15, pp. 102–137. Essay based on DOBBS (1975).

FONTENELLE, BERNARD LE BOVIER DE (1790). *Oeuvres de Fontenelle*. New ed. 8 vols. Paris: Chez Jean-François Bastien.

FORBES, ERIC G. (1971). *The Euler-Mayer correspondence (1751–1755), a new perspective on eighteenth-century advances in the lunar theory*. London: Macmillan.

FOUCAULT, MICHEL (1970). *The order of things: An archaeology of the human sciences*. New York: Pantheon Books. Translation of *Les mots et les choses: Une archéologie des sciences humaines*, Paris: Gallimard, 1966.

(1972). *The archaeology of knowledge*. Translated by A. M. Sheridan Smith. New York: Pantheon Books. Translation of *L'archéologie du savoir*, Paris: Gallimard, 1969.

(1973). *The birth of the clinic: An archaeology of medical perception*. Translated by A. M. Sheridan Smith. New York: Pantheon Books. Translation of *Nais-*

sance de la clinique: Une archéologie du regard médical, Paris: Presses Universitaires de France, 1963, 1972.

FRANKLIN, BENJAMIN (1941). *Benjamin Franklin's experiments*. A new edition of Franklin's *Experiments and observations on electricity*, edited with a critical and historical introduction by I. Bernard Cohen. Cambridge (Mass.): Harvard University Press.

FREUD, SIGMUND (1920). 'A note on the prehistory of the technique on analysis'. Vol. 18, pp. 263–265, of FREUD (1966–1974).

(1923). 'Josef Popper-Lynkeus and the theory of dreams'. Vol. 19, pp. 260–263, of FREUD (1966–1974).

(1932). 'My contact with Josef Popper-Lynkeus'. Vol. 22, pp. 219–224, of FREUD (1966–1974).

(1937). 'Analysis terminable and interminable'. Vol. 23, pp. 209–253, of (1966–1974).

(1966–1974). *The standard edition of the complete psychological works of Sigmund Freud*. Translated from the German under the general editorship of James Strachey. 24 vols. London: The Hogarth Press and The Institute of Psychoanalysis.

GABBEY, ALAN [W. ALLAN] (1971). 'Force and inertia in seventeenth-century dynamics'. *Studies in History and Philosophy of Science*, vol. 2, pp. 1–67.

(1976). [Essay review of Newton (1972).] *Historia Mathematica*, vol. 3, pp. 237–243.

GAGNEBIN, BERNARD. *See* FATIO DE DUILLIER (1949).

GALILEO GALILEI (1890–1909). *Le opere di Galileo Galilei*. 20 vols. Florence: tipografia di G. Barbèra. Editor, Antonio Favoro; *coadiutore letterario*, Isidoro del Lungo; *assistente per la cura del testo*, Umberto Marchesini (Vittorio Lami for vol. 3[1]). A reprint was made in 1929–1939 and in 1964–1966 by Barbèra. Vol. 8 (1898) contains the *Discorsi e dimostrazioni matematiche intorno a due nuove scienze*; vol. 7 (1897) contains the *Dialogo sopra i due massimi sistemi del mondo*.

(1953). *Dialogue concerning the two chief world systems – Ptolemaic and Copernican*. Translated by Stillman Drake, foreword by Albert Einstein. Berkeley, Los Angeles: University of California Press (2d, rev. ed. 1962).

(1957). *Discoveries and opinions of Galileo, including: The starry messenger (1610), Letter to the Grand Duchess Christina (1615), and excerpts from Letters on sunspots (1613), The assayer (1623)*. Translated with an introduction and notes by Stillman Drake. Garden City, N.Y.: Doubleday & Co., Doubleday Anchor Books.

(1974). *Two new sciences, including Centers of gravity and force of percussion*. Translated with introduction and notes by Stillman Drake. Madison: The University of Wisconsin Press.

GALISON, PETER (1978). 'Model and reality in Descartes' theory of light'. *Synthesis: the University Journal in the History and Philosophy of Science, vol. 4* (no. 4), pp. 2–23 [actually pub. 1979].

GELL-MANN, MURRAY (1977). 'The search for unity in particle physics'. Communication presented at the 1580th stated meeting of the American Academy of Arts and Sciences, 13 April.

GERHARDT, C. I. *See* LEIBNIZ (1849–1863).

GEYMONAT, LUDOVICO (1965). *Galileo Galilei: A biography and inquiry into his philosophy of science*. Foreword by Giorgio de Santillana. Text translated from the Italian with additional notes and an appendix by Stillman Drake. New York, Toronto, London: McGraw-Hill Book Company. Translation of *Galileo Galilei*, Milan: Giulio Einaudi, 1957.

GILBERT, WILLIAM (1900). *On the magnet, magnetick bodies also, and on the great magnet the earth; a new physiology, demonstrated by many arguments & experiments*. London: at the Chiswick Press. Translation of *De magnete* (1600) by a group of British scholars, chiefly Silvanus Phillips Thompson. See also Thompson's *Notes on the De magnete of Dr. William Gilbert* (London: privately printed, 1901). Translation and *Notes* reprinted in facsimile (1958), New York: Basic Books [The Collector's Series in Science, (Derek J. Price, ed.)].

GILLISPIE, CHARLES COULSTON (1960). *The edge of objectivity, an essay in the history of scientific ideas*. Princeton: Princeton University Press.

GINGERICH, OWEN (1972). 'Johannes Kepler and the new astronomy'. *Quarterly Journal of the Royal Astronomical Society*, vol. 13, pp. 346–373.

(1977a). 'The origins of Kepler's third law'. *Vistas in Astronomy*, vol. 18, pp. 595–601.

(1977b). 'Was Ptolemy a fraud?' Cambridge (Mass.): Center for Astrophysics [Harvard College Observatory, Smithsonian Astrophysical Observatory, preprint series, no. 751].

GINZBURG, BENJAMIN (1933). 'Newton, Sir Isaac'. Pp. 369a–370b of vol. 11 of *Encyclopaedia of the Social Sciences* (New York: The Macmillan Company, reprinted 1937.)

GLAISHER, JAMES WHITBREAD LEE (1888). 'The bicentenary of Newton's *Principia*'. *Cambridge Chronicle*, 20 April, pp. 7–8. Address given on 19 April 1888 in the ante-chapel of Trinity College.

GOCLENIUS, RUDOLPH (1613). *Lexicon philosophicum, quo tanquam clave philosophiae fores aperiuntur*. Frankfurt: typis viduae Matthiae Beckeri, impensis Petri Musculi & Ruperti Pistorii. Facsimile reprint (1964), Hildesheim: Georg Olms Verlagsbuchhandlung.

GOLDSTEIN, BERNARD R. (1967). *The Arabic version of Ptolemy's Planetary hypotheses*. Philadelphia: The American Philosophical Society [*Transactions of the American Philosophical Society*, vol. 57, pt. 4].

(1969). 'Some medieval reports of Venus and Mercury transits'. *Centaurus*, vol. 14, pp. 45–59.

GOMBRICH, E. H. (1960). *Art and illusion*. New York: Pantheon Books [Bollingen Series, 35–5]. *See also* GREGORY & GOMBRICH (1973).

GRANT, EDWARD (1971). *Physical science in the Middle Ages*. New York, London, Sydney, Toronto: John Wiley & Sons.

ed. (1974). *A source book in medieval science*. Cambridge (Mass.): Harvard University Press.

GRAY, GEORGE J. (1907). *A bibliography of the works of Sir Isaac Newton, together with a list of books illustrating his works*. 2d ed., rev. and enl. Cambridge (England): Bowes and Bowes.

GREENSTREET, W. J., ed. (1927). *Isaac Newton, 1642–1727. A memorial volume edited for the Mathematical Association*. London: G. Bell and Sons.

GREGORY, DAVID (1702). *Astronomiae physicae & geometricae elementa*. Oxford: e theatro Sheldoniano.

(1715). *The elements of astronomy, physical and geometrical*. Done into English, with additions and corrections. To which is annex'd, Dr. Halley's Synopsis of the astronomy of comets. In two volumes. London: printed for J. Nicholson.

(1726). *The elements of physical and geometrical astronomy*. Done into English, with additions and corrections. The second edition. To which is annex'd Dr. Halley's Synopsis of the astronomy of comets. The whole newly revised, and compared with the Latin and corrected throughout, by Edmund Stone. 2 vols. London: printed for D. Midwinter. Reprinted (1972), New York, London: Johnson Reprint Corporation.

(1937). *David Gregory, Isaac Newton and their circle, extracts from David Gregory's memoranda 1677–1708*. Edited by W. G. Hiscock. Oxford: printed for the editor.

GREGORY, R. L. (1966). *Eye and brain, the psychology of seeing*. New York: McGraw-Hill Book Co.

(1970). *The intelligent eye*. London: Weidenfeld and Nicolson; New York: McGraw-Hill Book Co.

GREGORY, R. L., & E. H. GOMBRICH, eds. (1973). *Illusion in nature and in art*. New York: Charles Scribner's Sons.

GRIGORIAN, ANCHOTE T. (1964). 'Les études newtoniennes de A. N. Krylov'. Pp. 198–207 of COHEN & TATON, eds. (1964, vol. 1).

GRMEK, M. D. (1975). 'Santorio, Santorio'. *Dictionary of scientific biography*, vol. 12, pp. 101–104.

GROSSER, MORTON (1962). *The discovery of Neptune*. Cambridge (Mass.): Harvard University Press.

GUERICKE, OTTO VON (1962). *Experimenta nova (ut vocantur) Magdeburgica de vacuo spatio*. . . . Aalen: Otto Zellers Verlagsbuchhandlung [Miliaria: Faksimile-drucke zur Dokumentation der Geistesentwicklung, herausgegeben von Hellmut Rosenfeld und Otto Zeller, 1]. Facsimile ed. of a work published in Amsterdam in 1672, 'apud Joannem Janssonium à Waesberge', with supplementary notes by Hans Schimank.

GUERLAC, HENRY (1963a). 'Francis Hauksbee: Expérimentateur au profit de Newton', *Archives Internationales d'Histoire des Sciences*, vol. 16, pp. 113–128.

(1963b). *Newton et Epicure*. Paris: Palais de la Découverte [Histoire des Sciences: D-91].

(1964). 'Sir Isaac and the ingenious Mr. Hauksbee'. Pp. 228–253 of COHEN & TATON, eds. (1964, vol. 1).

(1967). 'Newton's optical aether. His draft of a proposed addition to his *Opticks*'. *Notes and Records of the Royal Society of London*, vol. 22, pp. 45–57.

(1972). 'Hales, Stephen'. *Dictionary of scientific biography*, vol. 6, pp. 35–48.

(1973). 'Newton and the method of analysis'. Pp. 378–391 of WIENER, ed. (1973, vol. 3).

(1975). *Antoine-Laurent Lavoisier, chemist and revolutionary*. New York: Charles Scribner's Sons. Based on his article on Lavoisier in the *Dictionary of scientific biography*, vol. 7, pp. 66–91.

(1977). 'The Newtonianism of Dortous de Mairan'. Pp. 131–141 of Jean Macary, ed., *Essays on the age of enlightenment in honor of Ira O. Wade*. Geneva: Librairie Droz.

HAHN, ROGER (1971). *The anatomy of a scientific institution: The Paris Academy of Sciences, 1666–1803*. Berkeley, Los Angeles, London: University of California Press.

HALES, STEPHEN (1961). *Vegetable staticks*. Foreword by Michael Hoskin. London: Oldbourne. Reprint of Hales's *Vegetable staticks: or, an account of some statical experiments on the sap in vegetables*. . . . *Also an attempt to analyse the air, by a great variety of chymio-statical experiments*, London, 1727.

(1964). *Statical essays: containing Haemastaticks*. With an introduction by

André Cournand. Published under the auspices of the Library of the New York Academy of Medicine [The History of Medicine Series, no. 22]. New York, London: Hafner. Facsimile reprint of the 1733 ed.

    (1969). *Vegetable staticks*. Foreword by M. A. Hoskins. London: Macdonald; New York: American Elsevier. Reprint of the 1727 ed.

HALL, A. RUPERT (1948). 'Sir Isaac Newton's notebook, 1661–1665'. *Cambridge Historical Journal*, vol. 9, pp. 239–250.

    (1957). 'Newton on the calculation of central forces'. *Annals of Science*, vol. 13, pp. 62–71.

    (1963). *From Galileo to Newton 1630–1720*. New York, Evanston: Harper & Row.

    (1966). 'Mechanics and the Royal Society, 1668–70'. *The British Journal for the History of Science*, vol. 3, pp. 24–38.

    (1975). 'Magic, metaphysics and mysticism in the scientific revolution'. Pp. 275–290 of RIGHINI BONELLI & SHEA, eds. (1975).

HALL, A. R., & MARIE BOAS HALL (1959a). 'Newton's electric spirit: four oddities'. *Isis*, vol. 50, pp. 473–476.

    (1959b). 'Newton's "mechanical principles" '. *Journal of the History of Ideas*, vol. 20, pp. 167–178.

    (1960). 'Newton's theory of matter'. *Isis*, vol. 51, pp. 131–144.

    (1961). 'Clarke and Newton'. *Isis*, vol. 52, pp. 583–585.

    eds. (1962). *Unpublished scientific papers of Isaac Newton. A selection from the Portsmouth Collection in the University Library, Cambridge*. Cambridge (England): at the University Press.

HALL, MARIE BOAS (1975). 'Newton's voyage in the strange seas of alchemy'. Pp. 239–246 of RIGHINI BONELLI & SHEA, eds. (1975). *See also* BOAS, MARIE.

HALLEY, EDMOND (1932). *Correspondence and papers of Edmond Halley*. Preceded by an unpublished memoir of his life by one of his contemporaries and the 'Eloge' by D'Ortous de Mairan. Arranged and edited by Eugene Fairfield MacPike. Oxford: at the Clarendon Press.

HANKINS, THOMAS L. (1967), 'The reception of Newton's second law of motion in the eighteenth century'. *Archives Internationales d'Histoire des Sciences*, vol. 20, pp. 43–65.

    (1970). *Jean d'Alembert: Science and the enlightenment*. Oxford: Clarendon Press.

HANSON, NORWOOD RUSSELL (1958). *Patterns of discovery: An inquiry into the conceptual foundations of science*. Cambridge: at the University Press.

    (1962). 'Leverrier: The zenith and nadir of Newtonian mechanics'. *Isis*, vol. 53, pp. 359–378.

HARRIS, JOHN (1704, 1710). *Lexicon technicum: or, an universal English dictionary of arts and sciences: explaining not only the terms of art, but the arts themselves*. London: printed for Dan. Brown Goodwin. Vol. 2 (1710) has its own alphabetical sequence, A–Z. Reprinted (1966), New York, London: Johnson Reprint Corporation.

HARTLEY, SIR HAROLD, ed. (1960). *The Royal Society, its origins and founders*. London: The Royal Society.

HARTNER, WILLY (1964). 'Mediaeval views on cosmic dimensions'. Pp. 254–282 of COHEN & TATON, eds. (1964, vol. 1).

HARVEY, WILLIAM (1928). *The anatomical exercises of Dr. William Harvey (De motu cordis 1628: De circulatione sanguinis 1649: the first English text of 1653 now newly edited by Geoffrey Keynes)*. London: The Nonesuch Press.

HAWES, JOAN L. (1968a). 'Newton and the "electrical attraction unexcited" '. *Annals of Science*, vol. 24, pp. 121–130.

(1968b). 'Newton's revival of the aether hypothesis and the explanation of gravitational attraction'. *Notes and Records of the Royal Society of London*, vol. 23, pp. 200–212.

(1971). 'Newton's two electricities'. *Annals of Science*, vol. 27, pp. 95–103.

HERBERT, SANDRA (1971). 'Darwin, Malthus, and selection'. *Journal of the History of Biology*, vol. 4, pp. 209–217.

HERIVEL, J. W. (1960a). 'Halley's first visit to Newton'. *Archives Internationales d'Histoire des Sciences*, vol. 13, pp. 63–65.

(1960b). 'Newton's discovery of the law of centrifugal force'. *Isis*, vol. 51, pp. 546–553.

(1961a). 'Interpretation of an early Newton manuscript'. *Isis*, vol. 52, pp. 410–416.

(1961b). 'The originals of the two propositions discovered by Newton in December 1679?' *Archives Internationales d'Histoire des Sciences*, vol. 14, pp. 23–33.

(1964). 'Galileo's influence on Newton in dynamics'. Pp. 294–302 of COHEN & TATON, eds. (1964, vol. 1).

(1965a). *The background to Newton's Principia. A study of Newton's dynamical researches in the years 1664–84*. Oxford: at the Clarendon Press.

(1965b). 'Newton's first solution to the problem of Kepler motion'. *British Journal for the History of Science*, vol. 2, pp. 350–354.

HERMANN, JACOB (1716). *Phoronomia, sive de viribus et motibus corporum solidorum et fluidorum libri duo*. Amsterdam: R. & G. Wetstenios.

HESSE, MARY B. (1961). *Forces and fields. The concept of action at a distance in the history of physics*. London: Thomas Nelson and Sons.

(1966). *Models and analogies in science*. Notre Dame (Ind.): University of Notre Dame Press. 'The introduction and first three chapters were first published in England in 1963 by Sheed and Ward Ltd., in a volume of the same title'.

(1967). 'Models and analogy in science'. Vol. 5, pp. 354–359, of Paul Edwards, ed., *Encyclopedia of philosophy*.

HESSEN, BORIS (1931). 'The social and economic roots of Newton's "Principia". Pp. 149–229 of N. I. Bukharin et al., *Science at the cross roads*. Papers presented to the International Congress of the History of Science and Technology, London, 29 June–3 July 1931, by the delegates of the U.S.S.R. London: Kniga. Reprinted (1971), with a new foreword by Joseph Needham and a new introduction by P. G. Werskey, London: Frank Cass & Co.

(1971). *The social and economic roots of Newton's 'Principia'*. With a new introduction by Robert S. Cohen. New York: Howard Fertig (facsimile reprint of HESSEN, 1931).

HIRSCH, E. D., JR. (1967). *Validity in interpretation*. New Haven: Yale University Press.

(1976). *The aims of interpretation*. Chicago, London: The University of Chicago Press.

HOBSON, E. W. (1912). *Mathematics, from the points of view of the mathematician and of the physicist*. An address delivered to the Mathematical and Physical Society of University College, London. Cambridge: at the University Press.

(1923). *The domain of natural science*. The Gifford Lectures delivered in the University of Aberdeen in 1921 and 1922. Cambridge: at the University Press.

HOLTON, GERALD (1965). 'The thematic imagination in science'. Pp. 88–108 of Gerald

Holton, ed., *Science and culture: A study of cohesive and disjunctive forces.*
Boston: Houghton Mifflin Company. Reprinted, with some revisions, from
Harry Woolf, ed., *Science as a cultural force*, Baltimore: The Johns Hopkins
Press, 1964.

(1973). *Thematic origins of scientific thought, Kepler to Einstein.* Cambridge
(Mass.): Harvard University Press.

(1975). 'On the role of themata in scientific thought'. *Science*, vol. 188, pp. 328–344.
See MERTON (1975).

(1977). 'Analysis and synthesis as methodological themata'. *Methodology and
Science*, vol. 10, pp. 3–33.

(1978). *The scientific imagination: case studies.* Cambridge (England): at the
University Press. See especially chap. 4, 'Analysis and synthesis as methodological
themata', pp. 111–151.

HOME, RODERICK W. (1968). 'The third law in Newton's mechanics'. *The British
Journal for the History of Science*, vol. 4, pp. 39–51.

HOPPE, EDMUND (1926[?]). *Geschichte der Optik.* Leipzig: Verlagsbuchhandlung
J. J. Weber.

HORROX, JEREMIAH (1673). *Opera posthuma; viz. Astronomia Kepleriana, defensa &
promota.* . . . London: typis Gulielmi Godbid, impensis J. Martin.

HUSSERL, EDMUND (1970). *The crisis of European sciences and transcendental
phenomenology: an introduction to phenomenological philosophy.* Translated
with an introduction by David Carr. Evanston: Northwestern University Press.
The original work, ed. by Walter Biemel, was entitled *Die Krisis der
europäischen Wissenschaften.* . . , The Hague: Martinus Nijhoff, 1954, 1962.

HUTCHINS, ROBERT MAYNARD, ed. (1952). *Great books of the Western world.*
Vol. 16. Chicago, London, Toronto: Encyclopaedia Britannica. Contains English
translations of *The almagest* by Ptolemy, *On the revolutions of the heavenly
spheres* by Nicolaus Copernicus, *Epitome of Copernican astronomy* (bks. 4 and
5) and *The harmonies of the world* (bk. 5) by Johannes Kepler.

HUXLEY, THOMAS H. (1894). *Discourses biological and geological.* New York:
D. Appleton and Company [Collected Essays, vol. 8].

HUYGENS, CHRISTIAAN (1673). *Horologium oscillatorium sive de motu pendulorum
ad horologia aptato demonstrationes geometricae.* Paris: chez F. Muguet.

(1888–1950). *Oeuvres complètes de Christiaan Huygens.* Publiées par la Société
Hollandaise des Sciences. The Hague: Martinus Nijhoff.

JAMMER, MAX (1957). *Concepts of force.* Cambridge (Mass.): Harvard University
Press.

JONES, HOWARD MUMFORD, I. B. COHEN, & EVERETT MENDELSOHN, eds. (1963).
A treasury of scientific prose, a nineteenth-century anthology. Boston: Little,
Brown and Co. Reprint (1977), Westport (Conn.): Greenwood Press, Publishers;
published also as *Science before Darwin, an anthology of British scientific
writing in the early nineteenth century*, London: Andre Deutsch, 1963.

JOUGET, E. (1924). *Lectures de mécanique.* La mécanique enseignée par les auteurs
originaux. 2 vols. New printing with notes and additions. Paris: Gauthier- Villars
et Cie, éditeurs.

KEELE, K. D. (1952). *Leonardo da Vinci on movement of the heart and blood.*
Foreword by Charles Singer. London: Harvey and Blythe.

KEPLER, JOHANNES (1611). *Dioptrice seu demonstratio eorum quae visui & visibilibus
propter conspicilla non ita pridem inventa accidunt.* . . . Augsburg: typis
Davidis Franci (facsimile reprint, Cambridge [England]: W. Heffer & Sons, 1962).

(1929). *Neue Astronomie.* Translated and edited by Max Casper. Munich, Berlin: Verlag R. Oldenbourg.

(1937–). *Gesammelte Werke.* Herausgegeben in Auftrag der Deutschen Forschungsgemeinschaft und der Bayerischen Akademie der Wissenschaften. Munich. C. H. Beck'sche Verlagsbuchhandlung. Originally planned by Walther von Dyck and Max Caspar. Caspar and Franz Hammer have served, successively, as editors. The major works used in the present study are *Mysterium cosmographicum* (vols. 1, 8), *Astronomia nova* (vol. 3), *Harmonice mundi* (vol. 6), and *Epitome astronomiae Copernicanae* (vol. 7).

(1952a). *Epitome of Copernican astronomy.* Books 4 & 5, translated by C. G. Wallis. Pp. 839–1004 of HUTCHINS, ed. (1952).

(1952b). *The harmonies of the world.* Book 5, translated by C. G. Wallis. Pp. 1005–1085 of HUTCHINS, ed. (1952).

(1965). *Kepler's conversation with Galileo's sidereal messenger.* First complete translation, with an introduction and notes, by Edward Rosen. New York, London: Johnson Reprint Corporation.

(1967). *Kepler's Somnium: The dream, or posthumous work on lunar astronomy.* Translated with a commentary by Edward Rosen. Madison, Milwaukee, London: The University of Wisconsin Press.

(in press). *Mysterium cosmographicum, The secret of the universe.* Translated by Alistair M. Duncan, with an introduction and commentary by Eric J. Aiton. New York: Abaris Books.

See also BEER & BEER, eds. (1975) and BRASCH, ed. (1931).

KILGOUR, FREDRICK G. (1954). 'William Harvey's use of the quantitative method'. *Yale Journal of Biology and Medicine,* vol. 26, pp. 410–421.

KLINE, MORRIS (1972). *Mathematical thought from ancient to modern times.* New York: Oxford University Press.

KNIGHT, DAVID (1975). *Sources for the history of science, 1660–1914.* Ithaca: Cornell University Press.

KOYRÉ, ALEXANDRE (1939). *Etudes galiléennes.* Paris: Hermann & Cie (reprint ed., 1966).

(1943). 'Galileo and Plato'. *Journal of the History of Ideas,* vol. 4, pp. 400–428 (reprinted in KOYRÉ, 1968).

(1950a). 'La gravitation universelle de Kepler à Newton'. *Actes du VIe Congrès International d'Histoire des Sciences* (Amsterdam), pp. 196–211; *Archives Internationales d'Histoire des Sciences,* vol. 4 (1951), pp. 638–653.

(1950b). 'The significance of the Newtonian synthesis'. *Archives Internationales d'Histoire des Sciences,* vol. 3 [29], pp. 291–311.

(1952a). 'La mécanique céleste de J. A. Borelli'. *Revue d'Histoire des Sciences et de leurs Applications,* vol. 5, pp. 101–138.

1952b). 'An unpublished letter of Robert Hooke to Isaac Newton'. *Isis,* vol. 43, pp. 312–337.

(1955a). 'A documentary history of the problem of fall from Kepler to Newton. De motu gravium naturaliter cadentium in hypothesi terrae motae'. *Transactions of the American Philosophical Society,* vol. 45, pp. 329–395.

(1955b). 'Pour une édition critique des oeuvres de Newton'. *Revue d'Histoire des Sciences,* vol. 8, pp. 19–37.

(1956). 'L'hypothèse et l'expérience chez Newton'. *Bulletin de la Société Française de Philosophie,* vol. 50, pp. 59–79.

(1957). *From the closed world to the infinite universe.* Baltimore: The Johns Hopkins Press.

(1960a). 'Le "*De motu gravium* de Galilée": de l'expérience imaginaire et de son abus'. *Revue d'Histoire des Sciences*, vol. 13, pp. 197–245. Translated by R. E. W. Maddison in KOYRÉ (1968).

(1960b). 'Newton, Galileo, and Plato'. *Actes du IXe Congrès International d'Histoire des Sciences* (Barcelona–Madrid, 1959), pp. 165–197 (reprinted in *Annales: Economies, Sociétés, Civilisations*, vol. 6, pp. 1041–1059). Reprinted in KOYRÉ (1965), pp. 201–220.

(1960c). 'Les queries de l'Optique'. *Archives Internationales d'Histoire des Sciences*, vol. 13, pp. 15–29.

(1960d). 'Les Regulae philosophandi'. *Archives Internationales d'Histoire des Sciences*, vol. 13, pp. 3–14.

(1961a). 'Attraction, Newton, and Cotes'. *Archives Internationales d'Histoire des Sciences*, vol. 14, pp. 225–236.

(1961b). *La révolution astronomique: Copernic, Kepler, Borelli*. Paris: Hermann [Histoire de la Pensée, 3].

(1965). *Newtonian studies*. Cambridge (Mass.): Harvard University Press; London: Chapman & Hall. More than half the volume consists of a previously unpublished study on 'Newton and Descartes', pp. 53–200.

(1966). *Etudes d'histoire de la pensée scientifique*. Foreword by René Taton. Paris: Presses Universitaires de France.

(1968). *Metaphysics and measurement: Essays in scientific revolution*. London: Chapman & Hall; Cambridge (Mass.): Harvard University Press.

(1973). *The astronomical revolution: Copernicus-Kepler-Borelli*. Translated by R. E. Maddison. Paris: Hermann; London: Methuen; Ithaca: Cornell University Press. Translated from KOYRÉ, 1961b.

KOYRÉ, ALEXANDRE, & I. BERNARD COHEN (1960). 'Newton's "electric & elastic spirit"'. *Isis*, vol. 51, p. 337.

(1961). 'The case of the missing *tanquam*: Leibniz, Newton, and Clarke'. *Isis*, vol. 52, pp. 555–566.

(1962). 'Newton & the Leibniz-Clarke correspondence, with notes on Newton, Conti, and Des Maizeaux'. *Archives Internationales d'Histoire des Sciences*, vol. 15, pp. 63–126.

KRAFFT, FRITZ (1973). 'Johannes Keplers Beitrag zur Himmelsphysik'. Pp. 55–139 of KRAFFT, MEYER, & STICKER, eds. (1973).

(1977). 'Kepler's contribution to celestial physics'. *Vistas in Astronomy*, vol. 18, pp. 567–572.

KRAFFT, FRITZ, KARL MEYER, & BERNHARD STICKER, eds. (1973). *Internationales Kepler-Symposium, Weil der Stadt 1971, Referate und Diskussionen*. Hildesheim: Verlag Dr. H. A. Gerstenberg.

KRILOFF, A. N. (1924). 'On a theorem of Sir Isaac Newton'. *Monthly Notices of the Royal Astronomical Society*, vol. 84, pp. 392–395.

KUHN, THOMAS S. (1955). 'La Mer's version of "Carnot's cycle"'. *American Journal of Physics*, vol. 23, pp. 387–389.

(1962). *The structure of scientific revolutions*. Chicago: The University of Chicago Press [International Encyclopedia of Unified Science, vol. 2, no. 2]. Second ed., enlarged, 1970.

(1974). 'Second thoughts on paradigms'. Pp. 459–517 of SUPPE (1974).

(1976). 'Mathematical versus experimental traditions in the development of physical science'. *The Journal of Interdisciplinary History*, vol. 7, pp. 1–31.

(1977). *The essential tension*. Chicago: University of Chicago Press.

See LAKATOS & MUSGRAVE (1970).

LAGRANGE, JOSEPH LOUIS (1788). *Méchanique analytique.* Paris: chez la veuve Desaint, Libraire.

(1797). *Théorie des fonctions analytiques, contenant les principes du calcul différentiel, dégagés de toute considération d'infiniment petits ou d'évanouissans de limites ou de fluxions, et réduits à l'analyse algébrique des quantités finies.* Paris: Impr. de la République [prairial an V].

LAKATOS, IMRE (1971). 'History of science and its rational reconstructions'. *Boston Studies in the Philosophy of Science,* vol. 8, pp. 91–136. Comments by Thomas S. Kuhn, Herbert Feigl, Richard J. Hall, and Noretta Koertge, with 'Replies to critics' by Lakatos, 137–182.

LAKATOS, IMRE, & ALAN MUSGRAVE, eds. (1970). *Criticism and the growth of knowledge.* Proceedings of the International Colloquium in the Philosophy of Science, London, 1965, vol. 4. Cambridge (England): at the University Press. Contents: T. S. Kuhn, 'Logic of discovery or psychology of research?': discussions by J. W. N. Watkins, S. E. Toulmin, L. Pearce Williams, K. R. Popper, Margaret Masterman, I. Lakatos, P. K. Feyerabend; T. S. Kuhn, 'Reflections on my critics'.

LAKATOS, IMRE, & ELIE ZAHAR (1975). 'Why did Copernicus' research program supersede Ptolemy's'. Pp. 354–383 of WESTMAN, ed. (1975).

LALANDE, JOSEPH JEROME LE FRANÇAIS DE (1764). *Astronomie.* 2 vols. Paris: chez Desaint & Saillant.

LA MER, VICTOR K. (1949). 'Some current misconceptions of N. L. Sadi Carnot's memoir and cycle'. *Science,* vol. 109, p. 598.

(1954, 1955). 'Some current misinterpretations of Sadi Carnot's memoir and cycle'. *American Journal of Physics,* vol. 22, pp. 20–27; vol. 23, pp. 95–102.

LAPLACE, PIERRE SIMON, MARQUIS DE (1966). *Celestial mechanics.* Translated with a commentary by Nathaniel Bowditch. 4 vols. New York: Chelsea Publishing Company. Corrected facsimile reprint of the volumes published in Boston in 1829, 1832, 1834, 1839.

LAUDAN, LARRY [LAURENS] (1977). *Progress and its problems: Towards a theory of scientific growth.* Berkeley, Los Angeles, London: University of California Press. *See also* BUCHDAHL (1970).

LEATHERDALE, W. H. (1974). *The role of analogy, model and metaphor in science.* Amsterdam, Oxford: North Holland Publishing Company; New York: American Elsevier Publishing Co.

LECKY, WILLIAM E. H. (1879). *History of the rise and influence of the spirit of rationalism in Europe.* Rev. ed. New York: D. Appleton and Company.

LEFORT, F., and J.-B. BIOT. *See* COLLINS ET AL. (1856).

LEIBNIZ, G. W. (1689*a*). 'De lineis opticis, et alia'. *Acta Eruditorum,* January 1689, pp. 36–38.

(1689*b*). 'Schediasma de resistentia medii, & motu projectorum gravium in medio resistente'. *Acta Eruditorum,* February 1689, pp. 38–47. Reprinted in *Leibnizens Mathematische Schriften,* vol. 6, pp. 135–144.

(1689*c*). 'Tentamen de motuum coelestium causis'. *Acta Eruditorum,* February 1689, pp. 82–96. Reprinted in *Leibnizens Mathematische Schriften,* vol. 6, pp. 144–161.

(1849–1863). *Leibnizens Mathematische Schriften.* Herausgegeben von C. I. Gerhardt. 2 parts: 7 vols. [Leibnizens gesammelte Werke aus den Handschriften der Königlichen Bibliothek zu Hannover herausgegeben von Georg Heinrich Pertz. Dritte Folge.] Berlin: Verlag von A. Asher & Comp.; Halle: Druck und Verlag von H. W. Schmidt. Photo-reprint (1962), Hildesheim: Georg Olms Verlagsbuchhandlung.

(1952). *Theodicy*. Edited with introduction by Austin Farrer. Translated by E. M. Huggard. New Haven: Yale University Press.

(1956). *Philosophical papers and letters*. A selection translated and edited with an introduction by Leroy E. Loemker. 2 vols. Chicago: The University of Chicago Press.

LEIBNIZ, G. W., & JOHANN BERNOULLI (1745). *Virorum celeberr. Got. Gul. Leibnitii et Johan. Bernoulli Commercium philosophicum et mathematicum*. Lausanne and Geneva: sumpt. Marci-Michaelis Bousquet & Socior. 2 vols.

LEIBNIZ, G. W., & SAMUEL CLARKE (1956). *The Leibniz-Clarke correspondence, together with extracts from Newton's 'Principia' and 'Opticks'*. Edited with introduction and notes by H. G. Alexander. Manchester: Manchester University Press.

(1957). *Correspondance Leibnitz-Clarke présentée d'après les manuscrits originaux des bibliothèques de Hanovre et de Londres*. Edited by André Robinet. Paris: Presses Universitaires.

LENOBLE, ROBERT (1943). *Mersenne, ou la naissance du mécanisme*. Paris: Librairie Philosophique J. Vrin.

LEVIN, HARRY (1969). *The myth of the golden age in the Renaissance*. Bloomington, London: Indiana University Press.

LEVY, JACQUES R. (1973). 'Le Verrier, Urbain Jean Joseph'. *Dictionary of scientific biography*, vol. 8, pp. 276–279.

L'HOSPITAL [L'HÔPITAL], GUILLAUME-FRANÇOIS-ANTOINE DE (1696). *Analyse des infiniment petits pour l'intelligence des lignes courbes*. Paris: de l'Imprimerie Royale (2d ed. Paris, 1715).

LÖWY, H. (1932). 'Die Erkenntnistheorie von Popper-Lynkeus und ihre Beziehung zur Machschen Philosophie'. *Die Naturwissenschaften*, vol. 20, pp. 767–770.

LOHNE, JOHANNES A. (1960). 'Hooke versus Newton. An analysis of the documents in the case of free fall and planetary motion'. *Centaurus*, vol. 7, pp. 6–52.

(1961). 'Newton's "proof" of the sine law'. *Archive for History of Exact Sciences*, vol. 1, pp. 389–405.

(1965). 'Isaac Newton: The rise of a scientist 1661–1671'. *Notes and Records of the Royal Society of London*, vol. 20, pp. 125–139.

(1967). 'The increasing corruption of Newton's diagrams'. *History of Science*, vol. 6, pp. 69–89.

(1968). 'Experimentum crucis'. *Notes and Records of the Royal Society of London*, vol. 23, pp. 169–199.

LOHNE, JOHANNES A., & BERNHARD STICKER (1969). *Newtons Theorie des Prismenfarben mit Übersetzung und Erläuterung der Abhandlung von 1672*. Munich: Werner Fritsch.

LOSEE, JOHN (1972). *A historical introduction to the philosophy of science*. London, Oxford, New York: Oxford University Press.

LOWES, JOHN LIVINGSTON (1927). *The road to Xanadu: A study in the ways of the imagination*. Boston, New York: Houghton Mifflin Company.

LUKACS, GEORG (1948). *Karl Marx und Friedrich Engels als Literaturhistoriker*. Berlin: Aufbau-Verlag.

(1952). *Balzac und der französische Realismus*. Berlin: Aufbau-Verlag.

LYONS, SIR HENRY (1944). *The Royal Society, 1660–1940: A history of its administration under its charters*. Cambridge: at the University Press.

MACH, ERNST (1898). *Popular scientific lectures*. Translated by Thomas J. McCormack. La Salle (Ill.): The Open Court Publishing Company (5th ed., 1943).

(1926). *The principles of physical optics, an historical and philosophical treatment.* Translated by John S. Anderson and A. F. A. Young. London: Methuen & Co. Reprint (1953), New York: Dover Publications.

(1960). *The science of mechanics: A critical and historical account of its development.* Translated by Thomas J. McCormack, new introduction by Karl Menger. 6th ed., with revisions through the 9th German ed. La Salle (Ill.): The Open Court Publishing Company.

MacLachlan, James (1973). 'A test of an "imaginary" experiment of Galileo's'. *Isis*, vol. 64, pp. 374–379.

Macomber, Henry P. *See* Babson Collection (1950).

Macpike, Eugene Fairfield (1937). *Hevelius, Flamsteed and Halley, three contemporary astronomers and their mutual relations.* London: Taylor & Francis.

(1939). *Dr. Edmond Halley (1656–1742).* A bibliographical guide to his life and work arranged chronologically; preceded by a list of sources, including references to the history of the Halley family. London: Taylor & Francis.

See also Halley (1932).

Maeyama, Yasukatsu (1971). *Hypothesen zur Planetentheorie des 17. Jahrhunderts.* Frankfurt am Main: Institut für Geschichte der Naturwissenschaften, Johann Wolfgang Goethe–Universität.

Maeyama, Yasukatsu, & W. Saltzer, eds. (1977). ΠΡΙΣΜΑΤΑ: *Festschrift für Willy Hartner.* Wiesbaden: Steiner Verlag.

Magirus, Johann (1642). *Physiologiae peripateticae libri sex, cum commentariis. . . .* Cambridge: ex officina R. Daniels.

Maier, Anneliese (1949). *Die Vorläufer Galileis im 14. Jahrhundert.* Rome: Edizioni di Storia e Letteratura.

Mandelbaum, Maurice (1964). *Philosophy, science and sense perception: Historical and critical studies.* Baltimore: The Johns Hopkins Press.

Manuel, Frank E. (1968). *A portrait of Isaac Newton.* Cambridge (Mass.): The Belknap Press of Harvard University Press.

(1974). *The religion of Isaac Newton, the Freemantle Lectures 1973.* Oxford: at the Clarendon Press.

Marcolongo, R. (1919). *Il problema dei tre corpi da Newton (1686) ai nostri giorni.* Milan: Ulrico Hoepli.

Marichal, Juan (1970). 'From Pistoia to Cadiz: A generation's itinerary'. Pp. 97–110 of Alfred O. Aldridge, ed., *The Ibero-American Enlightenment.* Urbana: The University of Illinois Press.

Marx, Karl, & Friedrich Engels (1937). *Briefwechsel.* vol. 3. Moscow: Verlagsgenossenschaft Ausländischer Arbeiter in der UdSSR. Reprinted. 1950, Berlin: Dietz Verlag.

(1947). *Literature and art. Selections from their writings.* New York: International Publishers.

(1949). *Über Kunst und Literatur. Eine Sammlung aus ihren Schriften.* Edited by Michail Lifschitz. Berlin: Verlag Bruno Henschel und Sohn.

(1954). *Sur la littérature et l'art.* Textes choisis, précédés d'une introduction de Maurice Thorez et d'une étude de Jean Fréville. Paris: Editions Sociales.

(1967). *Über Kunst und Literatur.* Auswahl und Redaktion: Manfred Kliem. 2 vols. Berlin: Dietz Verlag.

Maupertuis, Pierre Louis Moreau de (1736). 'Sur les loix de l'attraction'. *Suite des Mémoires de mathématique et de physique, tirés des registres de l'Académie Royale des Sciences de l'année M.DCCXXXII* (Amsterdam: chez Pierre Mortier), vol. 2, pp. 473–505. Vol. 1 has a different title page: *Histoire de l'Académie*

Royale des Sciences, année M.DCCXXXII, avec les mémoires de mathématique & de physique pour la même année. In the section "histoire", pp. 158–165 are devoted to an essay 'Sur l'attraction newtonienne'.

(1756). 'Discours sur les différentes figures des astres, où l'on essaye d'expliquer les principaux phénomènes du ciel'. *Oeuvres de Mr de Maupertuis*, nouvelle édition, corrigée & augmentée, vol. 1, pp. 79–270. Lyon: chez Jean-Marie Bruyset. This essay was first published in Paris in 1732.

McGUIRE, J. E. (1966). 'Body and void and Newton's *De mundi systemate*: Some new sources'. *Archive for History of Exact Sciences*, vol. 3, pp. 206–248.

(1967). 'Transmutation and immutability: Newton's doctrine of physical qualities'. *Ambix*, vol. 14, pp. 69–95.

(1968). 'The origin of Newton's doctrine of essential qualities'. *Centaurus*, vol. 12, pp. 233–260.

See WESTMAN & McGUIRE (1977).

McGUIRE, J. E., & P. M. RATTANSI (1966). 'Newton and the "pipes of Pan"'. *Notes and Records of the Royal Society of London*, vol. 21, pp. 108–143.

McKIE, DOUGLAS, & G. R. DE BEER (1952). 'Newton's apple'. *Notes and Records of the Royal Society of London*, vol. 9, pp. 46–54, 333–335.

McMULLIN, ERNAN, ed. (1967). *Galileo, man of science*. New York, London: Basic Books.

McMURRICH, J. PLAYFAIR (1930). *Leonardo the anatomist (1452–1519)*. Foreword by George Sarton. Baltimore: The Williams & Wilkins Company [for The Carnegie Institute of Washington].

MEDAWAR, PETER B. (1967). *The art of the soluble: Creativity and originality in science*. London: Methuen & Co. Reprinted (1969), Harmondsworth (Middlesex, England): Penguin Books.

(1969). *Induction and intuition in scientific thought*. Philadelphia: The American Philosophical Society.

MERTON, ROBERT K. (1938). 'Science, technology and society in seventeenth century England'. *Osiris*, vol. 4, pp. 360–632. Reprinted (1970), with a new introduction by the author, New York: Howard Fertig, 1970; New York: Harper Torchbooks.

(1957). 'Priorities in scientific discovery'. *American Sociological Review*, vol. 22, pp. 635–659. Reprinted in MERTON (1973).

(1961). 'Singletons and multiples in science'. *Proceedings of the American Philosophical Society*, vol. 105, pp. 470–486 (reprinted in MERTON, 1973).

(1973). *The sociology of science, theoretical and empirical investigations*. Edited and with an introduction by Norman W. Storer. Chicago, London: The University of Chicago Press.

(1975). 'Thematic analysis in science: Notes on Holton's concept'. *Science*, vol. 188, pp. 335–338.

MIDDLETON, W. E. KNOWLES (1971). *The experimenters, a study of the Accademia del Cimento*. Baltimore, London: The Johns Hopkins Press.

MILLER, PERRY (1958). 'Bentley and Newton'. Pp. 271–278 of NEWTON (1958).

MILLER, SAMUEL (1803). *A brief retrospect of the eighteenth century*. Part first; in two volumes; containing a sketch of the revolutions and improvements in science, arts, and literature, during that period. 2 vols. New York: printed by T. and J. Swords. No 'part second' was ever published.

MILLIKAN, ROBERT ANDREWS (1947). *Electrons (+ and −), protons, photons, neutrons, mesotrons, and cosmic rays*. Chicago: The University of Chicago Press.

MITTON, SIMON (1978). *The Crab Nebula*. New York: Charles Scribner's Sons.

MORE, HENRY (1659). *The immortality of the soul, so farre forth as it is demonstrable from the knowledge of nature and the light of reason.* London: printed by J. Flescher for William Morden.

(1679). *Enchiridion metaphysicum: sive, de rebus incorporeis succincta & luculenta dissertatio.* Pars prima: De existentia & natura rerum incorporearum in genere. London: typis impressa J. Macock, sumptibus J. Martyn & Gualt. Kettilby. *Opera omnia* [vol. 1].

MORE, L. T. (1934). *Isaac Newton: A biography.* New York, London: Charles Scribner's Sons (Reprint ed., New York: Dover Publications, 1962).

MORRISON, PHILIP (1976). 'Dissonance in the heavens'. *Bulletin of the American Academy of Arts and Sciences*, vol. 30 (no. 2), pp. 26–36.

MOSCOVICI, SERGE (1958). 'Recherches de Giovanni-Battista Baliani sur le choc des corps élastiques'. Pp. 98–115 of *Actes du Symposium International des sciences physiques et mathématiques dans la première moitié du XVIIe siècle (Pise-Vince, 16–18 juin 1958).* Paris: Hermann & Cie.

(1967). *L'expérience du mouvement: Jean-Baptiste Baliani, disciple et critique de Galilée.* Paris: Hermann.

MOTTE, ANDREW. *See* NEWTON (1729*b*), (1934).

MOUY, PAUL (1934). *Le développement de la physique cartésienne, 1646–1712.* Paris: Librairie philosophique J. Vrin.

(1938). 'Malebranche et Newton'. *Revue de Métaphysique et de Morale*, vol. 45, pp. 411–435.

MOVIUS, HALLAM L., JR. (1948). 'The lower palaeolithic cultures of southern and eastern Asia'. *Transactions of the American Philosophical Society*, vol. 38, pp. 329–420.

NAGEL, ERNEST (1956). *Logic without metaphysics, and other essays in the philosophy of science.* Glencoe (Ill.): The Free Press.

(1961). *The structure of science: Problems in the logic of scientific explanation.* New York & Burlingame: Harcourt, Brace & World.

NASH, LEONARD K. (1963). *The nature of the natural sciences.* Boston: Little, Brown and Co.

NEEDHAM, JOSEPH (1959). *Science and civilisation in China.* vol. 3. *Mathematics and the sciences of the heavens and the earth.* With the collaboration of Wang Ling. Cambridge: at the University Press.

NEEDHAM, JOSEPH, & WALTER PAGEL, eds. (1938). *Background to modern science.* New York: The Macmillan Company; Cambridge (England): at the University Press. Reprint (1975), New York: Arno Press (from a 1940 reprint).

NEUGEBAUER, OTTO E. (1946). 'The history of ancient astronomy: Problems and methods'. *Publications of the Astronomical Society of the Pacific*, vol. 58, pp. 17–43, 104–142. An expanded and revised version of an article with the same title, which appeared in *Journal of Near Eastern Studies*, vol. 4 (1945), pp. 1–38.

(1948). 'Mathematical methods in ancient astronomy'. *Bulletin of the American Mathematical Society*, vol. 54, pp. 1013–1041.

(1957). *The exact sciences in antiquity.* 2d ed. Providence: Brown University Press.

NEWTON, ISAAC (1672). 'A letter of Mr. Isaac Newton . . . containing his new theory about light and colors'. *Philosophical Transactions*, vol. 6, pp. 3075–3087.

(1702). *A new and most accurate theory of the moon's motion; whereby all her irregularities may be solved, and her place truly calculated to two minutes.* Written by that incomparable mathematician Mr. Isaac Newton, and published in Latin by Mr. David Gregory in his excellent astronomy. London: printed and sold by A. Baldwin. Reprinted in NEWTON (1975).

(1715). '[Recensio libri =] An account of the book entituled *Commercium epistolicum Collinii & aliorum, De analysi promota*, published by order of the Royal-Society, in relation to the dispute between Mr. Leibnits and Dr. Keill, about the right of invention of the new geometry of fluxions, otherwise call'd the differential method'. *Philosophical Transactions*, vol. 29, pp. 173–224. The title of the book given in the *Philosophical Transactions* is not the exact title of the *Commercium epistolicum* [first edition] itself. A French translation appeared in the *Journal Littéraire*, vol. 6 (Nov./Dec. 1715), pp. 13sqq, 345sqq.

(1728*a*). *De mundi systemate liber*. London: impensis J. Tonson, J. Osborn, & T. Longman.

(1728*b*). *Optical lectures read in the publick schools of the University of Cambridge, Anno Domini, 1669*. By the late Sir Isaac Newton, then Lucasian Professor of the Mathematicks. Never before printed. Translated into English out of the original Latin. London: printed for Francis Fayram.

(1728*c*). *A treatise of the system of the world*. Translated into English. London: printed for F. Fayram.

(1729*a*). *Lectiones opticae, annis* MDCLXIX, MDCLXX, & MDCLXXI. In scholis publicis habitae: et nunc primum ex MSS. in lucem editae. London: apud Guil. Innys.

(1729*b*). *The mathematical principles of natural philosophy*. Translated into English by Andrew Motte. To which are added, *The laws of the moon's motion, according to gravity*. By John Machin. In two volumes. London: printed for Benjamin Motte (Facsimile reprint (1968), London: Dawsons of Pall Mall.

(1736). *The method of fluxions and infinite series; with its application to the geometry of curve-lines*. Translated from the author's Latin original not yet made publick. To which is subjoin'd a perpetual comment . . . by John Colson. London: printed by Henry Woodfall and sold by John Nourse.

(1737). *A treatise of the method of fluxions and infinite series, with its application to the geometry of curve lines*. Translated from the Latin original not yet published. London: printed for T. Woodman and J. Millan (reprinted in vol. 1 of NEWTON, 1964–1966).

(1759). *Principes mathématiques de la philosophie naturelle*. Translated 'par feue Madame la Marquise du Chastellet'. 2 vols. Paris: chez Desaint & Saillant [& chez] Lambert.

(1934). *Sir Isaac Newton's Mathematical principles of natural philosophy and his System of the world*. Translated into English by Andrew Motte in 1729. The translations revised, and supplied with an historical and explanatory appendix, by Florian Cajori. Berkeley: University of California Press. Often reprinted.

(1952). *Opticks or a treatise of the reflections, refractions, inflections & colours of light*. Based on the fourth edition: London, 1730. With a foreword by Albert Einstein; an introduction by Sir Edmund Whittaker; a preface by I. Bernard Cohen; and an analytical table of contents prepared by Duane H. D. Roller. New York: Dover Publications.

(1958). *Isaac Newton's papers & letters on natural philosophy and related documents*. Edited, with a general introduction, by I. Bernard Cohen assisted by Robert E. Schofield. Cambridge (Mass.): Harvard University Press. 2d, rev. ed., Harvard University Press, 1978.

(1959–1977). *The correspondence of Isaac Newton*. Vol. 1, 1661–1675 (1959), vol. 2, 1676–1687 (1960), vol. 3, 1688–1694 (1961), edited by H. W. Turnbull; vol. 4, 1694–1709 (1967), edited by J. F. Scott; vol. 5, 1709–1713 (1975), vol. 6, 1713–1718 (1976), vol. 7, 1718–1727 (1977) edited by A. Rupert Hall and Laura Tilling. Cambridge (England): at the University Press [published for the Royal Society].

(1964–1967). *The mathematical works of Isaac Newton*. Assembled with an introduction by Dr. Derek T. Whiteside. 2 vols. New York, London: Johnson Reprint Corporation [The Sources of Science].

(1967–). *The mathematical papers of Isaac Newton*. Vol. 1, 1664–1666 (1967); vol. 2, 1667–1670 (1968); vol. 3, 1670–1673 (1969); vol. 4, 1674–1684 (1971); vol. 5, 1683–1684 (1972); vol. 6, 1684–1691 (1974); vol. 7, 1691–1695 (1976); edited by D. T. Whiteside, with the assistance in publication of M. A. Hoskin and A. Prag. Cambridge: at the University Press. To be complete in 8 vols.

(1969). *A treatise of the system of the world*. Translated into English. With an introduction by I. B. Cohen. London: Dawsons of Pall Mall. Photo-reprint of the 2d ed. (1731), plus the front matter of the 1st ed. (1728).

(1972). *Isaac Newton's Philosophiae naturalis principia mathematica*. The third edition (1726) with variant readings assembled by Alexandre Koyré, I. Bernard Cohen, & Anne Whitman. 2 vols. Cambridge (England): at the University Press; Cambridge (Mass.): Harvard University Press.

(1973). *The unpublished first version of Isaac Newton's Cambridge lectures on optics 1670–1672*. A facsimile of the autograph, now Cambridge University Library MS. Add. 4002, with an introduction by D. T. Whiteside. Cambridge (England): The University Library.

(1975). *Isaac Newton's 'Theory of the moon's motion' (1702)*. With a bibliographical and historical introduction by I. Bernard Cohen. London: Dawson.

NIETO, MICHAEL MARTIN (1972). *The Titius-Bode law of planetary distances: its history and theory*. Oxford, New York: Pergamon Press.

OLSCHKI, LEONARDO (1937). *Storia letteraria delle scoperte geografiche*. Florence: Leo S. Olschki.

(1941). 'What Columbus saw on landing in the West Indies'. *Proceedings of the American Philosophical Society*, vol. 84, pp. 633–659.

O'MALLEY, C. D. (1964). *Andreas Vesalius of Brussels 1514–1564*. Berkeley, Los Angeles: University of California Press.

ORNSTEIN, MARTHA (1928). *The role of scientific societies in the seventeenth century*. Chicago: The University of Chicago Press. Reprint edition (1975), New York: Arno Press; author's name given as Martha Ornstein [Bronfenbrenner].

OVENDEN, MICHAEL W. (1977). 'Bode's law – truth or consequences'. *Vistas in Astronomy*, vol. 18, pp. 473–496.

PAGEL, WALTER (1935). 'Religious motives in the medical biology of the XVIIth century'. *Bulletin of the History of Medicine*, vol. 3, pp. 97–128, 213–231, 265–312.

(1944). 'The religious and philosophical aspects of Van Helmont's science and medicine'. *Bulletin of the History of Medicine*, suppl. 2, pp. 1–44.

(1948). 'J. B. Van Helmont, *De tempore*, and biological time'. *Osiris*, vol. 8, pp. 346–417.

(1951). 'Giordano Bruno, the philosophy of circles and the circular movement of the blood'. *Journal of the History of Medicine*, vol. 6, pp. 116–124.

(1958). *Paracelsus, an introduction to philosophical medicine in the era of the Renaissance*. Basel, New York: S. Karger.

(1967). *William Harvey's biological ideas: Selected aspects and historical background*. Basel: S. Karger; New York: Hafner Publishing Company.

(1969–1970). 'William Harvey revisited'. *History of Science*, vol. 8, pp. 1–31, vol. 9, pp. 1–41.

PALTER, ROBERT (1970). 'Newton and the inductive method'. Pp. 244–257 of PALTER, ed. (1970).

ed. (1970). *The annus mirabilis of Sir Isaac Newton 1666–1966*. Cambridge (Mass.), London: M.I.T. Press.

PARTINGTON, J. R. (1961). *A history of chemistry*. Vol. 2. London: Macmillan & Co.

PASCAL, BLAISE (1937). *The physical treatises of Pascal. The equilibrium of liquids and the weight of the mass of the air*. Translated by I. H. B. and G. H. Spiers, with introduction and notes by Frederick Barry. New York: Columbia University Press, Records of Civilization, Sources and Studies, No. 28.

(1956). *Traités de l'équilibre des liqueurs et de la pesanteur de la masse de l'air*. Paris: Gauthier-Villars, Éditeur-Imprimeur-Librairie, Les maîtres de la pensée scientifique, collection de mémoires et d'ouvrages. Facsimile reprint of 1819 ed. with new introduction by Maurice Solovine.

PATTERSON, LOUISE DIEHL (1949, 1950). 'Hooke's gravitation theory and its influence on Newton. I: Hooke's gravitation theory. II: The insufficiency of the traditional estimate'. *Isis*, vol. 40, pp. 327–341; vol. 41, pp. 32–45.

PEACOCK, GEORGE (1855). *Life of Thomas Young*. London: John Murray.

PEARCE, GLENN, & PATRICK MAYNARD, eds. (1973). *Conceptual change*. Dordrecht: D. Reidel Publishing Company.

PEMBERTON, HENRY (1728). *A view of Sir Isaac Newton's philosophy*. London: S. Palmer.

PHYSICS SURVEY COMMITTEE, NATIONAL RESEARCH COUNCIL (1973). *Physics in perspective: The nature of physics and the subfields of physics*. Student ed. Washington, D.C.: National Academy of Sciences.

PIGHETTI, CLELIA (1960). 'Cinquant' anni di studi newtoniani (1908–1959)'. *Rivista Critica di Storia della Filosofia*, fascicoli 2–3, pp. 181–203, 295–318.

(1961). 'Per la storia del newtonianesimo in Italia. I: Newton e la cultura italiana: Aspetti e problemi di un importante incontro'. *Rivista Critica di Storia della Filosofia*, fascicolo 4, pp. 425–434.

(1962). 'A proposito delle ipotesi nella metodologia newtoniana'. *Archives Internationales d'Histoire des Sciences*, vol. 15, pp. 291–302.

See also SIZI (1964).

POINCARÉ, HENRI (1912). 'Sur un théorème de géométrie'. *Rendiconti del Circolo Matematico di Palermo*, vol. 33, pp. 1–34.

POPPER, KARL (1957). 'The aim of science'. *Ratio*, vol. 1, pp. 24–35. Reprinted in POPPER (1972), pp. 191–205.

(1972). *Objective knowledge, an evolutionary approach*. Oxford: at the Clarendon Press.

PORTSMOUTH COLLECTION (1888). *A catalogue of the Portsmouth Collection of books and papers written by or belonging to Sir Isaac Newton, the scientific portion of which has been presented by the Earl of Portsmouth to the University of Cambridge*. Prepared by H. R. Luard, G. G. Stokes, J. C. Adams, and G. D. Liveing. Cambridge: at the University Press.

PURVER, MARGERY (1967). *The Royal Society: Concept and creation*. London: Routledge and Kegan Paul.

RANDALL, JOHN HERMAN, JR. (1942). 'Newton's natural philosophy, its problems and consequences.' Pp. 335–357 of F. P. Clarke & M. C. Nahm, eds., *Philosophical essays in honor of Edgar Singer.*

RATTANSI, P. M. (1972). 'Newton's alchemical studies'. Pp. 167–182 of DEBUS, ed. (1972), vol. 2.

(1973). 'Some evaluations of reason in sixteenth- and seventeenth-century natural philosophy'. Pp. 148–166 of TEICH & YOUNG, eds. (1973).

RICHTER, IRMA A., ed. (1952). *Selections from the notebooks of Leonardo da Vinci.* Edited with commentaries. London, New York, Toronto: Oxford University Press.

RIGAUD, STEPHEN PETER (1838). *Historical essay on the first publication of Sir Isaac Newton's Principia.* Oxford: at the University Press. Reprinted (1972), New York, London: Johnson Reprint Corporation.

ed. (1841). *Correspondence of scientific men of the seventeenth century . . . in the collection of . . . the Earl of Macclesfield.* 2 vols. Oxford: at the University Press.

RIGHINI BONELLI, M. L., & WILLIAM R. SHEA, eds. (1975). *Reason, experiment, and mysticism in the scientific revolution.* New York: Science History Publications.

ROBERTS, MICHAEL, & E. R. THOMAS (1934). *Newton and the origin of colours: A study of one of the earliest examples of scientific method.* London: G. Bell & Sons.

ROBINSON, RICHARD (1962). *Plato's earlier dialectic.* 2d ed. Oxford: at the Clarendon Press.

ROMANES, GEORGE JOHN (1895). *Darwin, and after Darwin.* An exposition of the Darwinian theory and a discussion of post-Darwinian questions. Vol. 2: Post-Darwinian questions: Heredity and utility. ed. 2. Chicago: The Open Court Publishing Company.

RONCHI, VASCO (1970). *The nature of light: An historical survey.* Translated by V. Barocas. Cambridge (Mass.): Harvard University Press; London: Heinemann Educational Books.

ROSCOE, HENRY E., & ARTHUR HARDEN (1896). *A new view of Dalton's atomic theory: A contribution to chemical history, together with letters and documents.* . . . London, New York: Macmillan and Co.

ROSEN, EDWARD (1947). *The naming of the telescope.* New York: Henry Schuman.

(1971). *Three Copernican treatises.* The *Commentariolus* of Copernicus, the *Letter against Werner,* the *Narratio prima* of Rheticus. Translated with introduction and notes. 3d ed., rev., with a biography of Copernicus, and Copernicus bibliographies, 1939–1958 and 1959–1970. New York: Octagon Books. 1st ed., New York: Columbia University Press, 1939.

(1973). 'Cosmology from antiquity to 1850'. Pp. 535–554 of WIENER, ed. (1973), vol. 1.

ROSENBERGER, FERDINAND (1882–1890). *Die Geschichte der Physik in Grundzügen mit synchronistischen Tabellen der Mathematik, der Chemie und beschreibenden Naturwissenschaften sowie der allgemeinen Geschichte.* 3 vols. Brunswick: Druck und Verlag von Friedrich Vieweg und Sohn.

(1895). *Isaac Newton und seine physikalischen Principien.* Ein Hauptstück aus der Entwickelungsgeschichte der modernen Physik. Leipzig: Johann Ambrosius Barth (Arthur Meiner).

ROSENFELD, LEON (1965). 'Newton and the law of gravitation'. *Archive for History of Exact Sciences,* vol. 2, pp. 365–386.

ROSSI, PAOLO (1968). *Francis Bacon, from magic to science.* Translated by Sacha Rabinovitch. Chicago: The University of Chicago Press.

(1975). 'Hermeticism, rationality, and the scientific revolution'. Pp. 247–273 of RIGHINI BONELLI & SHEA (1975).

ROUTH, EDWARD JOHN (1896–1902). *A treatise on analytical statics, with numerous examples.* 2d ed. 2 vols. Cambridge (England): at the University Press.

ROYAL SOCIETY (1947). *Newton tercentenary celebrations.* Cambridge (England): at the University Press. Contains E. N. da C. Andrade, 'Newton'; Lord Keynes, 'Newton, the man'; J. Hadamard, 'Newton and the infinitesimal calculus'; S. I. Vavilov, 'Newton and the atomic theory'; N. Bohr, 'Newton's principles and modern atomic mechanics'; H. W. Turnbull, 'Newton: the algebraist and geometer'; W. Adams, 'Newton's contributions to observational astronomy'; J. C. Hunsaker, 'Newton and fluid mechanics'.

RUFFNER, JAMES ALAN (1966). The background and early development of Newton's theory of comets. Bloomington: [doctoral dissertation, Indiana University].

(1971). 'The curved and the straight: Cometary theory from Kepler to Hevelius'. *Journal for the History of Astronomy,* vol. 2, pp. 178–195.

RUFUS, W. CARL (1931). 'Kepler as an astronomer'. Pp. 1–38 of BRASCH, ed. (1931).

RUSSELL, HENRY NORRIS, RAYMOND SMITH DUGAN, & JOHN QUINCY STEWART (1926). *Astronomy.* Vol. 1: *The solar system.* Boston: Ginn and Company.

RUSSELL, J. L. (1964). 'Kepler's laws of planetary motion: 1609–1666'. *The British Journal for the History of Science,* vol. 2, pp. 1–24.

RUTHERFORD, LORD (1938). 'Forty years of physics'. Pp. 47–74 of NEEDHAM & PAGEL, eds. (1938).

SABRA, A. I. (1967). *Theories of light from Descartes to Newton.* London: Oldbourne.

(1976). 'The scientific enterprise'. Pp. 181–192 of Bernard Lewis et al., *The world of Islam: Faith, people, culture.* London: Thames and Hudson.

SCHWEBER, SILVAN S. (1977). 'The origins of the *Origin* revisited'. *Journal of the History of Biology,* vol. 10, pp. 229–316.

SCOTT, J. F. *See* NEWTON (1959–1977).

SETTLE, THOMAS B. (1961). 'An experiment in the history of science'. *Science,* vol. 133, pp. 19–23.

(1967). 'Galileo's use of experiment as a tool of investigation'. Pp. 315–337 of McMULLIN, ed. (1967).

SHAPERE, DUDLEY (1974). 'Scientific theories and their domains'. Pp. 518–599 of SUPPE (1974).

SHAPLEY, HARLOW, & HELEN E. HOWARTH (1929). *A source book in astronomy.* New York: McGraw-Hill Book Company.

SHEA, WILLIAM R. (1972). *Galileo's intellectual revolution, middle period, 1610–1632.* New York: Science History Publications; London: Macmillan.

(1975). 'Trends in the interpretation of seventeenth century science'. Pp. 1–18 of RIGHINI BONELLI & SHEA, eds. (1975).

SINGER, CHARLES (1922). *The discovery of the circulation of the blood.* London: G. Bell and Sons. Reprinted (1956), London: Wm. Dawson & Sons.

SIZI, FRANCESCO (1611). ΔΙΑΝΟΙΑ: *Astronomica, optica, physica, qua syderei nuncii rumor de quatuor planetis a Galilaeo Galilaeo . . . recens perspicilli cuiusdam ope conspectus redditur.* Venice: apud Mariam Bertanum.

(1964). *Dianoia: Astronomica, ottica, fisica. Per mezzo della quale si rende vano il rumore sollevato dal "Sidereus nuncius" a proposito di quattro pianeti osservati recentemente per mezzo di un certo perspicillo da Galileo Galilei, matematico celeberrimo.* Traduzione e nota critica di Clelia Pighetti. Florence: G. Barbèra.

SKINNER, QUENTIN (1966). 'The limits of historical explanations'. *Philosophy: The Journal of the Royal Institute of Philosophy*, vol. 41, pp. 199–215.

SOTHEBY & CO. (1936). *Catalogue of the Newton papers, sold by order of the Viscount Lymington to whom they have descended from Catherine Conduitt, Viscountess Lymington, great-niece of Sir Isaac Newton*. London: Sotheby & Co.

SPINOZA, BENEDICTUS DE (1905). *The principles of Descartes' philosophy*. Translated from the Latin, with an introduction by Halbert Hains Britan. Chicago: The Open Court Publishing Company.

———— (1930). *Spinoza's Ethics and De intellectus emendatione*. Translated by A. Boyle. London & Toronto: J. M. Dent & Sons; New York: E. P. Dutton.

STEINER, GEORGE (1959). *Tolstoy or Dostoevsky: An Essay in the old criticism*. New York: Alfred A. Knopf (reprint ed., New York: E. P. Dutton & Company, (1971).

———— (1970). *Language and silence: Essays on language, literature, and the inhuman*. New York: Atheneum.

STEPHENSON, F. RICHARD, & DAVID H. CLARK (1976). 'Historical supernovas'. *Scientific American*, vol. 234 (no. 6), pp. 100–107. See also CLARK AND STEPHENSON (1977).

STEWART, DUGALD (1877). *Elements of . . . human mind*, 3 vols., edited by Sir William Hamilton, Edinburgh: T. & T. Clark. These are vols. 2–4 of *The collected works. . .* , edited by Sir William Hamilton. An earlier ed. (1854) was published in Edinburgh by Thomas Constable and Co. (Boston: Little, Brown, and Co.).

STRAKER, STEPHEN MORY (1970). Kepler's optics: A study in the development of seventeenth-century natural philosophy. Bloomington: [doctoral dissertation, Indiana University].

STREETE, THOMAS (1661). *Astronomia Carolina: A new theorie of the celestial motions*. London: for Lodowick Lloyd.

STRONG, E. W. (1951). 'Newton's "mathematical way" '. *Journal of the History of Ideas*, vol. 12, pp. 90–110. Reprinted (1957) in WIENER & NOLAND, eds., pp. 412–432.

STRUIK, D. J., ed. (1969). *A source book in mathematics, 1200–1800*. Cambridge (Mass.): Harvard University Press.

STUEWER, ROGER H., ed. (1970). *Historical and philosophical perspectives of science*. Minneapolis: University of Minnesota Press [Minnesota Studies in the Philosophy of Science, Herbert Feigl and Grover Maxwell, eds.], vol. 5.

STUKELEY, WILLIAM (1936). *Memoirs of Sir Isaac Newton's life, 1752: Being some account of his family and chiefly of the junior part of his life*. Edited by A. Hastings White. London: Taylor & Francis.

SUPPE, FREDERICK, ed. (1974). *The structure of scientific theories*. Urbana: University of Illinois Press.

TABARRONI, G. (1971). 'J. D. Cassini et la deuxième loi de Kepler'. *Publications of the XIIIth International Congress of the History of Science*, sect. 6, p. 56. English summary in *Vistas in Astronomy*, vol. 18, p. 593.

TATON, RENÉ (1975). 'The mathematical revolution of the seventeenth century'. Pp. 283–290 of RIGHINI BONELLI & SHEA, eds. (1975). See COHEN & TATON, eds. (1964).

TAYLOR, A. M. (1966). *Imagination and the growth of science*. Being the Tallman Lectures 1964–5 delivered at Bowdoin College, Brunswick, Maine, U.S.A. London: John Murray.

TEICH, MIKULAŠ, & ROBERT YOUNG, eds. (1973). *Changing perspectives in the history of science: Essays in honour of Joseph Needham.* London: Heinemann.

THOMAS, E. R. *See* ROBERTS & THOMAS (1934).

THORNTON, JOHN L., & R. I. J. TULLY (1971). *Scientific books, libraries and collectors. A study of bibliography and the book trade in relation to science.* 3d, rev. ed. London: The Library Association.

TISSERAND, FRANÇOIS FELIX (1882). 'Les planètes intra-mercurielles'. *Annuaire du Bureau des Longitudes*, pp. 729–772.

TOULMIN, STEPHEN (1972). *Human understanding.* Vol. 1: General introduction and part 1. Princeton: Princeton University Press.

TRUESDELL, CLIFFORD (1960). 'A program toward rediscovering the rational mechanics of the Age of Reason'. *Archive for History of Exact Sciences*, vol. 1, pp. 3–36.

——— (1968). *Essays in the history of mechanics.* New York: Springer-Verlag.

——— (1970). 'Reactions of late baroque mechanics to success, conjecture, error, and failure in Newton's *Principia*'. Pp. 192–232 of PALTER, ed. (1970).

——— (1972). 'Leonard Euler, supreme geometer (1707–1783)'. Pp. 51–95 of *Studies in eighteenth-century culture.* Vol. 2: *Irrationalism in the eighteenth century.* London, Cleveland: The Press of Case Western Reserve University.

——— (1973). 'The scholar's workshop and tools'. *Centaurus*, vol. 17, pp. 1–10.

TULLY, R. I. J. *See* THORNTON & TULLY (1971).

TURBAYNE, COLIN MURRAY (1962). *The myth of metaphor.* New Haven, London: Yale University Press.

TURGOT, ANNE-ROBERT-JACQUES, BARON DE L'AULNE (1808–1811). *Oeuvres de M.* Turgot, Ministre d'Etat. 9 vols. Paris: Impression de Delance.

——— (1913–1923). *Oeuvres de Turgot et documents le concernant.* 5 vols. Paris: F. Alcan.

——— (1973). *Turgot on progress, sociology and economics. A philosophical review of the successive advances of the human mind, On universal history, Reflections on the formation and the distribution of wealth.* Translated, edited and with an introduction by Ronald L. Meek. Cambridge: at the University Press.

TURNBULL, HERBERT WESTREN, ed. (1939). *James Gregory: Tercentenary memorial volume. Containing his correspondence with John Collins and his hitherto unpublished mathematical manuscripts, together with addresses and essays communicated to the Royal Society of Edinburgh, July 4, 1938.* London: G. Bell and Sons [published for the Royal Society of Edinburgh].

——— (1945). *The mathematical discoveries of Newton.* London, Glasgow: Blackie & Son.

——— (1951). 'The discovery of the infinitesimal calculus'. *Nature*, vol. 167, pp. 1048–1050.

——— (1953). 'Isaac Newton's letters: Some discoveries'. *Manchester Guardian*, 3 October, p. 4. Reprinted, *Manchester Guardian Weekly*, 8 October 1953, p. 11. *See also* NEWTON (1959–1977).

UNAMUNO, MIGUEL DE (1951). *Ensayos.* 2 vols. Madrid: Aguilar.

——— (1968). *Obras completas.* Vol. 3: *Nuevos ensayos.* Madrid: Escelicer.

VALLERY-RADOT, RENÉ (1911). *La vie de Pasteur.* 13th ed. Paris: Librairie Hachette.

VAN HELDEN, ALBERT (1977). 'The invention of the telescope'. *Transactions of the American Philosophical Society*, vol. 67, pt. 4.

VARIGNON, PIERRE (1700). 'Des forces centrales ou des pesanteurs nécessaires aux planètes pour leur faire décrire les orbes qu'on leur a supposez jusqu'ici'. *Histoire de l'Académie Royale des Sciences, année M.DCC., avec les mémoires de mathématique & de physique pour la même année* (seconde édition, revue, corrigée & augmentée; Paris, 1761); *mémoires*, pp. 224–243.

VESALIUS, ANDREAS (1543). *De humani corporis fabrica.* Basel: [ex officina Joannis Oporini]. Facsimile (1964) ed., Brussels: Culture et Civilisation.

(1950). *The illustrations from the works of Andreas Vesalius of Brussels.* With annotations, and translations, a discussion of the plates and their background, authorship and influence, and a biographical sketch of Vesalius by J. B. de C. M. Saunders and Charles D. O'Malley. Cleveland, New York: The World Publishing Company.

VIETS, HENRY (1922). '*De staticis experimentis* of Nicolaus Cusanus'. *Annals of Medical History,* vol. 4, pp. 115–135.

WAFF, CRAIG (1975). Universal gravitation and the motion of the moon's apogee: The establishment and reception of Newton's inverse-square law, 1687–1749. Baltimore: [doctoral dissertation. The Johns Hopkins University].

(1976). 'Isaac Newton, the motion of the lunar apogee, and the establishment of the inverse square law'. *Vistas in Astronomy,* vol. 20, pp. 99–103.

WALLACE, WILLIAM A. (1968). 'The enigma of Domingo de Soto: *Uniformiter difformis* and falling bodies in late medieval physics'. *Isis,* vol. 59, pp. 384–401.

(1975). 'Soto, Domingo de'. *Dictionary of Scientific Biography,* vol. 12, pp. 547–548.

WALLIS, JOHN (1659). *Tractatus duo. Prior, de cycloide et corporibus inde genitis. Posterior, epistolaris; in qua agitur, de cissoide, et corporibus inde genitis. . . .* Oxford: typis academicis Lichfieldianis. Contains Wren's rectification of the cycloid (pp. 62–68), Wren's account of curtate and prolate cycloids (pp. 69–72), and Wren's use of the cycloid in solving Kepler's problem (pp. 72–73) [In the copy examined, these pages are misnumbered 70–76, 77–80, 80–; p. 73 is correctly numbered.]

(1669). 'A summary account given by Dr. John Wallis, of the General Laws of Motion . . .'. *Philosophical Transactions,* vol. 3, pp. 864–866.

(1670). *Mechanica: Sive, de motu, tractatus geometricus.* London: typis Gulielmi Godbid, impensis Mosis Pitt.

WEBSTER, CHARLES (1975). *The great instauration: Science, medicine and reform 1626–1660.* London: Duckworth. For a major review of this book by Quentin Skinner, see *Times Literary Supplement,* 2 July 1976, no. 3877, pp. 810–812.

WEINBERG, STEVEN (1977). 'The search for unity: notes for a history of quantum field theory'. *Daedalus,* vol. 106, number 4, pp. 17–35.

WESTFALL, RICHARD S. (1967). 'Hooke and the law of universal gravitation'. *The British Journal for the History of Science,* vol. 3, pp. 245–261.

(1970). 'Uneasily fitful reflections on fits of easy transmission'. Pp. 88–104 of PALTER, ed. (1970). First published in *The Texas Quarterly,* vol. 10 (no. 3), pp. 86–102.

(1971). *Force in Newton's physics: The science of dynamics in the seventeenth century.* London: Macdonald; New York: American Elsevier.

(1972). 'Newton and the hermetic tradition'. Pp. 183–198 of DEBUS, ed. (1972), vol. 2.

(1973). 'Newton and the fudge factor'. *Science,* vol. 179, pp. 751–758.

(1975). 'The role of alchemy in Newton's career'. Pp. 189–232 of RIGHINI BONELLI & SHEA, eds. (1975).

WESTMAN, ROBERT S. (1975a). 'Kepler's theory of hypothesis'. *Vistas in Astronomy,* vol. 18, pp. 713–720.

(1975b). 'Kepler's theory of hypothesis and the "realist dilemma"'. *Vistas in Astronomy,* vol. 18, pp. 721–724.

ed. (1975). *The Copernican achievement.* Berkeley, Los Angeles, London:

University of California Press, UCLA Center for Medieval and Renaissance studies – Contributions.

WESTMAN, ROBERT S., & J. E. McGUIRE (1977). *Hermeticism and the scientific revolution.* Los Angeles: University of California, William Andrews Clark Memorial Library.

WEYL, HERMANN (1940). 'The mathematical way of thinking'. *Science*, vol. 92, pp. 437–446.

(1949). *Philosophy of mathematics and natural science.* Rev. and augmented English ed., based on a translation by Olaf Helmer. Princeton: Princeton University Press.

WHEWELL, WILLIAM (1832). *On the free motion of points, and on universal gravitation, including the principal propositions of books I. and III. of the Principia; the first part of a new edition of a treatise on dynamics.* Cambridge (England): printed for J. & J. Deighton [and Whittaker, Treacher & Arnot, London].

(1847). *The philosophy of the inductive sciences, founded upon their history.* A new edition, with corrections and additions, and an appendix, containing philosophical essays previously published. 2 vols. London: John W. Parker. Reprint ed. (1967), with a new introduction by John Herivel, New York, London: Johnson Reprint Corporation.

(1856). 'Of the transformation of hypotheses in the history of science'. *Transactions of the Cambridge Philosophical Society*, vol. 9, pp. 139–146. Reprinted (1860) as Appendix G in Whewell's *On the philosophy of discovery*, London: John W. Parker and Son.

(1858). *Novum organon renovatum*: Being the second part of the philosophy of the inductive sciences. 3d ed. London: John W. Parker and Son.

(1860). *On the philosophy of discovery, chapters historical and critical.* Including the completion of the third edition of *The philosophy of the inductive sciences.* London: John W. Parker and Son.

(1865). *History of the inductive sciences, from the earliest to the present time.* 3d ed., with additions. 2 vols. New York: D. Appleton and Company.

WHISTON, WILLIAM (1716). *Sir Isaac Newton's mathematick philosophy more easily demonstrated: With Dr. Halley's account of comets illustrated.* Being forty lectures read in the Public Schools at Cambridge. By William Whiston, M.A. Mr. Lucas's Professor of the Mathematicks in that University. For the use of the young students there. In this English edition the whole is corrected and improved by the author. London: printed for J. Senex and W. Taylor. Reprinted (1972), New York, London: Johnson Reprint Corporation.

(1728). *Astronomical lectures, read in the Publick Schools at Cambridge.* Whereunto is added a collection of astronomical tables; being those of Mr. Flamsteed, corrected; Dr. Halley; Monsieur Cassini; and Mr. Street. For the use of young students in the university. The second edition corrected. London: printed for J. Senex, W. and J. Innys, J. Osborne and T. Longman. Reprinted (1972), New York, London: Johnson Reprint Corporation.

(1749). *Memoirs of the life of Mr. William Whiston by himself.* 2 vols. London: printed for the author and sold by Mr. Whiston and Mr. Bishop.

WHITE, HAYDEN V. (1973). 'Foucault decoded: Notes from the underground'. *History and Theory*, vol. 12, pp. 23–54.

WHITEHEAD, ALFRED NORTH (1922). 'The first physical synthesis'. Pp. 161–178 of F. S. Marvin, ed., *Science and civilization.* London: Humphrey Milford, Oxford University Press.

(1925). *Science and the modern world.* Lowell Lectures, 1925. New York: The Macmillan Company. Many printings and reprintings in England and the United States.

WHITESIDE, DEREK T. (1960). 'Wren the mathematician'. Pp. 107–112 of HARTLEY, ed. (1960).

(1961*a*). 'Newton's discovery of the general binomial theorem'. *Mathematical Gazette,* vol. 45, pp. 175–180.

(1961*b*). 'Patterns of mathematical thought in the later seventeenth century'. *Archive for History of Exact Sciences,* vol. 1, pp. 179–388.

(1962). 'The expanding world of Newtonian research'. *History of Science,* vol. 1, pp. 16–29.

(1964*a*). 'Isaac Newton: Birth of a mathematician'. *Notes and Records of the Royal Society of London.* vol. 19, pp. 53–62.

(1964*b*). 'Newton's early thoughts on planetary motion: A fresh look'. *British Journal for the History of Science,* vol. 2, pp. 117–137.

(1966*a*). 'Newtonian dynamics'. *History of Science,* vol. 5, pp. 104–117. A review-article on HERIVEL (1965*a*).

(1966*b*). 'Newton's marvellous year: 1666 and all that'. *Notes and Records of the Royal Society of London,* vol. 21, pp. 32–41.

(1970*a*). 'Before the *Principia*: The maturing of Newton's thoughts on dynamical astronomy, 1664–84'. *Journal for the History of Astronomy,* vol. 1, pp. 5–19.

(1970*b*). 'The mathematical principles underlying Newton's *Principia*'. *Journal for the History of Astronomy,* vol. 1, pp. 116–138.

(1974*a*). 'Keplerian planetary eggs, laid and unlaid, 1600–1605'. *Journal for the History of Astronomy,* vol. 5, pp. 1–21.

(1974*b*). 'Mercator, Nicolaus'. *Dictionary of scientific biography,* vol. 9, pp. 310–312.

(1976). 'Newton's lunar theory: From high hope to disenchantment'. *Vistas in Astronomy,* vol. 19, pp. 317–328.

(1977). 'From his claw the Greene lyon'. *Isis,* vol. 68, pp. 116–121. An essay-review of DOBBS (1975).

See also NEWTON (1964–1967), (1967–).

WIENER, PHILIP P., ed. (1973). *Dictionary of the history of ideas, studies of pivotal ideas.* 4 vols. New York: Charles Scribner's Sons.

WIENER, PHILIP P., & AARON NOLAND, eds. (1957). *Roots of scientific thought, a cultural perspective.* New York: Basic Books.

WILSON, CURTIS (1968). 'Kepler's derivation of the elliptical path'. *Isis,* vol. 59, pp. 5–25.

(1970). 'From Kepler's laws, so-called, to universal gravitation: empirical factors'. *Archive for History of Exact Sciences,* vol. 6, pp. 89–170.

(1972). 'How did Kepler discover his first two laws?' *Scientific American,* vol. 226 (March), pp. 92–106.

(1974). 'Newton and some philosophers on Kepler's "laws"'. *Journal of the History of Ideas,* vol. 35, pp. 231–258.

(1977). 'Kepler's ellipse and area rule – their derivation from fact and conjecture'. *Vistas in Astronomy,* vol. 18, pp. 587–591.

WING, VINCENT (1651). *Harmonicon coeleste: Or, the coelestial harmony of the visible world: Containing an absolute and entire piece of astronomie* . . . London: printed by Robert Leybourn, for the Company of Stationers.

(1669). *Astronomia Britannica: In qua per novam, concinnioremque methodum, hi quinque tractatus traduntur* . . . 2 vols. London: typis Johannis Macock, impensis Georgii Sawbridge.

WISAN, WINIFRED L. (1974). 'Galileo's science of motion'. *Archive for History of Exact Science*, vol. 13, pp. 103–306.

WOLFSON, HARRY AUSTRYN (1929). *Crescas' critique of Aristotle*. Problems of Aristotle's *Physics* in Jewish and Arabic philosophy. Cambridge (Mass.): Harvard University Press.

(1934). *The philosophy of Spinoza*. 2 vols. Cambridge (Mass.): Harvard University Press. Reprinted (1958), without the 'index of references', New York: Meridian Books.

WORTIS, JOSEPH (1954). *Fragments of an analysis with Freud*. New York: Simon and Schuster (reprint ed., New York: McGraw Hill Book Company, 1975).

WREN, CHRISTOPHER (1669). 'Generatio corporis cylindroidis hyperbolici, elaborandis lentibus hyperbolicis accommodati'. *Philosophical Transactions*, vol. 4, pp. 961–962.

See also WALLIS (1659).

YATES, FRANCES A. (1964). *Giordano Bruno and the hermetic tradition*. London: Routledge and Kegan Paul; Chicago: University of Chicago Press.

(1967). 'The hermetic tradition in Renaissance science'. Pp. 255–274 of Charles S. Singleton, ed., *Art, science, and history in the Renaissance*. Baltimore: The Johns Hopkins Press.

(1972). *The Rosicrucian enlightenment*. London and Boston: Routledge & Kegan Paul.

YOUNG, THOMAS (1855). *Miscellaneous works*. Edited by George Peacock. Vol. I. London: John Murray.

YOUSCHKEVITCH, A. P. (1968). 'Sur la révolution en mathématiques des temps modernes'. *Acta historiae rerum naturalium necnon technicarum*, Prague, special issue no. 4, pp. 5–31.

ZEITLINGER, H. (1927). 'A Newton bibliography'. Pp. 148–170 of W. J. Greenstreet, ed., *Isaac Newton, 1642–1727, a memorial volume*, edited for the Mathematical Association. London: G. Bell and Sons.

INDEX

QC
7
.C66
c.2

35,193

35,193

QC   Cohen. I. Bernard,

Date Due

BJJH

PRINTED IN U.S.A.   CAT. NO. 24 161   BRODART